STRUCTURED SYSTEMS ANALYSIS:
tools & techniques

by

Chris Gane and Trish Sarson

IST databooks **Improved System Technologies Inc.**
888 seventh avenue, new york, n.y. 10019

FIRST EDITION, JULY 1977

Second impression, October 1977

© Improved System Technologies, Inc., 1977

All rights reserved. No part of this book may be reproduced in any form or by any means without permission in writing from the publisher.

ISBN 0-931096-00-7

Contents

CHAPTER 1   THE NEED FOR BETTER TOOLS ... 1

    1.1   WHAT GOES WRONG IN ANALYSIS? ... 2

    1.2   HOW MUCH CAN WE BLAME OUR TOOLS? ... 5
        1.2.1   No "model" in DP ... 5
        1.2.2   English narrative is too vague ... 5
        1.2.3   Flowcharts do more harm than good ... 6
        1.2.4   We have no systematic way of recording user preferences ... 8

    1.3   HOW MUCH DOES THE FUNCTIONAL SPECIFICATION MATTER? ... 9

    REFERENCES ... 10

CHAPTER 2   WHAT THE TOOLS ARE AND HOW THEY FIT TOGETHER ... 11

    2.1   FIRST, DRAW A LOGICAL DATA FLOW DIAGRAM ... 12
        2.1.1   Error conditions ... 18
        2.1.2   Alternative physical implementations ... 18
        2.1.3   The general system class ... 21

    2.2   NEXT, PUT THE DETAIL IN A DATA DICTIONARY ... 21

    2.3   DEFINE THE LOGIC OF THE PROCESSES ... 24

    2.4   DEFINE THE DATA STORES: CONTENTS AND IMMEDIATE ACCESS ... 27
        2.4.1   Are the logical data stores the simplest possible? ... 28
        2.4.2   What immediate accesses will be needed? ... 29

    2.5   USING THE TOOLS TO CREATE A FUNCTIONAL SPECIFICATION ... 33

    EXERCISES AND DISCUSSION POINTS ... 36

CONTENTS

CHAPTER 3 DRAWING UP DATA FLOW DIAGRAMS   37

    3.1   SYMBOL CONVENTIONS   38
          3.1.1 External entity   38
          3.1.2 Data flow   39
          3.1.3 Process   43
          3.1.4 Data store   45

    3.2   EXPLOSION CONVENTIONS   46

    3.3   ERROR AND EXCEPTION HANDLING   49

    3.4   GUIDELINES FOR DRAWING DATA FLOW DIAGRAMS   50

    3.5   EXAMPLE: Distribution with inventory   52

    3.6   MATERIALS FLOW AND DATA FLOW   66

    REFERENCES   68

    EXERCISES AND DISCUSSION POINTS   69

CHAPTER 4 BUILDING AND USING A DATA DICTIONARY   71

    4.1   THE PROBLEM OF DESCRIBING DATA   71

    4.2   WHAT WE MIGHT WANT TO HOLD IN A DATA DICTIONARY   75
          4.2.1 Describing a data element   76
          4.2.2 Describing data structures   83
          4.2.3 Describing data flows   86
          4.2.4 Describing data stores   87
          4.2.5 Describing processes   87
          4.2.6 Describing external entities   89
          4.2.7 Describing glossary entries   90

    4.3   MANUAL vs AUTOMATED DATA DICTIONARIES   90

    4.4   WHAT WE MIGHT WANT TO GET OUT OF A DATA DICTIONARY   92
          4.4.1 Ordered listings of all entries   93
          4.4.2 Composite reports   93
          4.4.3 Cross-referencing ability   95
          4.4.4 Finding a name from a description   95
          4.4.5 Consistency and completeness checking   96
          4.4.6 Generation of machine-readable data definitions   96
          4.4.7 Extraction of data dictionary entries from
                    existing programs   98

CONTENTS

4.5   AN EXAMPLE OF AN AUTOMATED DATA DICTIONARY   98

4.6   CROSS-PROJECT OR ORGANIZATION-WIDE DATA DICTIONARIES   108

4.7   DATA DICTIONARIES AND DISTRIBUTED PROCESSING   109

APPENDIX - Commercially available data dictionary packages   112

REFERENCES   113

EXERCISES AND DISCUSSION POINTS   114

CHAPTER 5   ANALYZING AND PRESENTING PROCESS LOGIC   115

5.1   THE PROBLEMS OF EXPRESSING LOGIC   115
    5.1.1   Not only but notwithstanding, and/or unless...   115
    5.1.2   Greater than, less than   118
    5.1.3   And/or ambiguity   119
    5.1.4   Undefined adjectives   121
    5.1.5   Handling combinations of conditions   121

5.2   DECISION TREES   126

5.3   DECISION TABLES   135
    5.3.1   Conditions, actions, and rules   136
    5.3.2   Building up the rule matrix   137
    5.3.3   Indifference   139
    5.3.4   Extended entry; the freight rate problem   141
    5.3.5   Decision tables vs decision trees   145

5.4   STRUCTURED ENGLISH, PSEUDOCODE, AND "TIGHT ENGLISH"   146
    5.4.1   The "structures" of Structured Programming   146
    5.4.2   Conventions for Structured English   152
    5.4.3   Pseudocode   156
    5.4.4   Logically "Tight English"   158
    5.4.5   Pros and cons of the four tools   161
    5.4.6   Who does what?   164

REFERENCES   165

EXERCISES AND DISCUSSION POINTS   166

CONTENTS

CHAPTER 6   DEFINING THE CONTENTS OF DATA STORES   169

    6.1   WHAT COMES OUT MUST GO IN   169

    6.2   SIMPLIFYING DATA STORE CONTENTS BY INSPECTION   174

    6.3   SIMPLIFYING DATA STORE CONTENTS BY NORMALIZATION   176
         6.3.1   The vocabulary of normalization   178

    6.4   SOME NORMALIZED FORMS ARE SIMPLER THAN OTHERS   180
         6.4.1   First normal form (1NF)   180
         6.4.2   Second normal form (2NF)   182
         6.4.3   Third normal form (3NF)   182

    6.5   MAKING RELATIONS OUT OF RELATIONS--PROJECTION AND JOIN   185
         6.5.1   Projection   186
         6.5.2   Join   186

    6.6   THE IMPORTANCE OF THIRD NORMAL FORM   189

    6.7   A PRACTICAL EXAMPLE OF 3NF   190
         6.7.1   Normalization of the "customers" data store   192
         6.7.2   Normalizations of the "books" data store   194
         6.7.3   Normalization of the "accounts receivable" data store   196
         6.7.4   Normalization of the "inventory" data store   197
         6.7.5   Putting the relations together   197

    REFERENCES   201

    EXERCISES AND DISCUSSION POINTS   202

CHAPTER 7   ANALYZING RESPONSE REQUIREMENTS   203

    7.1   DESCRIBING THE WAYS DATA IS USED   203

    7.2   PHYSICAL TECHNIQUES FOR IMMEDIATE ACCESS   206
         7.2.1   Indexes   206
         7.2.2   Hierarchical records   209

    7.3   GENERAL INQUIRY LANGUAGE CAPABILITY   213

    7.4   TYPES OF QUERY   216
         7.4.1   Entities and attributes   216
         7.4.2   Six basic query types   218
         7.4.3   Variations on the basic types of queries   222

CONTENTS

7.5	FINDING OUT WHAT THE USERS NEEDS AND PREFERENCES ARE	222
	7.5.1 Operational access versus informational access	223
	7.5.2 Getting a composite "wish-list"	224
	7.5.3 Refining the "wish-list"	231
7.6	SECURITY CONSIDERATIONS	234
	APPENDIX - General Inquiry Packages	235
	REFERENCES	236
	EXERCISES AND DISCUSSION POINTS	237

CHAPTER 8 USING THE TOOLS: a structured methodology   239

8.1	THE INITIAL STUDY	239
8.2	THE DETAILED STUDY	244
	8.2.1 Defining in more detail who the users of a new system would be	244
	8.2.2 Building a logical model of the current system	246
	8.2.3 Refining the estimates of IRACIS	247
8.3	DEFINING A "MENU" OF ALTERNATIVES	250
	8.3.1 Deriving objectives for the new system from the limitations of the current system	251
	8.3.2 Developing a logical model of the new system	252
	8.3.3 Producing tentative alternative physical designs	254
8.4	USING THE "MENU" TO GET COMMITMENT FROM USER DECISION MAKERS	259
8.5	REFINING THE PHYSICAL DESIGN OF THE NEW SYSTEM	261
	8.5.1 Refining the logical model	261
	8.5.2 Designing the physical data base	262
	8.5.3 Deriving the hierarchy of modular functions that will be programmed	263
	8.5.4 Defining the new clerical tasks that will interface with the new system	263
	8.5.5 A note on estimating	264
8.6	LATER PHASES OF THE PROJECT	268
	REFERENCES	269
	EXERCISES AND DISCUSSION POINTS	270

CONTENTS

CHAPTER 9 DERIVING A STRUCTURED DESIGN FROM THE LOGICAL MODEL — 273

- 9.1 THE OBJECTIVES OF DESIGN — 274
  - 9.1.1 Performance considerations — 275
  - 9.1.2 Control considerations — 281
  - 9.1.3 Changeability considerations — 283

- 9.2 STRUCTURED DESIGN FOR CHANGEABILITY — 286
  - 9.2.1 What makes for a changeable system? — 286
  - 9.2.2 Deriving a changeable system from the data flow diagram — 288
  - 9.2.3 Module coupling — 295
  - 9.2.4 Well-formed modules: cohesiveness, cohesion, binding — 298
  - 9.2.5 Scope of effect/scope of control — 300

- 9.3 THE TRADE-OFF BETWEEN CHANGEABILITY AND PERFORMANCE — 304

- 9.4 AN EXAMPLE OF STRUCTURED DESIGN — 307
  - 9.4.1 The boundaries of the design — 307
  - 9.4.2 Physical file considerations — 308
  - 9.4.3 Locating the central transform — 320
  - 9.4.4 Refining the design from the top down — 322

- 9.5 TOP-DOWN DEVELOPMENT — 334
  - 9.5.1 Possible top-down versions of the CBM system — 334
  - 9.5.2 Why develop top-down? — 336
  - 9.5.3 The role of the analyst — 338
  - 9.5.4 Summary — 342

REFERENCES — 343

EXERCISES AND DISCUSSION POINTS — 344

CHAPTER 10 INTRODUCING STRUCTURED SYSTEMS ANALYSIS INTO YOUR ORGANIZATION — 347

- 10.1 STEPS IN IMPLEMENTATION OF STRUCTURED SYSTEMS ANALYSIS — 347
  - 10.1.1 Reviewing the groundrules for conducting projects — 347
  - 10.1.2 Establishing standards and procedures for the use of the data dictionary and other software — 352
  - 10.1.3 Training analysts in the use of the tools and techniques — 352
  - 10.1.4 Orienting users to the new approaches — 353

10.2	BENEFITS AND PROBLEMS	355
	10.2.1 Benefits from using Structured Systems Analysis	355
	10.2.2 Potential problems	357
REFERENCES		360
GLOSSARY		361
INDEX		371

Preface

We are excited about the techniques described in this book. They are proving their worth in a troublesome area of data processing: the analysis and definition of what a new system should do if it is to be of most value to the people who are paying for it.

The discipline consists of an evolving set of tools and techniques which have grown out of the success of Structured Programming and Structured Design. The underlying concept is the building of a logical (non-physical) model of a system, using graphical techniques which enable users, analysts, and designers to get a clear and common picture of the system and how its parts fit together to meet the user's needs. Until the development of the Structured Systems Analysis tools, there was no way of showing the underlying logical functions and requirements of a system; one very quickly got bogged down in the details of the current or proposed physical implementation.

The book starts with a discussion of some of the problems we face in analysis, and then reviews the graphical tools and how they fit together to make a logical model. We then take each tool in turn and treat them in detail in Chapters 3 through 7, starting with the key tool, the logical data flow diagram. Since we are using tools which build a logical model, the approach to system development which results is somewhat different from traditional approaches; in Chapter 8 we sketch out a structured systems development methodology, which takes advantage of the new tools. The methodology involves building a system top-down by successive refinement; first producing an overall system data flow, then developing detailed data flows, next defining the detail of data structure and process logic, then moving into the design of a modular structure, and so on. We analyze top-down, we design top-down, we develop top-down, we test top-down. Further we recognize that good development involves iteration; one has to be prepared to refine the logical model and the physical design in the light of information resulting from the use of an early version of that model or design.

We distinguish the work of analysis (defining "what" the system will do) from the work of design (defining "how" it will do it), recognizing that analysts often do design and designers often do analysis. Part of the value of Structured Systems Analysis is that it provides the designer with the inputs needed to define the programs for maximum changeability using Structured Design. In Chapter 9, we review the importance of changeability and the techniques and concepts of Structured Design, taking a realistic system, analyzing it and designing it down to the module level.

Finally, in Chapter 10, we discuss the issues that arise in changing over to these new techniques from the traditional approaches, with their implications for management control of projects, and the benefits that one can expect.

We have tried to avoid introducing new terms as far as possible; since the discipline draws on Structured Design (which has its own vocabulary), and relational data base theory (which has its own vocabulary) there may be some unfamiliar terminology. Each such term is explained where it first appears, and is also defined in the Glossary at the back of the book.

We hope *you* will find these tools and techniques useful, whether you are a systems analyst, or a designer, or a manager, or a user of data processing services. We would like to hear about your experiences in using Structured Systems Analysis, particularly if you are willing to share those experiences with others. Please write us at Improved System Technologies Inc., 888 Seventh Avenue, New York, N.Y. 10019, or call us on (212) 586 1098, and tell us the type of system you have used the techniques on, the size of effort involved, the phase that the project has reached, and the effect that using the Structured Systems Analysis techniques has had on the project. We will be building a store of this data, to put people working on similar systems or in the same area in touch with one another if they want to share information.

We gratefully acknowledge the help of those who have given us permission to reproduce their copyright material, and the contributions made to the development of these ideas by our former colleagues at Yourdon inc, Tom de Marco, Victor Weinberg, and Ed Yourdon.

Chris Gane   Trish Sarson

Chapter 1

The need for better tools

In many ways, systems analysis is the toughest part of the development of a data processing system. It's not simply the technical difficulty of the work, though many projects demand that the analyst have deep knowledge of current D.P. technology. It's not simply the political difficulties that arise, especially in larger projects where the new system will serve several, possibly conflicting, interest groups. It's not simply the communication problems that arise in any situation where people of different backgrounds, with different views of the world and different vocabularies, have to work together. It's the compounding of these difficulties together that makes systems analysis so hard and demanding; the fact that the analyst must play the middleman between the user community, who have a gut-feel for their problems, but find it hard to explain them and are vague about what computers can do to help, and the programming community, who are anxious that the organization have a sharp data processing function, but do not have the information to know what is best for the business. The analyst must make a match between what is currently *possible* in our onrushing technology (minis, micros, distributed processing, data base, data communications) and what is *worth doing* for the business, as run by the people in it.

Making the match in a way which is acceptable to all parties and will stand the test of time is the hardest part

1. THE NEED FOR BETTER TOOLS

of the effort; if it is done well, then no matter how difficult the design and programming, the system which is built will serve the needs of the business. If it is done poorly, then no matter how excellent the implementation, the system will not be what the organization needs and the costs will outweigh the benefits. In making that match, we need all the help we can get. This book presents some tools which have proved helpful.

1.1 WHAT GOES WRONG IN ANALYSIS?

The problems that the analyst faces are intertwined together; that's one reason why they are tough problems. We can distinguish five aspects which are worth commenting on:

Problem #1

> The analyst finds it hard to learn enough about the business to see the system requirements through the user's eyes. (When we use the term "business," by the way, we mean the enterprise of any organization, whether profit-making or not.) Again, and again, we hear it said, "We built a technically excellent system, but it wasn't what the users wanted." Why should this be? Why can't the analyst simply study the business and gather enough facts to specify the right system? At the heart of this problem is the fact that many user managers are "doers" rather than "explainers." They acquire and handle the information they need on an intuitive basis, without thinking in terms of information flow, or decision logic. This is natural; one becomes a manager by *making* the right decision and *doing* a superior job, not necessarily by explaining how the job is done, and how the decisions are taken. But it means that the analyst has no right to expect a lucid explanation of the system requirements from the users; he has to help them work out their needs. At the same time, analysts do not have the gift of telepathy; they do not know what they have not been told. This painful fact shows up particularly in terms of the relative importance that users give to various features of the system. Suppose a particular manager wants a cash report each morning. Which is more important, that he have it by 8.30, even if there are some items not yet resolved, or that he

have it accurate to the penny, even if it takes until
11 A.M. some days? The manager knows very well, and
might say, "Heck, any dummy who knows anything about
the business would know that!" But getting that level
of intuitive feel for the trade-offs in the business is
tough.

Problem #2

People in the user community do not yet know enough
about data processing to know what is feasible and what
isn't. The propaganda about computers has, in general,
not left people with any specific or accurate ideas
about what they can or can't do. Many people have no
idea of the capability of an on-line CRT; why should
they? The technology is still too young for people to
have the background knowledge and exposure which would
enable them to imagine the way a new system would affect them. The popular media haven't helped; the image
of computers is either one of expensive and senseless
mistakes, or one of science-fiction where boxes with a
mind of their own take over the world.

Compare this situation with people's ideas about, say,
the construction industry. Even though a businessman
may never have commissioned a factory before, he has
been in and out of factories all his working life, and
has formed a whole background which enables him to make
sense of the things his architect will say to him. Then
again, one factory is much like another; at least they
will have much more in common than, say, a batch system
and an on-line system. Our problems in data processing
are much worse than those of the construction industry,
not least because we have had no way to make a *model* of
what we are going to build. In a construction or engineering project of any size, the architect will discuss the requirements of his clients, and then produce
a model of what the finished structure will look like.
Everyone who has an interest can look at the model,
relate it to their previous experience with such structures, and form a clear idea of what they will be getting for their money. The tools of Structured Systems
Analysis enable us to produce a pictorial model of a
system, which can play much the same role as the model
of a building or oil refinery.

1. THE NEED FOR BETTER TOOLS

Problem #3

The analyst can quickly get overwhelmed with detail, both the detail of the business and the technical detail of the new system. A large part of the time in the analysis phase of the project is spent in acquiring detailed information about the current situation, the clerical procedures, the input documents, the reports produced and required, the policies in effect, and the myriad of facts which are thrown up by such a complex thing as a real business. Unless there is some scheme or structure to organize these details, the analyst (or even a whole team of analysts) can become overloaded with facts and paper. The details are needed, they must be available when required, but the analyst must have tools to control the detail, or he will find he "can't see the wood for the trees." Part of the value of a top-down approach, as we shall see, is that it enables one to look at the big picture, and then home in on the detail of each piece, as and when required.

Problem #4

The document setting out the details of a new system (which may be called variously the System Specification, or General Design, or Functional Specification, or some equivalent name), effectively forms a contract between the user department and the systems development group, yet it is frequently impossible for the users to understand because of its sheer bulk and the technical concepts built into it. It somewhat resembles an old-style insurance policy; the things that are really going to matter in the end are buried in the fine print. Users often make a valiant effort to master these documents, and end up mentally shrugging their shoulders and signing off, saying to themselves, "Well, I guess these computer people know what they're doing." Only when the finished system is delivered do they have something which they can understand and react to, and of course, by then it's too late.

Problem #5

If the specification document *can* be written in such a way as to make sense to users, it may not be very useful to the physical designers and programmers who have to build the system. Often a considerable amount of re-analysis goes on, essentially duplicating the work that the analyst has done, but redefining data and process logic in terms which the programmers can use. Even if the analyst has a technical background, and so writes the specification with an eye to the subsequent ease of programming, he may end up limiting the programmer's freedom of action to implement the system in the best way. The physical design of the files, programs, and input/output methods should be done by someone with up-to-date technical knowledge, based on an understanding of the complete logical requirements of the system. To begin to specify physical design before the logical model of the system has been built is to be "prematurely physical," and too often results in an inferior design.

1.2 HOW MUCH CAN WE BLAME OUR TOOLS?

Even with the best possible analytical tools, some of the problems just discussed will always be with us. No analytical tool will enable analysts to know what is in a user's mind without being told, for instance. Nonetheless, it is the theme of this book that the problems of analysis can be significantly eased with the logical tools we describe; and we identify four limitations of our present analytical tools.

1.2.1 NO "MODEL" IN DP

We have no way of showing a *vivid tangible model* of the system to users. It's hard for users to imagine what the new system is going to do for them until it is actually in operation, by which time it's usually too late. "How do I know what I want till I see what I get?" is the disguised cry of many users. The pictorial tools in this book give the user a better "model" of the system than was possible until now.

1.2.2 ENGLISH NARRATIVE IS TOO VAGUE AND LONG-WINDED

Since we have had no way of showing a tangible model, we have had to do the next best thing, which is to use English narrative to describe the proposed system. Can you imagine spending five years' salary on a custom-built house on the

1. THE NEED FOR BETTER TOOLS

basis of an exhaustive narrative description of how the house will be built? No pictures, no plans, no visits to a similar house, just the 150 page narrative. "The living room, which faces South-South East, will be 27' x 16' at its greatest width, with the western half taking a trapezoidal form, the West wall being 13'4" long, (abutting the northern portion of the East wall of the kitchen)......"

Having spent the money on the basis of the narrative, and not being shown anything until the house is finished, would you be surprised if you were disappointed when you moved in? Is it surprising that users are disappointed with systems when they get them?

If you use English to describe a complex system (or building), the result takes up so much space that it's hard for the reader to grasp how the parts fit together. Worse than that, as we shall see in Chapter 5, English has some built-in problems that make it very difficult to use where precision is needed.

1.2.3 FLOWCHARTS DO MORE HARM THAN GOOD

If we can't make a model and English is too vague and wordy then what about a picture? Unfortunately, up till now the only picture we have had for a system has been the flowchart. Though one flowchart *can* be worth a thousand words, it traps the analyst into a commitment; to use the standard flowchart symbols, (see Figure 1.1), means inevitably that the analyst must commit to a *physical implementation* of the new system. The very act of drawing a flowchart means that a decision must be made as to whether the input will be on cards or through a CRT, which files will be on tape and which on disk, which programs will produce output, and so on. Yet these decisions are the essence of the *designer's* job. Once the analyst has drawn a systems flowchart, what is left for the designer to do? The designer has a choice between accepting the analyst's physical design, and dealing with the details of program and file structure, or (as too often happens), going back to the written specification and producing a new design from that. Neither course is satisfactory. In Fred Brooks' words:

1.2 HOW MUCH CAN WE BLAME OUR TOOLS?

> "The manual (specification) must not only describe
> everything the user does see, including all inter-
> faces; it must also refrain from *describing* what the
> user does *not* see. That is the implementer's busi-
> ness, and there his freedom must be unconstrained.
> The architect (analyst) must always be prepared to
> show an implementation for any feature he describes,
> but must not attempt to dictate the implementation." [1.1]

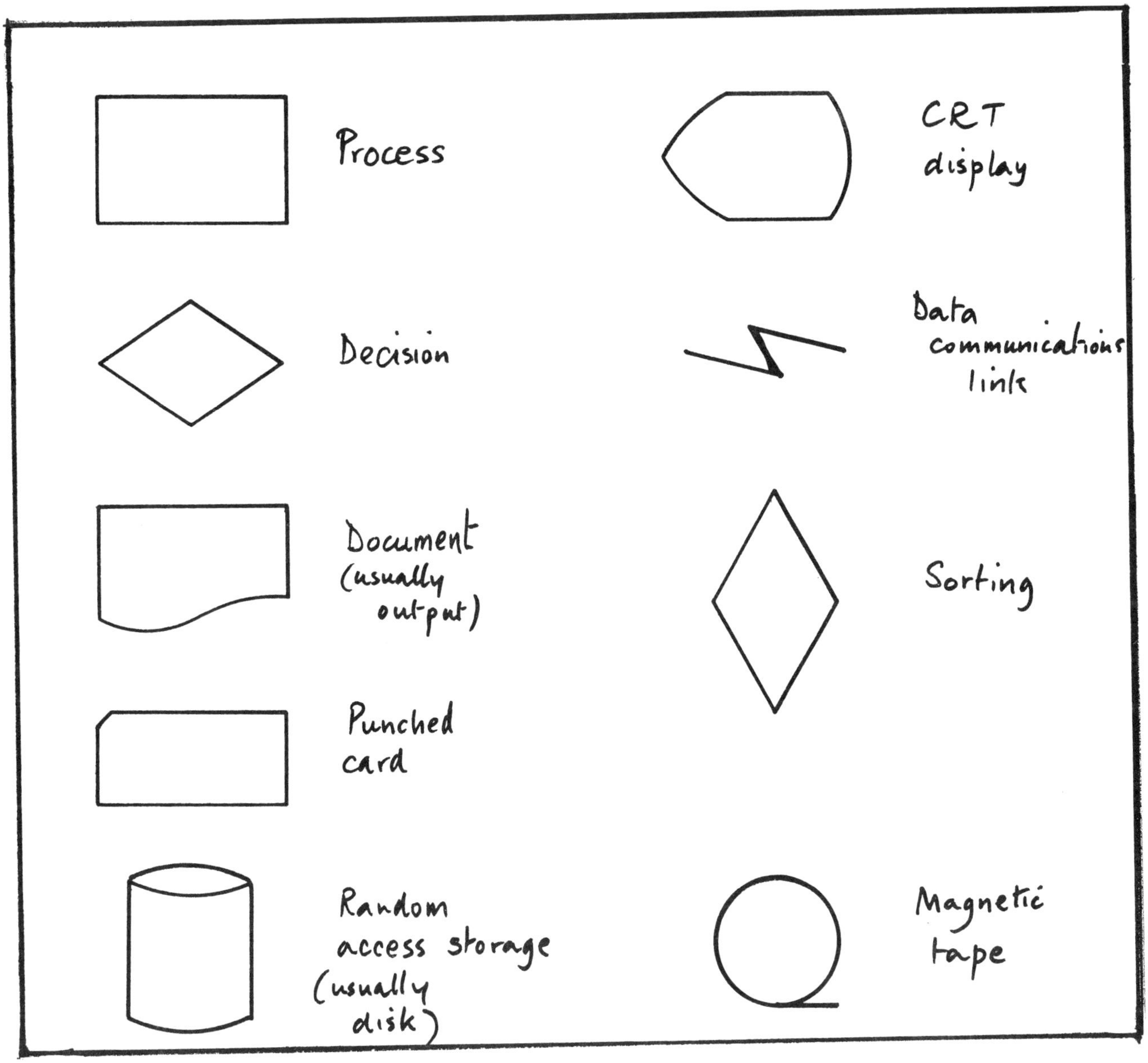

Figure 1.1
Common Conventional Flowchart Symbols (from IBM flowcharting template X20-8020)

1. THE NEED FOR BETTER TOOLS

If the analyst and designer are the same person, drawing the flowchart must be recognized as an act of design, not of analysis. There is a great temptation to sketch a physical design of the new system before one has a full understanding of all the logical requirements; this is what is meant by being "prematurely physical."

Also, once a design is in being, it is much harder to consider alternatives. The flowcharts for an on-line system, and for one which carries out the same logical functions but in batch mode, are very different. The underlying logical similarity is impossible to see. Even worse, the flowchart decision symbol encourages the creator to get into the details of decisions--e.g., what happens to invalid transactions?--and without knowing it, the system overview chart has become a detailed program flowchart. We need the detail, but not in an overview. The analyst desperately needs to be able to see "a map of the forest." It is not as though the flowchart was helpful in "modelling" the system to users. While it is true that some sophisticated users can learn to read flowcharts, to the majority they are "visual jargon."

1.2.4 WE HAVE NO SYSTEMATIC WAY OF RECORDING USER PREFERENCES AND TRADE-OFFS, ESPECIALLY IN TERMS OF IMMEDIATE ACCESS TO DATA.

As we noted, the analyst needs to draw out the user's (often intuitive) preferences for various aspects of the new system. To some users it will not matter that figures are not 100% accurate provided the report is available by 9 A.M. without fail, while other users will be prepared to wait, provided the figures are correct when they come. Obviously, we can make everybody happy, but only at a price. This data on user trade-offs is very important to the designer when weighing the cost-effectiveness of various designs. But unless the analyst has a tool for explicitly recording preferences, the information gets lost. This is a particularly difficult issue where data on file must be immediately accessed in several ways, for example, in the classic purchasing problem where any one vendor supplies many parts, and any one part can be supplied by many vendors. If we set up the file by vendor we can easily produce an immediate answer to the question, "What parts does Vendor A supply?" But what about the question "Who supplies Part 123456?" To give an immediate answer to this question may involve the construction of a secondary index, or some other data base technique. But how important is the second question? Is it vital to the way the user will do business? Or is it just a "nice-to-have," mentioned to the analyst by one manager at the end of an interview and built into the specification without question? In Chapter 7 we will discuss tools for recording and analyzing preferences.

1.3 HOW MUCH DOES THE FUNCTIONAL SPECIFICATION MATTER?

We can use the tools of Structured Systems Analysis described in this book to prepare a functional specification that:

1. is well-understood and fully agreed by the users

2. sets out the logical requirements of the system without dictating a physical implementation

3. expresses preferences and trade-offs.

But so what? What will it buy us? Why don't we just accept that analysis is an inexact art, and that since users are going to change their minds anyhow, we can work with the current methods of specification?

Some compelling evidence has been produced by Barry Boehm [1.2], showing how the cost of correcting an error grows out of all proportion, the later in the system development life cycle it is detected, (see Figure 1.2).

The graph plots the relative cost of fixing up an error --note the logarithmic scale!-- depending on the phase of system development where the error was detected. Thus an error which shows up in the requirements phase--perhaps due to the users examining our pictorial logical model-- might cost 0.2 units (say $200), whereas if that same error is not detected until the system goes into operation, on average it costs more than 10 units ($10,000). Thus the building of a logical model which clearly communicates to users what the systems will and will not do, is a crucially important exercise in terms of the cost of fixing errors later on. Making changes on a piece of paper is cheap, making changes in code is several times more expensive, making changes in a working system is several times more expensive again. We can't afford to wait for users to "see what they get," before they "know what they want."

1. THE NEED FOR BETTER TOOLS

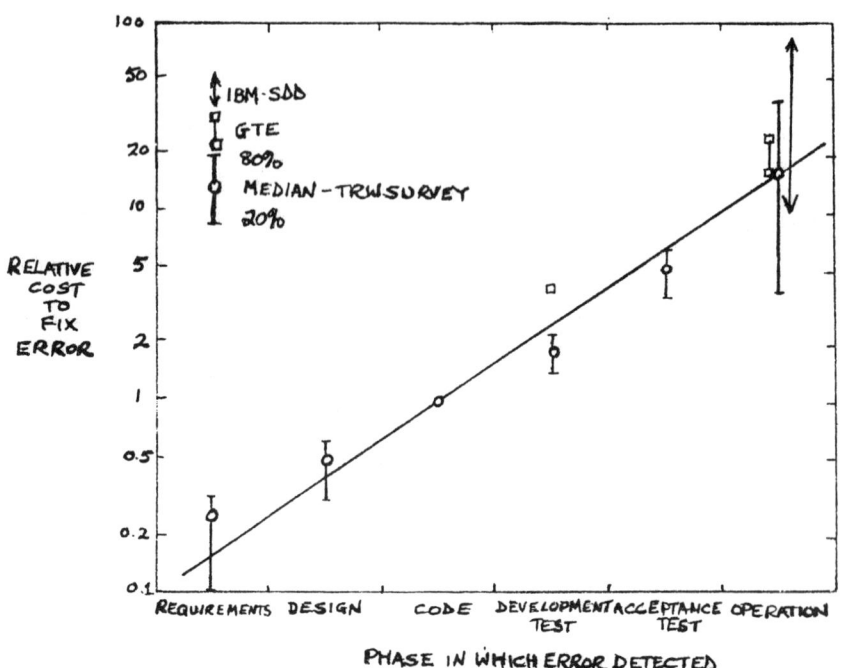

Figure 1.2

REFERENCES

1.1 F.P. Brooks, "*The Mythical Man-Month,*" Addison-Wesley, 1975.

1.2 B. Boehm, "Software Engineering," *IEEE Transactions on Computers*, Dec. 1976.

Chapter 2

What the tools are and how they fit together

Before examining each of the tools of Structured Analysis in detail, we will give an overview of each tool and show their relationship to one another, by seeing how they can be used in a relatively simple analysis situation.

The CBM Corporation (Computer Books by Mail) was recently acquired by a national holding corporation and is now a division. Established twelve years ago, the company's business has been to act as a "Book-jobber" receiving orders from librarians for books about computers, ordering the books from the appropriate publisher at a discount, and filling the order on receipt of the books from the publisher. Invoices are produced by a service bureau computer from forms filled out by CBM; business is currently running at about 100 invoices a day, each with an average of 4 book titles, and an average value per invoice of $50. The new management plans to expand the operation considerably, improving service levels by holding stocks of the 100 most frequently ordered books, and making it possible for all professionals (not only librarians) to order by calling a toll-free number 800-372-6657 (800-DP BOOKS), as well as by mail as at present. This will create problems of credit-checking, of course, and create the need for an inventory control system of some sort. The people who take orders off the phone will need rapid access to a catalog of books to verify authors and titles, and to be able to

2. WHAT THE TOOLS ARE AND HOW THEY FIT TOGETHER

advise callers what books are available on any given topic.

The volume of transactions in the new system will, of course, depend on the acceptance of this new method of ordering, but it is projected to grow to 1000 invoices per day or more, though with a lower average of books per invoice, (since librarians tend to order more at a time than professionals).

A systems analyst has been assigned to this newly acquired division, with the responsibility of investigating and specifying the new system on behalf of the Vice President of Marketing. How can he begin to build a logical model of the required system, without jumping to "prematurely physical" conclusions as to what will be automated, what will be manual, whether the automated systems will be on-line or batch, whether to use the bureau or have an in-house computer, and so on?

2.1 FIRST, DRAW A LOGICAL DATA FLOW DIAGRAM

At the most general level, we can say that, just like the present system, the new system will take book orders, check them against a file of books available, check against some file to see that the customer's credit is okay, and cause the book(s) ordered to be sent out with an invoice.

We can show this in a *logical data flow diagram (DFD)* thus:

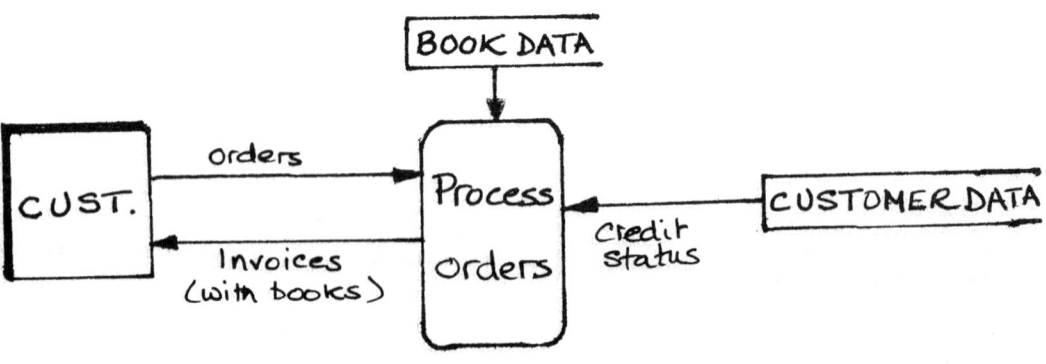

Figure 2.1

2.1 FIRST, DRAW A LOGICAL DATA FLOW DIAGRAM

In this DFD we have used four symbols:

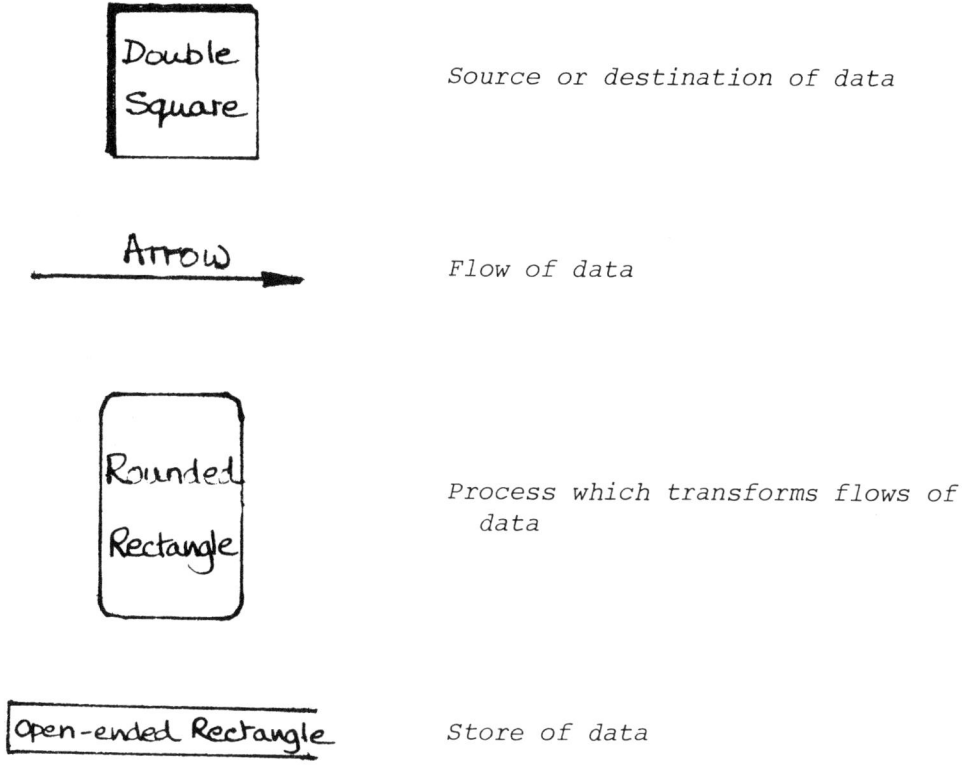

Figure 2.2

These symbols, and the concepts they stand for are at the *logical* level; a flow of data may physically be contained in a letter, or an invoice, in a telephone call, from program to program, via a satellite data link, anywhere data passes from one entity or process to another. A process may physically be a roomful of clerks calculating discounts, a JCL cataloged procedure, or a combination of manual and automated activities. A data store can be a rotary card file, a microfiche, a filing cabinet, a table in core, or a file on tape or disk. Using the four symbols enables us to draw a picture of the system without committing ourselves yet as to how it will be implemented.

Of course, the system pictured in Figure 2.1 is pretty general; at so high a level of abstraction as to be rather useless. Nonetheless, we can use just the four basic symbols to construct a "map of the forest" at any level of detail we choose. Let us expand "Process Orders" to show the logical functions that make up the present system. For starters, we

2. WHAT THE TOOLS ARE AND HOW THEY FIT TOGETHER

can note that incoming orders must be checked to make sure that the details are correctly stated, (that the title and author match, for example). Further, once we have a valid order, we need to batch it together with orders for other books from the same publisher, so that we can get the benefit of quantity discounts.

Figure 2.3 shows what the data flow looks like now:

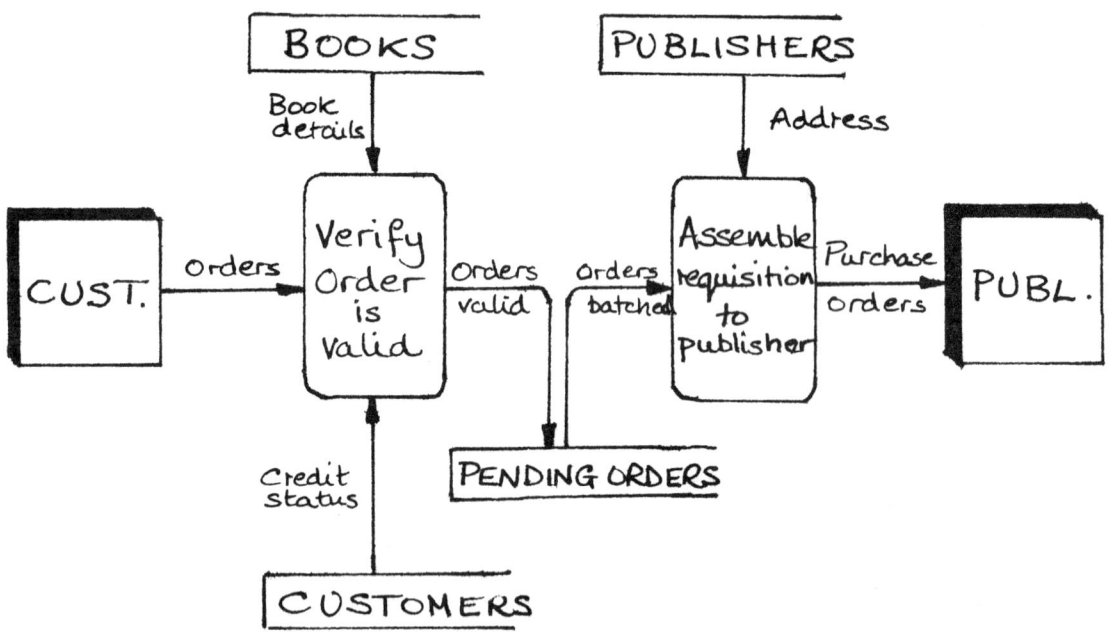

Figure 2.3

We see from this DFD that, as each order is checked, it is put in some store of pending orders, until (according to some logic we don't need to specify yet) a batch of orders can be assembled into a bulk order.

So far, so good: but what about filling the orders, and, hopefully, getting paid for them? Each publisher will send a shipping notice with each shipment, detailing its contents; this must be compared with the order that was placed to make sure that the right numbers of the right titles have been sent to us. Where do we find the details of the orders that we placed? Clearly there must be another data store, perhaps

2.1 FIRST, DRAW A LOGICAL DATA FLOW DIAGRAM

called "Publisher Orders" which can be interrogated. Once we have the right books available, we can assemble and ship orders to our individual customers. Figure 2.4 shows the system with these features added.

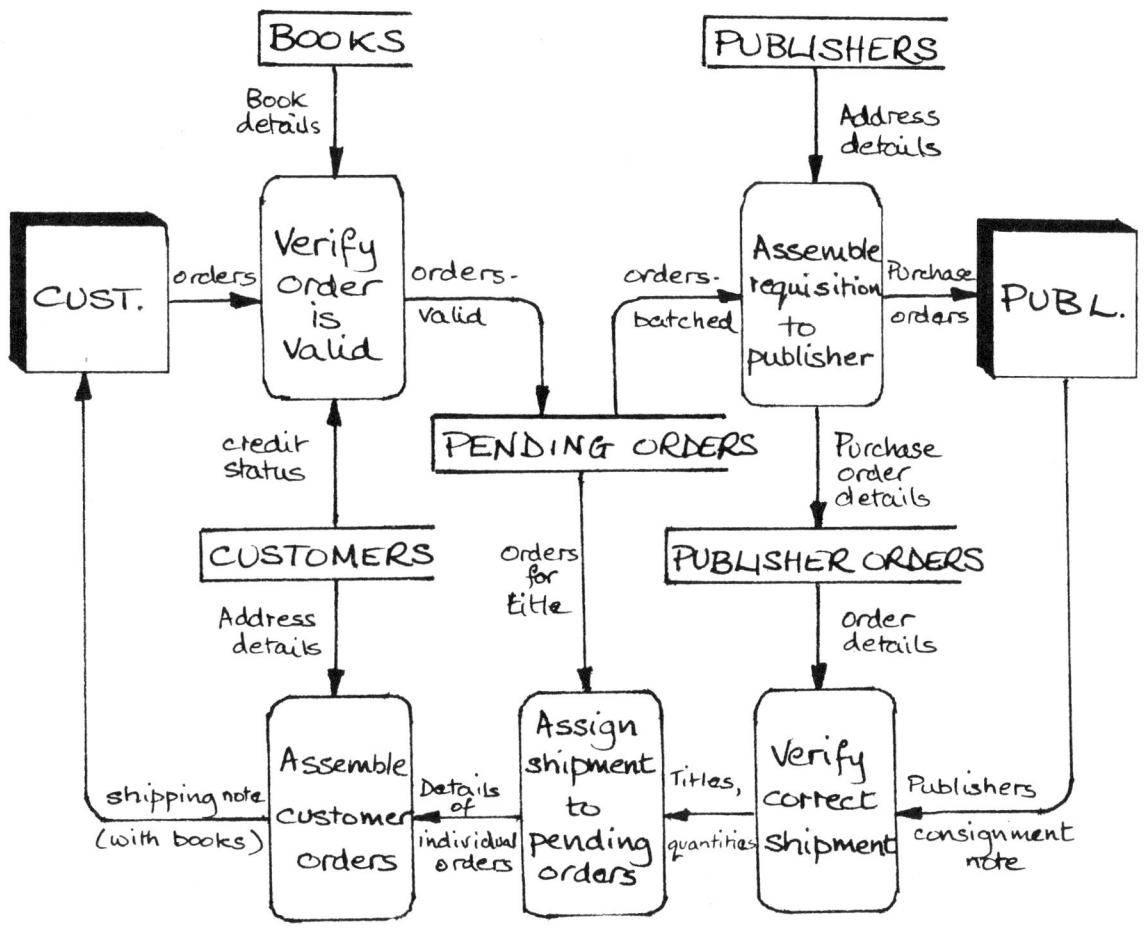

Figure 2.4

Note that we do not show the movement of books themselves; for our purposes, the books are not data and so are not included in a data flow diagram. We will discuss the relationship between a data flow diagram and a materials flow diagram in Chapter 3. For the moment, we are only interested in items, like the shipping notes, which *are* data about the books.

2. WHAT THE TOOLS ARE AND HOW THEY FIT TOGETHER

In Figure 2.4, still no one is getting paid for anything. We need to send out invoices to our customers (at present done by the service bureau, but that's a physical detail), and handle payments from our customers. There being no free lunch, we will in turn be billed by the publishers and have to pay them. Figure 2.5 shows the addition of these financial functions with their associated data stores, commonly known as Accounts Receivable and Accounts Payable.

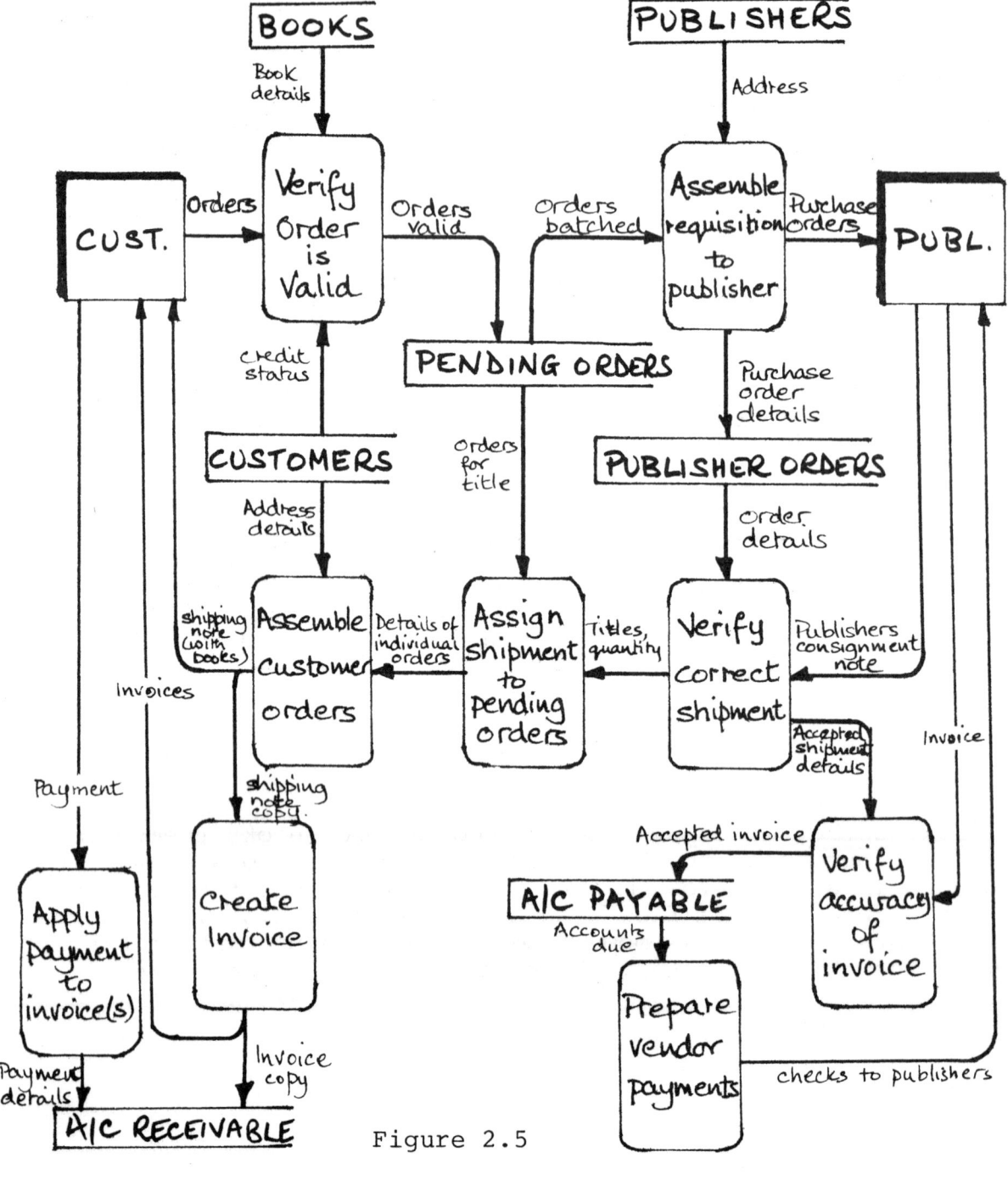

Figure 2.5

2.1 FIRST, DRAW A LOGICAL DATA FLOW DIAGRAM

For the sake of clarity, the logical functions for creation and maintenance of the customer file, the book file, and the publisher file have not been shown, nor do we handle any inquiries. We will deal with these functions in the next chapter.

Each of the process boxes that is shown of course summarizes a lot of detail. Each process box can be "exploded" into a lower level, more detailed, data flow diagram. Figure 2.6 shows the explosion of "Assemble requisition to publisher."

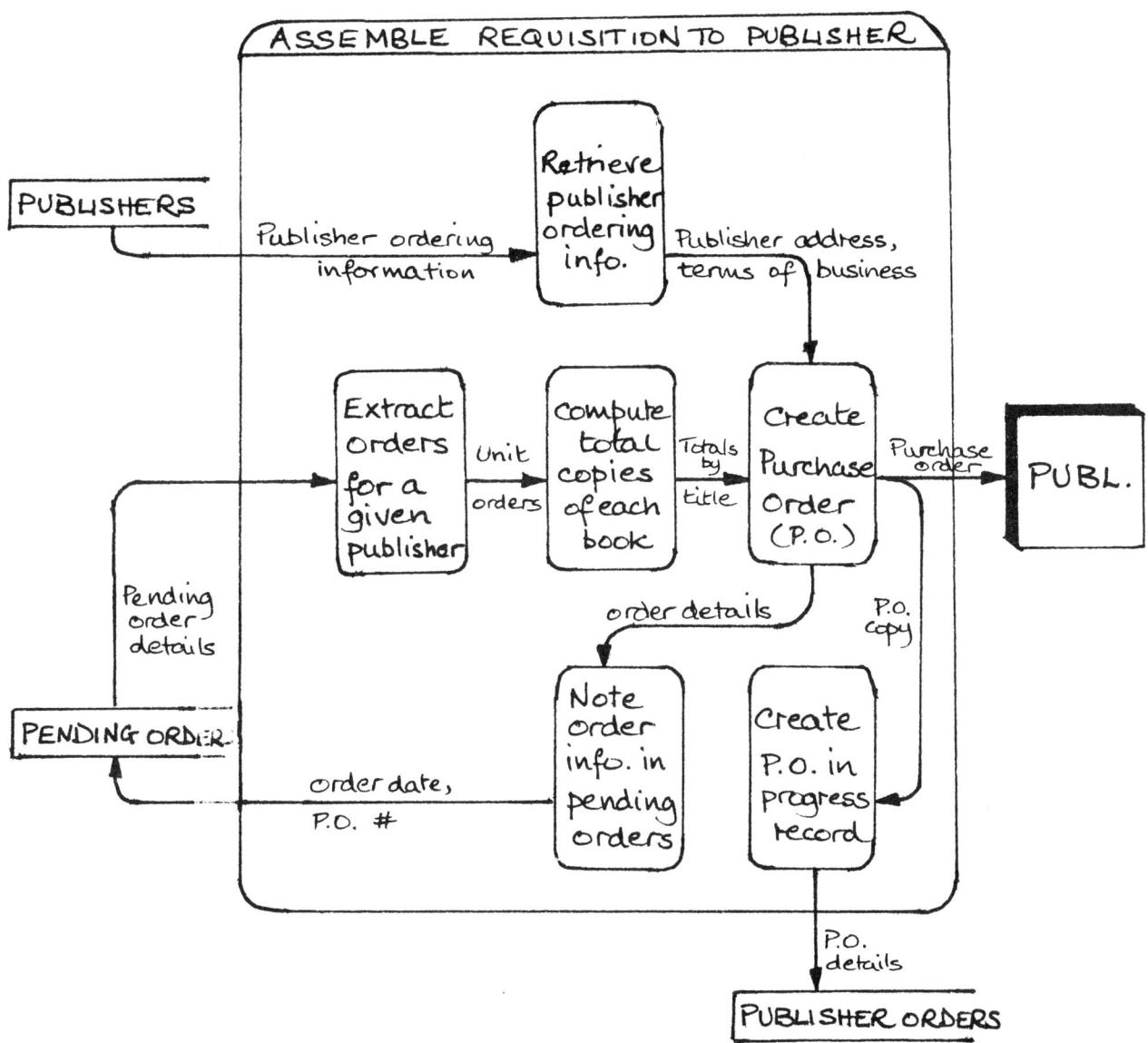

Figure 2.6

2. WHAT THE TOOLS ARE AND HOW THEY FIT TOGETHER

If necessary, each component process box can itself be broken down to a third level of detail. As we shall see in Chapter 3, this is not often necessary. In Chapter 3, we will deal in detail with guidelines for drawing these data flow diagrams, and conventions for handling explosion.

2.1.1 ERROR CONDITIONS

You will notice that we have taken no account of error conditions; we do not specify what happens to an order from a customer who is over his credit limit, or what happens to an invoice from a publisher for a shipment we never received. Of course, we must deal with these circumstances, but keep their treatment to the second and lower level diagrams so that they do not interfere with the "big picture." Without this simplifying rule, it is all too easy to get bogged down in error and exception handling and sink without a trace.

2.1.2 ALTERNATIVE PHYSICAL IMPLEMENTATIONS

Another important point about Figure 2.5 is that since it is a *logical* data flow diagram, it is very easy to envisage several physical implementations. As we know, in the present system, "Create Invoice" and "Apply Payment to Invoice" are automated on the service bureau, and all the other functions are manual. Now that we have a "map of the forest," we can read off different alternative solutions by drawing the system boundary around different processes and data stores. Figure 2.7 shows two such possibilities.

> BOUNDARY 1. Automating order validation, as well as invoicing and accounts receivable, while leaving the batching of orders and all purchasing functions to be done manually. Within this "automation region" we can envisage at least two physical solutions.
>
> > *Boundary 1:Batch.* Orders would be scanned for completeness and keypunched. After editing against the book file and the customer file, they would be added to a pending order file, a confirmation of order

18

2.1 FIRST, DRAW A LOGICAL DATA FLOW DIAGRAM

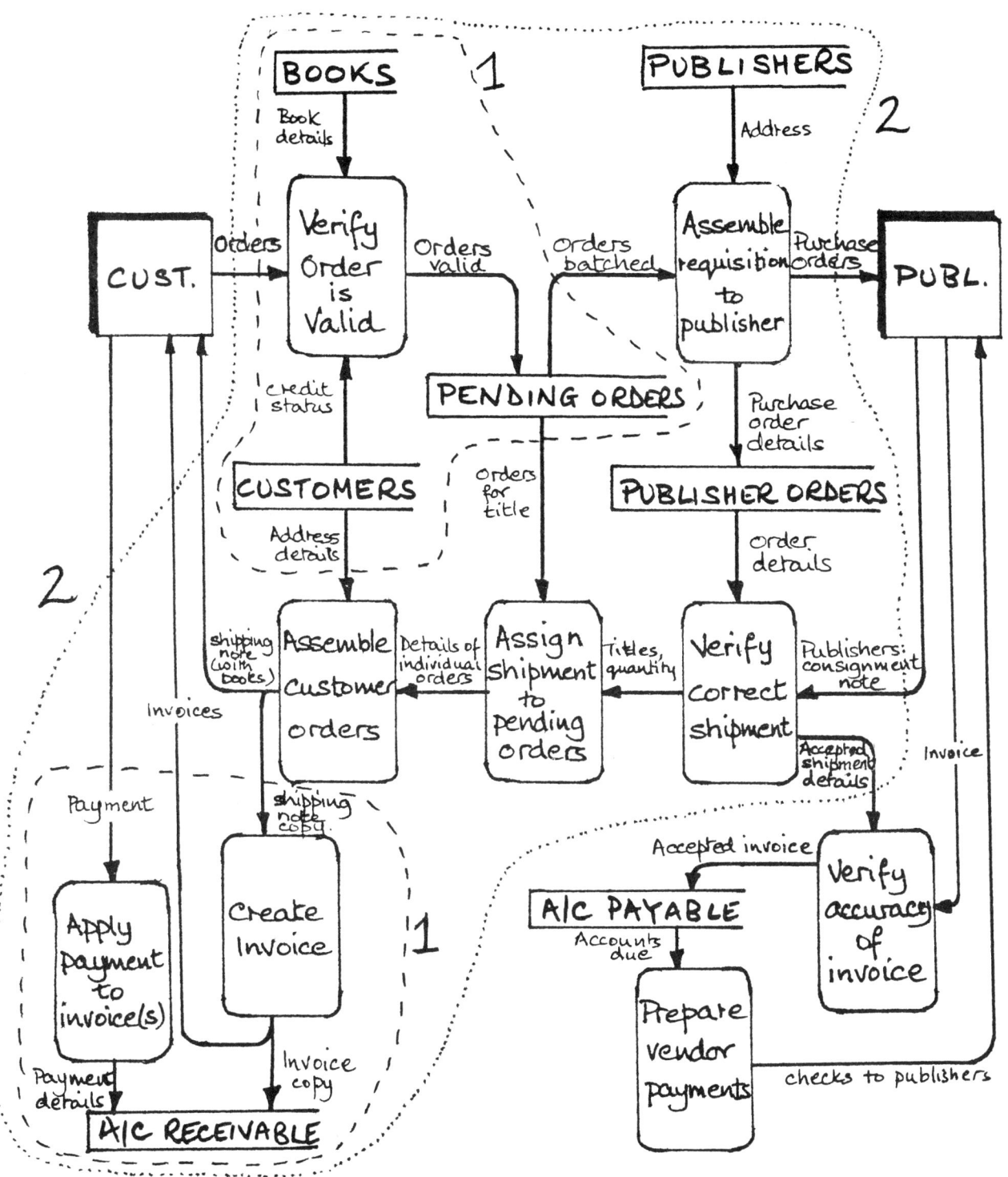

Figure 2.7

2. WHAT THE TOOLS ARE AND HOW THEY FIT TOGETHER

printed, and a printout produced sorted by title within publisher. This would be used as a basis for making up orders to publishers. Shipments and payments would be keypunched and used as input to a conventional accounts receivable system.

Boundary 1:On-line. As orders are received they would be entered via a CRT screen, with the ability to call up all books by a given author, or with a given word or phrase in the title, and with on-line edit against the customer file. Once an order is satisfactorily entered, a confirmation will be printed on the terminal printer, for eyeball checking against the order, and the pending orders file updated.

BOUNDARY 2. Automating the production of purchase orders, and the breaking of bulk shipments into individual orders, as well as order entry and accounts receivable.

Boundary 2:Batch. In addition to the system described for Boundary 1, this more extensive system runs through the pending orders tape each night, extracting those titles whose total orders exceed the economic order quantity, sorting by publisher, printing purchase orders and updating the publisher-orders-in-progress file. When shipments are received, their contents are keypunched and edited against the orders-in-progress file, then run against the pending orders file to create shipping notes for individual customer orders.

Boundary 2:On-line. The additional functions available would be that of immediately matching the contents of a shipment received from a publisher, and generating instructions for breaking that shipment down to individual orders on the spot.

As can be imagined, several families of systems are possible. Which system is the best compromise between cost and benefit depends on factors such as volume, size of bulk order discount, value put on fast turnaround to customers, and so on. The important point is that whatever physical implementation is chosen, the remaining manual functions can be clearly seen, and that the automated plus manual functions will always add up to the same logical data flow diagram.

2.1.3 THE GENERAL SYSTEM CLASS

It should be clear that Figure 2.7, with very little change, would apply to a business distributing autoparts, or cattle-feed, or hospital supplies. Business graduates will recognize it as a member of the class of "distribution operations without inventory;" in other words, any enterprise that takes small orders, combines them into bulk orders to a larger wholesaler or manufacturer, and then breaks bulk to supply customers, without holding stock for off-the-shelf supply.

In Chapter 3, we will develop the data flow diagram for CBM's new system, which is a distribution system *with* inventory.

The reader is invited to consider other *logically* different types of business systems. It is a useful exercise to sketch the data flow diagram for a different type of system, and mark in the automation boundaries for each of the implementations with which you are familiar.

2.2 NEXT, PUT THE DETAIL IN A DATA DICTIONARY

In the data flow diagrams of the previous section, we have given names to the data flows, data stores, and processes, which are as descriptive and meaningful as possible, being short enough to fit on the diagram. As soon as we start to look more closely, for instance to answer the question, "What exactly do you mean by "Orders?," we shall find ourselves getting quickly into great detail, perhaps specifying the layout of the order form, gathering samples, drawing card or tape record layouts, and so on. What we want to do is to stay at the logical level, identifying each of the data elements that

2. WHAT THE TOOLS ARE AND HOW THEY FIT TOGETHER

are present in a data flow, giving them meaningful names, defining each one, and organizing them so that we can easily look up the definition.

What *do* we mean by "Orders"? At the very least an order for a book will carry some identification of the order, the customer's name and address, and the details of at least one book, usually more. So we can begin by giving meaningful names, and showing this breakdown:

```
ORDER
   ORDER-IDENTIFICATION
   CUSTOMER-DETAILS
   BOOK-DETAILS
```

and we can expand these further, with notes:

```
ORDER
   ORDER-IDENTIFICATION
      ORDER-DATE
      CUSTOMER-ORDER-NUM......usually present
   CUSTOMER-DETAILS
      ORGANIZATION-NAME
      PERSON-AUTHORIZING......optional
         FIRST-NAME..........may be initial only
         LAST-NAME
      PHONE
         AREA-CODE
         EXCHANGE
         NUMBER
         EXTENSION...........optional
      SHIP-TO-ADDRESS
         STREET
         CITY-COUNTY
         STATE-ZIP
      BILL-TO-ADDRESS.........If not present, same as
         STREET                  SHIP-TO-ADDRESS
         CITY-COUNTY
         STATE-ZIP
   BOOK-DETAILS...............One or more iterations of this
                                 group of data elements
      AUTHOR-NAME.............One or more iterations of this
      TITLE                      data element

      ISBN....................International Standard Book
                                 Number, (optional)
      LOCN....................Library of Congress Number,
                                 (optional)
      PUBLISHER-NAME..........Optional
```

2.2 NEXT, PUT THE DETAIL IN A DATA DICTIONARY

Note that this data structure for an order tells us nothing about form layout, field size, whether the field is held in packed decimal, editing characteristics etc. While it omits this physical detail, it is precise enough for the analyst and the user to review for errors and omissions. For instance, the user might well say, "If a BILL-TO-ADDRESS is given, we need to record the name of the person to whose attention the invoice is sent," and the analyst can add in another data element.

Each of the data elements referred to in the data structure of ORDER may need to be defined separately. For instance, what do we mean by ISBN, the International Standard Book Number? We need to be able to quickly locate an explanation that the ISBN is a ten digit number, divided into four groups. The first digit, or pair of digits, represents the country, the next group the publisher, the third group the particular book, and the last digit is a check digit. For example:

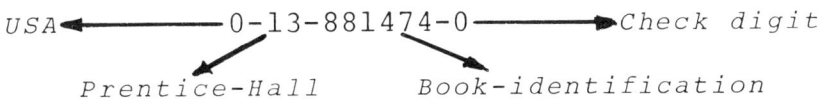

This is the ISBN for the book by Gary E. Whitehouse, "Systems Analysis and Design using Network Techniques" published by Prentice-Hall.

We examine a technique for documenting data element definitions in Chapter 4. For the moment let us note that if we had a definition and explanation of each data flow, the contents of each data store, and of each data element of which they were composed, and if we could arrange those definitions in alphabetical order for easy reference, we would have a *data dictionary* for the system. If anyone wanted to know, "What do you mean by Shipping Notice?" we would look up SHIPPING-NOTICE in the data dictionary under "S" and find its data structure. If, say, we further wanted to know, "What do you mean by METHOD-OF-TRANSPORT?," (one of the data elements of SHIPPING-NOTICE) we would look up METHOD-OF-TRANSPORT under "M" and find its definition and explanation there.

The techniques, implementations, and advantages of data dictionaries are discussed at length in Chapter 4. The most important benefit to us as analysts is that we can describe data flows and data stores by giving a single meaningful name, knowing that all the details for which that name stands are readily available if required.

2. WHAT THE TOOLS ARE AND HOW THEY FIT TOGETHER

2.3 DEFINE THE LOGIC OF THE PROCESSES

With each data element in the system defined, we can begin to explore what is going on inside the processes. What is meant by "Apply payment to invoice," in Figure 2.5, for example? We have seen that each such process can be exploded into lower level processes. Suppose one such lower level process, "Verify Discount," involves checking that the correct discount has been applied. If the analyst inquires, "What is the discount policy?," he may well be shown a memo, or a page in a procedures manual, saying something like:

> *"Trade discount (to established booksellers) is 20%. For private customers and libraries, 5% discount is allowed on orders for 6 books or more, 10% on orders for 20 books or more, and 15% on orders for 50 or more. Trade orders for 20 books and over receive the 10% discount in addition to the trade discount."*

This is a very simple piece of what we shall call "external logic," external in the sense that it concerns itself with business policy, procedures or clerical rules, as opposed to "internal logic" which specifies the way in which the computer will implement them.

Simple though this example may be, policy logic can very quickly get confusing, especially since exceptions and policy changes get dealt with by writing more memos, rather than rewriting the original policy document. The analyst needs tools to picture the structure of the logic and express policies in a comprehensive unambiguous form.

First of these tools is the *decision tree* as shown in Figure 2.8. The branches of the tree correspond to each of the logical possibilities; the way in which the amount of the discount depends on the combination of possibilities is self-evident. As a tool for sketching out logical structure, and for getting the user to confirm that the policy logic expressed is correct, the decision tree is excellent. It is possible to read off the combination of circumstances that leads to each action very directly and clearly.

2.3 DEFINE THE LOGIC OF THE PROCESSES

Type of customer   *Order size*   *Discount*

```
                          20 or more ─────── (20% + 10%)
           Trade ─────┤
                          less than 20 ───── 20%


                          50 or more ─────── 15%
                          20 - 49 ────────── 10%
         Private
           or    ─────┤   6 - 19 ─────────── 5%
         Librarian
                          less than 6 ────── nil
```

Figure 2.8

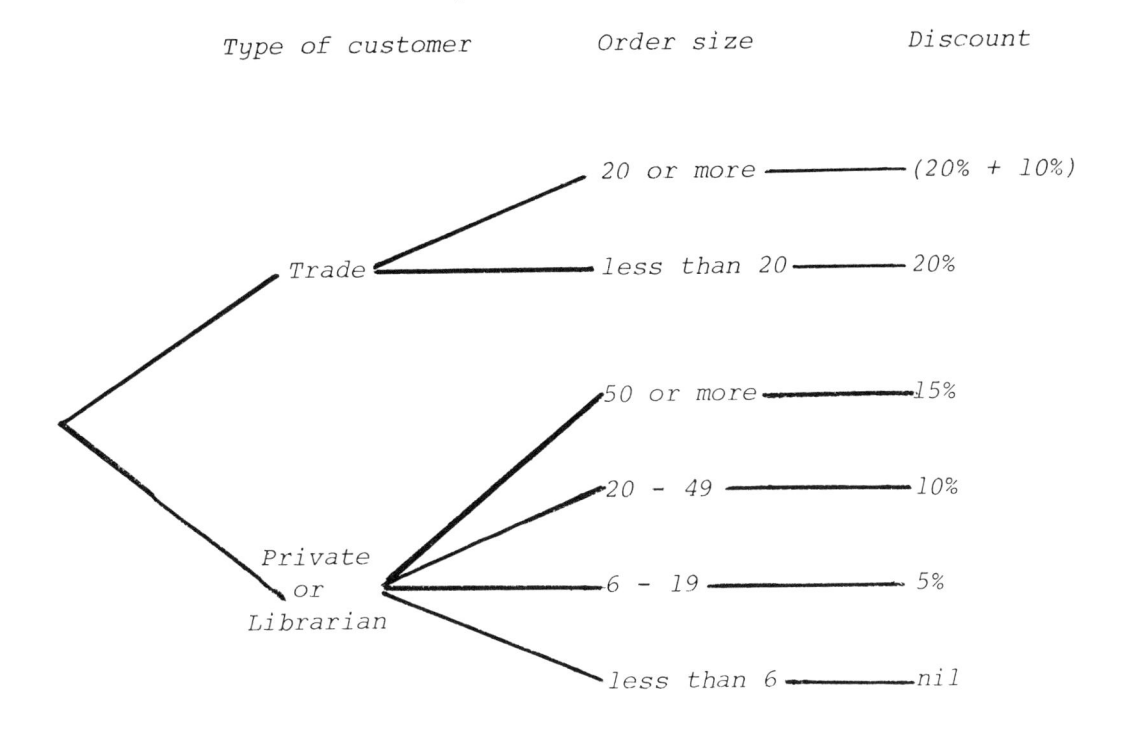

While the decision tree shows the "bare bones" of the decision structure very clearly, it does not lend itself easily to the incorporation of instructions or calculations. If we need to write down the logic as a step-by-step set of instructions, including the decision structure and intermediate calculations or actions (perhaps because we want a clerk to be able to follow it), we may prefer to use a strict logical form of English, as shown in Figure 2.9, on the next page. This type of English has been called *Structured English*, because it uses logical constructs similar to those of Structured Programming. Instructions to carry out actions which involve no decision are written out as imperative sentences ("Add up the total number of volumes...."). Where a **decision must be made**, it is expressed as a combination of IF, THEN, ELSE, SO, with IF and ELSE aligned appropriately to show the structure of the decision.

2. WHAT THE TOOLS ARE AND HOW THEY FIT TOGETHER

```
DISCOUNT POLICY

Add up the total number of volumes on the order (in the quantity column)

IF   the order is from a trade customer
   and-IF    the order calls for 20 or more volumes
        THEN: discount is 30%
      ELSE (order is for less than 20)
         SO: discount is 20%

ELSE (order is from a private customer or librarian)

   so-IF      the order calls for 50 or more volumes
              discount is 15%
      ELSE IF the order is for 20 to 49 volumes
              discount is 10%
      ELSE IF the order is for 6 to 19 volumes
              discount is 5%
      ELSE    (the order is for less than 6 volumes)
          SO:no discount is given
```

Figure 2.9

We can make the Structured English (and the decision tree) more precise and compact by using, where relevant, terms which are defined in the data dictionary. For instance, we could define ORDER-SIZE as a data element which can take up four values:

> SMALL - 5 or fewer volumes
> MEDIUM - 6 to 19 volumes
> LARGE - 20 to 49 volumes
> BULK   - 50 or more

The first part of Figure 2.9 would then read:

> *IF the order is from a trade customer*
> *and-IF ORDER-SIZE is LARGE or BULK*
> *THEN: discount is 30%*
> *ELSE (ORDER-SIZE is MEDIUM or SMALL)*
> *SO:............*

Conventions and notations for decision trees and Structured English, together with other techniques of representing policies and procedures are dealt with in detail in Chapter 5.

2.4 DEFINE THE DATA STORES: CONTENTS AND IMMEDIATE ACCESS

As we built up the data flow diagram, we identified places where data was held from one transaction to the next, or stored permanently because it described some aspect of the world outside the system (for example, the customer's address). The analyst must obviously specify the data elements that are held in each data store, in a similar way to the specification of each data flow as we saw in Section 2.2. Indeed, since data can only get into a data store via some data flow, and cannot get out unless it has been put in, the contents of a data store can be read off from the specification of the incoming and outgoing data flows as we shall show in Chapter 6. The logical contents of each data store are held in the data dictionary under the name of the data store. Just as with a data flow, the individual data elements in the data store are defined elsewhere in the data dictionary. For example, in Figure 2.5, we identify a data store called "BOOKS," which is used in order verification by entering with a title or an author and retrieving book details. What do we mean by "book details?" If we look up BOOK-DETAILS in the data dictionary, we will find the following:

```
BOOK-DETAILS
    AUTHOR-NAME........One or more iterations
    ORGANIZATION-AFFILIATION....University, corporation
    TITLE
    ISBN...............International Standard Book
                        Number, (optional)
    LOCN...............Library of Congress Number,
                        (optional)
    PUBLISHER-NAME.....Optional, see file of abbrev-
                        iations
    PRICE-HARDBACK
    PRICE-PAPER
    PUBLICATION-DATE....May be iterations if editions
```

Clearly, if all these data elements are present in the data flow, they must be present in the data store. The analyst can take this as a starting point, and as the system flow develops, ask himself, "Are there any other data elements which describe books, which should be stored here?" What about WEIGHT-FOR-MAILING, for instance? If the weight of an order could be computed from the weight of the component books the postage could be computed from a table of postage rates,

2. WHAT THE TOOLS ARE AND HOW THEY FIT TOGETHER

and entered as part of the invoice. Should this data element be stored? If so how (at the logical level) will the value of the data element be captured for each new book?

Having established the contents of each proposed logical data store on the data flow diagram, the analyst has two problems to resolve:

1. Are these logical data stores the simplest possible? Can they be combined? Should they be combined?

and

2. What immediate accesses to the data stores will be needed, and how valuable is each type of access?

2.4.1 ARE THE LOGICAL DATA STORES THE SIMPLEST POSSIBLE?

Once the contents of each data store have been identified, they must be compared for similarities and overlaps. Supposing we later identify an inventory data store, what will its contents be? The data elements must include the book's title, probably its author(s) and ISBN, for unique identification, and also elements describing:

> quantity-on-hand
> quantity-on-order
> orders-filled-in-the-last-30-days
> back-orders-during-the-last-30-days
> date-last-order-placed
> average-delivery-time

and other information needed for inventory management.
But this data store contains information about each book, just like the BOOKS data store. So why not combine the two, and keep the inventory information in BOOKS? Depending on the implementation chosen, there are arguments in favor of combining these files, and arguments in favor of keeping them separate, which we discuss in Chapter 9. For instance, if the analyst and designer decide that the BOOKS data store will be a printed catalog of books with sections organized by author, by title, and by subject, it is clearly inappropriate to include fast-changing inventory data. On the other hand, if BOOKS becomes an on-line data base accessible to the clerks taking orders over the telephone, it may make good

2.4 DEFINE THE DATA STORES: CONTENTS AND IMMEDIATE ACCESS

sense to have up-to-date inventory and delivery-date information within the BOOKS file, so that it can be retrieved at the same time as the data needed to verify an order.

Independent of physical considerations like this, it is important in the early stages of analysis that the analyst be able to define the details of each data store's contents, be sure that the definition is comprehensive, and be able to identify similarities and differences between the various data stores in the system. In doing this the "relational" approach--expressing a complex file as several simple "normalized relations"--is useful, and is discussed in detail in Chapter 6.

2.4.2 WHAT IMMEDIATE ACCESSES WILL BE NEEDED?

Much harder, and yet more critical than the task of specifying contents, is the task of deciding what immediate accesses to each data store are needed by the user. For example, let us suppose we have decided to have the BOOKs file available on-line via a CRT. The data element which uniquely identifies each book is the ISBN. But if we make that the primary key of the file, and have immediate access only by that key, it means that we have to know the ISBN of a book in order to display its title, subject-code, publisher, price, etc. This is comparable to having to know a person's insurance policy number in order to find their policy, or having to know the part-number of a part in order to know the price.

We much more frequently want immediate access based on the *name* of something, and this means that the analyst must specify *multiple* access to the file. In this case, the user might quite rightly decide that he wanted to display book details on his CRT screen as a result of keying in either the name of an author (in which case he might get several books), or a subject heading (in which case he would almost certainly get several screens full), or the actual title, or the ISBN. He might further specify that he wanted to have the response back *immediately*, say fast enough to answer a question to someone on the telephone. By immediate access we mean "faster than it is practicable to read the entire file from end to end or to sort it." Obviously the exact meaning of this will vary depending on the size of the file and the power of the computer; on a 370/168 it might be possible to read all through a 100,000 character file in less than a second, whereas on a slower machine some large customer data bases take all weekend to process from end to end.

2. WHAT THE TOOLS ARE AND HOW THEY FIT TOGETHER

The meaning of "immediately" is important to know in each context, because if the user requires to know the answer to a query, *and can wait while we pass or sort the entire file* the designer will not have to build any structure into the file. If the user insists on immediate response, faster than we can pass or sort the file, the designer will have to create some type of index or some structure, and provide some software to enable the user to access the file with the search arguments he wants. For instance, the designer might create a subsidiary author file, with, for each author, the ISBN's of all books he had written. Knowing these ISBN's, the retrieval software would access the main file and display to the user all the books written by the author whose name had been keyed in.

There are numerous techniques for achieving multiple access, and they are reviewed briefly in Chapter 7. No matter how it is done, immediate access costs money in terms of taking more time to develop and maintain, having more software overhead for accesses, and using more disk space. While it is hard to get exact figures, as a rule of thumb one might expect an immediate access to cost up to five times as much as having the same information provided by sorting the file at leisure.

It is therefore very important for the analyst to establish just which accesses are required to be immediate, and further, of those immediate accesses, which are the most valuable to the business. In designing the data base accesses the analyst and designer will go through a series of loops: first trying to satisfy everything the users have asked for, then, putting a cost on the system and if that puts the system over budget, deleting the most expensive or difficult facility, trying a tentative design again, and so on. Right in that last sentence is a key to one of the things that goes wrong, "Deleting the most expensive or difficult facility." If it is too expensive to provide *all* the facilities asked for, the first economy should be to delete the facility of *least value to the user, not* the most difficult or expensive. But how can this be done if the analyst does not know the relative importance of the various accesses?

In this simple case of the BOOKS file, it turns out that people either know the author but are vague about the title, or know the subject topic but are vague about both title and author. Ability to locate a book by ISBN, (while easy to provide since the ISBN is the primary key) is of little value. Location of a book by title alone is useful but of only moderate value. We can create a picture of these immediate accesses in a data immediate access diagram (DIAD) like this:

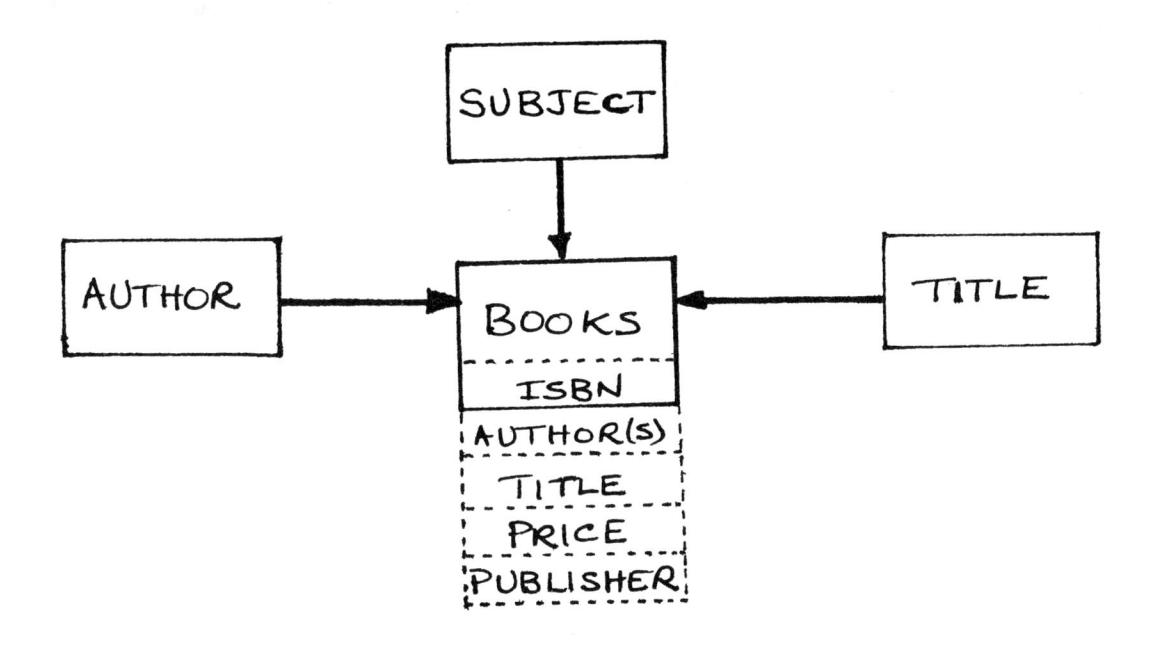

2.4 DEFINE THE DATA STORES: CONTENTS AND IMMEDIATE ACCESS

Figure 2.10

This diagram indicates that a file exists describing books with an ISBN as the primary key, and additional data elements describing attributes of each book such as author, price etc. The fact that ISBN is shown as being the primary key (within the solid rectangle) implies that, "Given an ISBN, show me the author(s) and title," can be answered immediately.

A query such as, "Given a publisher, show me all their titles," can *not* be answered immediately with the accesses defined in the DIAD above. However, the publisher does *appear* in each BOOK record, so it is always possible to sort the file on the publisher field, or to write a program which reads the entire file, extracting book records for any given publisher. (It is true that part of the ISBN is a publisher code, but the inquirer may not always know that code.)

On the other hand, the DIAD shows that the queries: "Given an author name, show me the books written by that author," and "Given a title, show me all the books with that title," *can* be answered immediately, even though "author" and "title" are

2. WHAT THE TOOLS ARE AND HOW THEY FIT TOGETHER

attributes of BOOKS and fields within the BOOKS record. The fact that we have shown separate rectangles in the DIAD, with arrows pointing to BOOKS, implies that we will allow immediate access to the file via separate "author" and "title" indexes, or by some rapid search mechanism.

The "subject" access is quite clear-cut; SUBJECT is not an attribute of BOOKS, it is a quite separate index into the file, and by including it in the immediate access diagram, we are telling the data base designer that, somehow, this facility must be designed in.

We can use the immediate access diagram to record the information that we need about user preferences and the value of each access in the business. Figure 2.11 shows the results of questioning three managers about the relative value of the various accesses to them and their people. The question asked was "Given that you could get this query answered tomorrow morning for $10, what would you pay to have it answered immediately (while a customer is on the phone)?" The three managers' preferences are coded in Figure 2.11 as "M1," "M2," and "M3."

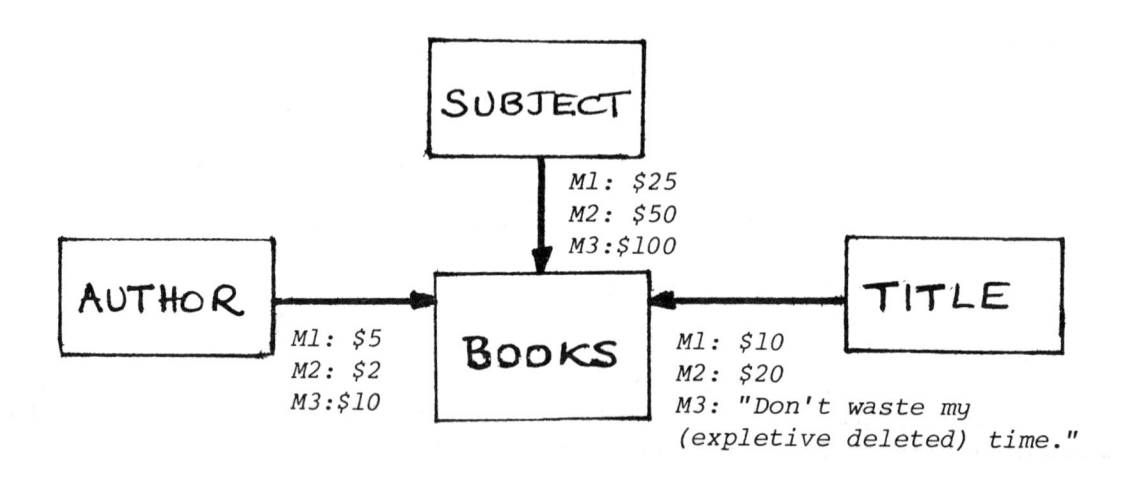

```
                    ┌─────────┐
                    │ SUBJECT │
                    └────┬────┘
                         │ M1: $25
                         │ M2: $50
                         ▼ M3: $100
┌────────┐          ┌─────────┐          ┌───────┐
│ AUTHOR │─────────▶│  BOOKS  │◀─────────│ TITLE │
└────────┘          └─────────┘          └───────┘
  M1: $5              M1: $10
  M2: $2              M2: $20
  M3: $10             M3: "Don't waste my
                           (expletive deleted) time."
```

Figure 2.11

2.5 USING THE TOOLS TO CREATE A FUNCTIONAL SPECIFICATION

If you were the analyst in conference with the designer, and you were told that only two of the three accesses were economically feasible, which one would you give up?

The conventions for representing a DIAD, and the interviewing techniques, are dealt with in more detail in Chapter 7. We also want to be able to provide the designer with information on the volume or frequency of these immediate accesses. We may want to gather security information as well, (e.g. who is allowed to have access to a person's salary history?).

The reader may find it useful to draw up a DIAD for some immediate access system with which he is familiar, and then draw the diagram which would, in his opinion, satisfy all the requirements of all the users, comparing the two.

2.5 USING THE TOOLS TO CREATE A FUNCTIONAL SPECIFICATION

To summarize this overview chapter, the logical data flow diagram shows the sources and destinations of data (and so by implication the boundaries of the system), identifies and names the logical functions, identifies and names the groups of data elements that connect one function to another, and the data stores which they access. Each data flow is analyzed, its structures and the definitions of its component data elements stored in the data dictionary. Each logical function may be broken down into a more detailed data flow diagram. When it is no longer useful to subdivide logical functions, their "external" business logic is expressed using decision trees, Structured English or other tools. The contents of each data store is analyzed and stored in the data dictionary. The relative value to users of various immediate access paths to data in data stores is analyzed and expressed in a data immediate access diagram. Figure 2.12 shows these relationships.

These documents make up a comprehensive account of a system, whether it be the existing system in an enterprise, or the "new" system that is to be built. When the complete package is prepared for a new system, we have a *logical functional specification*, a detailed statement of what the system is to do, which is as free as possible from physical considerations of how it will be implemented. This type of functional specification gives the designer a clear idea of the end results

2. WHAT THE TOOLS ARE AND HOW THEY FIT TOGETHER

OVERALL DATA FLOW DIAGRAM (Chapter 3)

"Explosion" of each process

DETAILED DATA FLOW DIAGRAM (Chapter 3)

Immediate access analysis

Content and structure of data flows

External logic of processes

IMMEDIATE ACCESS DIAGRAM (Chapter 7)

DATA DICTIONARY (Chapter 4)

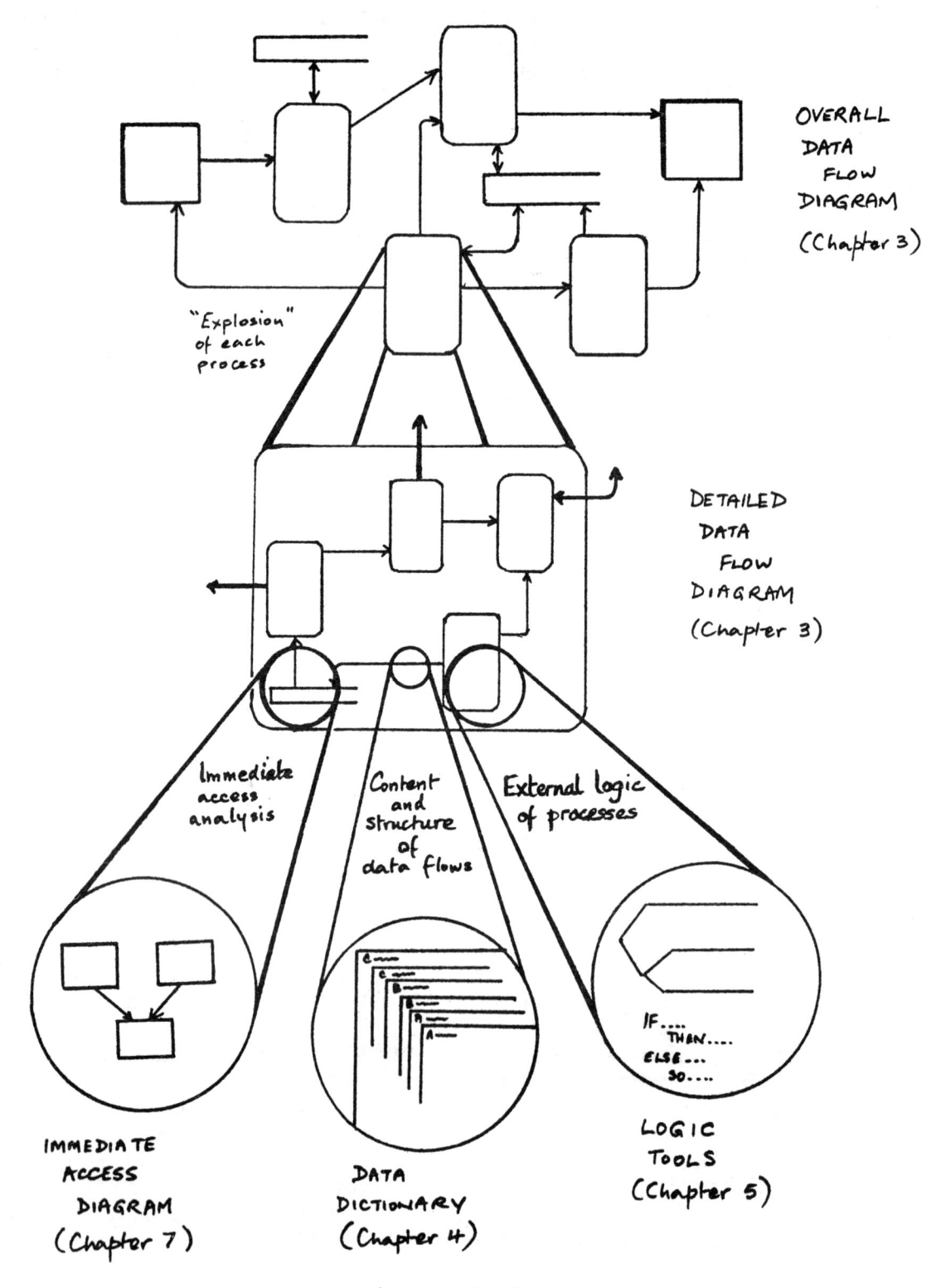

LOGIC TOOLS (Chapter 5)

Figure 2.12

2.5 USING THE TOOLS TO CREATE A FUNCTIONAL SPECIFICATION

he has to achieve, and written evidence of the user's preferences where trade-off judgements may have to be made. At this point the analyst may hand over prime responsibility for the technical work to a designer and continue with the other aspects of the project (such as planning the testing and acceptance of the system). If the analyst and designer are the same person, this person now "changes hats," and looks at the logical functional specification with the eye of a designer, secure in the knowledge that it represents a firm baseline of requirements from which to design the system.

Often, as we indicated, the analyst and designer will go through several iterations of setting up trial physical designs, working out their technical and cost implications, changing some features, and trying again. We will discuss this *tentative design* process, and the user's involvement in it, in Chapter 8.

2. WHAT THE TOOLS ARE AND HOW THEY FIT TOGETHER

EXERCISES AND DISCUSSION POINTS

1. List as many different physical methods of implementing a given data flow as you can.

2. List as many different physical methods of creating a data store as you can.

3. Explode each of the processes shown in Figure 2.5 to the same level of detail shown in Figure 2.6. Include any error and exception conditions that seem likely.

4. Adapt the data flow of Figure 2.5 to represent the Friendly Feed Company, which takes orders from farmers for cattle and hog feed, bulk orders from the mill, and then breaks the bulk shipment down for the individual farmers.

5. How many logically different types of enterprises (such as distribution without inventory, distribution with inventory, manufacturing to order, manufacturing with inventory) can you define?

6. Take a complex policy statement and draw a decision tree to indicate its logical structure.

7. What is the largest machine-readable file you personally are aware of? How long does it take:

    a) to search it

    b) to sort it?

8. What in your view, is the difference between analysis and design? Do the analysts you know only do analysis?

Chapter 3

Drawing up data flow diagrams

The idea of a chart for representing the flow of data through a system is not new; Martin and Estrin [3.1] proposed a "program graph" in 1967 in which the data flows were represented by arrows and the transforms by circles. A similar notation has been used in the data flow approach to program design which is part of the discipline of Structured Design [3.2]. Whitehouse [3.3] describes a similar flowgraph notation for modelling mathematical systems. Ross [3.4] has described a very general graphical approach to systems analysis which comprehends data flow as one of its aspects. What *is* new is the recognition that the data flow diagram at the logical level is the key tool for understanding and working with a system of any complexity, together with the refinement of the notation for use in analysis.

In analysis, as we saw in Chapter 2, we need to recognize external entities and data stores, as well as data flows and transforms or processes. In order to represent our logical system fully, we need to add symbols to the simple program graph. Also, since we need to describe our transforms or processes clearly and it is hard to get much legible writing inside a circle, we adopt a rectangle with rounded corners as the process symbol.

3. DRAWING UP DATA FLOW DIAGRAMS

3.1 SYMBOL CONVENTIONS

Let us examine the symbol conventions in detail.

3.1.1 EXTERNAL ENTITY

External entities are most usually logical classes of things or people which represent a source or destination of transactions, e.g., Customers, Employees, Aircraft, Tactical Units, Suppliers, Taxpayers, Policy-holders. They may also be a specific source or destination, e.g., Accounts Department, IRS, Office of the President, Warehouse. Where the system we are considering accepts data from another system or provides data to it, that other system is an external entity.

An external entity can be symbolized by a "solid" square, with the upper and left sides in double thickness to make the symbol stand out from the rest of the diagram. The entity can be identified by a lower-case letter in the upper left-hand corner for reference.

To avoid crossing data flow lines, the same entity can be drawn more than once on the same diagram; the two (or more) boxes per entity can be identified by an angled line in the bottom right-hand corner, as shown in Figure 3.1.

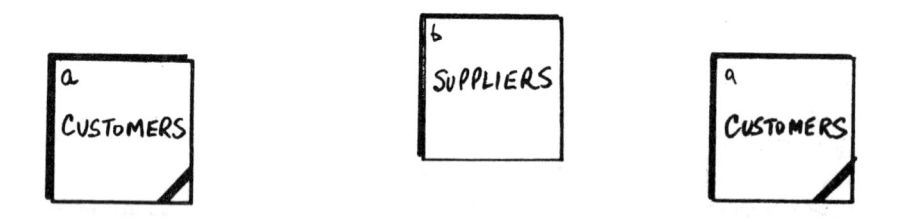

Figure 3.1

Where another entity is to be duplicated, the instances of it have two angled lines, and so on, see Figure 3.2.

38

3.1 SYMBOL CONVENTIONS

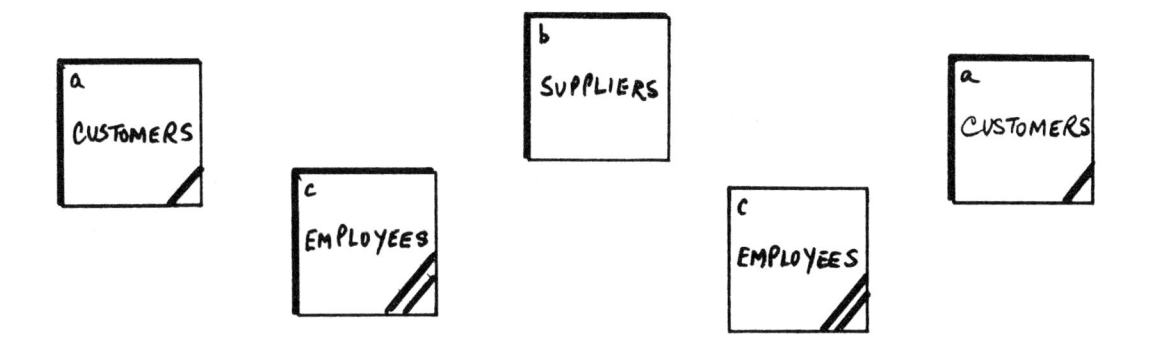

Figure 3.2

By designating some thing or some system as an external entity, we are implicitly stating it is outside the boundary of the system we are considering. As the analysis proceeds, and we learn more about the user's objectives, we may take some external entities and bring them into our system data flow diagram, or alternatively, take some part of our system function and remove it from consideration by designating it all as an external entity with data flows to and from it.

3.1.2 DATA FLOW

This is symbolized by an arrow, preferably horizontal and/or vertical, with an arrowhead showing the direction of flow. For clarity, and especially in early drafts of the diagram, a double-headed arrow may be used in place of two arrows, where data flows are paired.

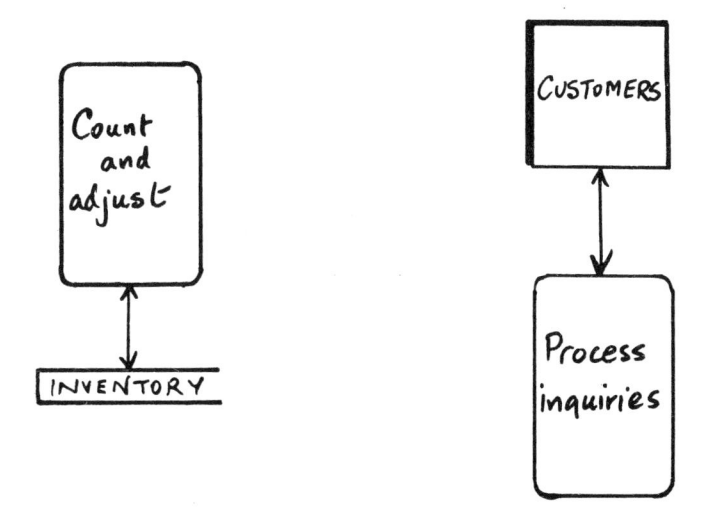

Figure 3.3

3. DRAWING UP DATA FLOW DIAGRAMS

Each data flow is to be thought of as a pipe, down which parcels of data are sent. The data flow pipe may be referenced by giving the processes, entities, or data stores, at its head and tail, but each data flow should have a description of its contents written alongside it. A description should be chosen which is as meaningful as possible to the users who will be reviewing the data flow diagram, (consistent with being a logical description). In early versions, the data flow description should be written on the diagram in upper and lower case letters.

Data flow reference: 29-c

Data flow description: Sales reports

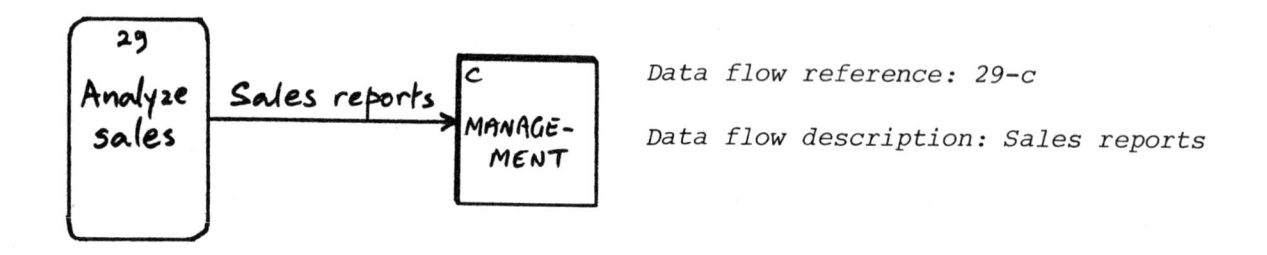

Figure 3.4

At a later stage of analysis, when the data dictionary contents have been defined, the description can be changed to all upper-case letters to show that it has been entered into the data dictionary. Frequently we find that several different "parcels" of data travel along the same data flow. For example, if we look up SALES-REPORTS in the data dictionary we may find its components listed as:

    SALES-REPORTS-BY-DAY
    SALES-TREND-ANALYSIS
    SALESMAN-PERFORMANCE-ANALYSIS
    SALES-BY-PRODUCT-ANALYSIS

Each of these components is in turn described in the data dictionary; each will contain a somewhat different assemblage of data elements.

Especially when drafting a data flow diagram, it is acceptable to leave out the description if it will be self-evident to the reviewer, but the creator of the diagram must be able to supply a description at all times. Figure 3.5 shows a data flow needing no description.

3.1 SYMBOL CONVENTIONS

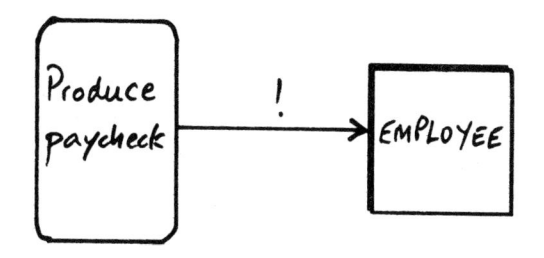

[Figure 3.5: "Produce paycheck" process connected to "EMPLOYEE" entity]

Figure 3.5

Occasionally it is difficult to find a description which adequately characterizes the contents of a data flow. For instance, customers may send in orders, payments, returns of damaged merchandise, inquiries, complaints, etc. It is unwieldy to draw:

[Figure 3.6: CUSTOMERS sending Orders, Payments, Returns, Inquiries, Complaints to "Route transactions" process, which outputs Orders, Payments, Returns, Inquiries, Complaints]

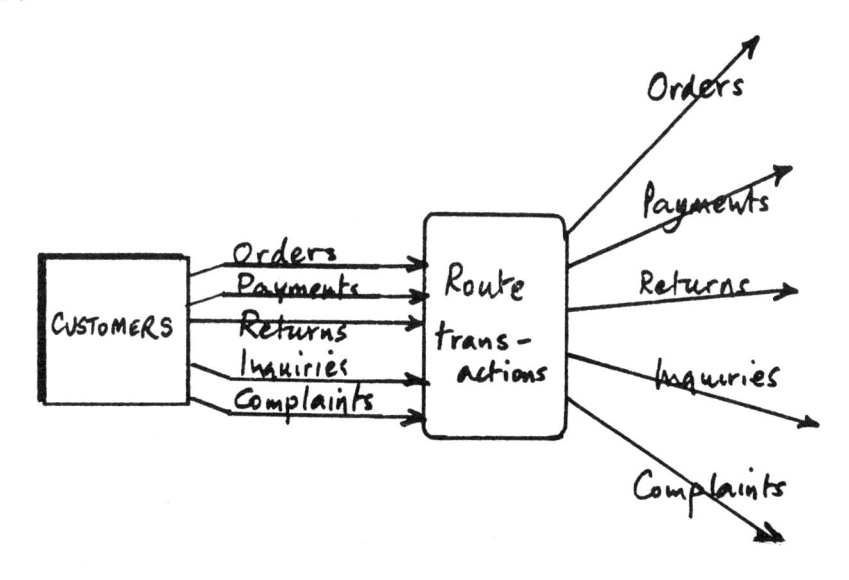

Figure 3.6

There are two ways out of such a situation. If the most important logical fact is that there is only a single data flow (perhaps to a sales office) and that the function "route transactions" is an important one, then the approach should be to lump the contents together under a necessarily vague term, such as "transactions from customers" or "management reports". The contents of this data flow can either be found in the data dictionary or by examining the output of the routing function. Figure 3.7 shows an example of this solution.

3.   DRAWING UP DATA FLOW DIAGRAMS

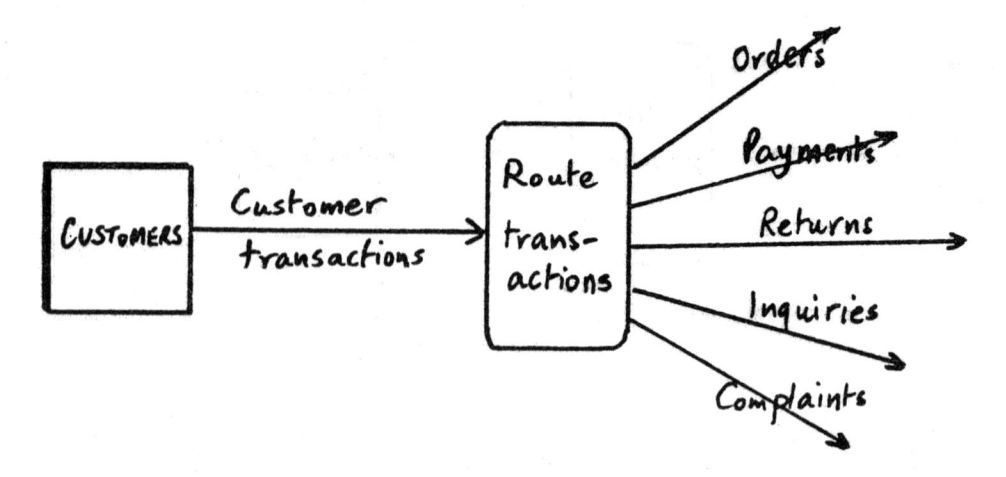

Figure 3.7

The second approach can be used where the routing function is trivial and each transaction gets processed in a different way (and indeed consists of different data elements). In this case, a different arrow can be drawn for each different type of transaction, each going to a different process box:

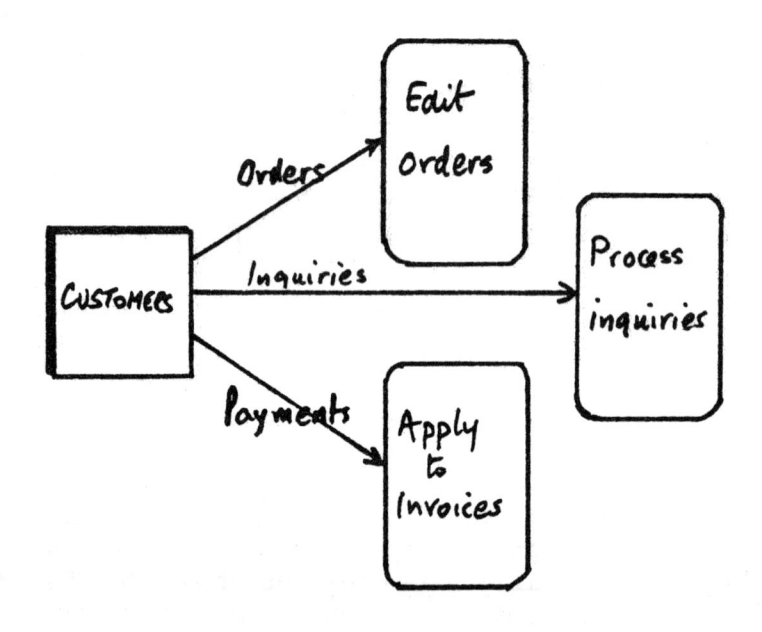

Figure 3.8

The description and contents of data flows in the data dictionary is dealt with in detail in Chapter 4.

3.1 SYMBOL CONVENTIONS

3.1.3 PROCESS

We need to describe the function of each process and, for easy reference, give each process a unique identifier, possibly tying it back to a physical system. Processes can be symbolized by an upright rectangle, with the corners rounded, optionally divided into three areas:

Identification

Description of Function

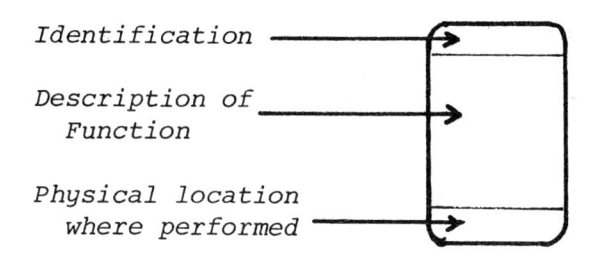

Physical location where performed

Figure 3.9

The *identification* can be a number, initially allocated approximately left to right across a data flow diagram, but having no meaning other than to identify the process. There is no point in assigning meaning to process numbers, since some processes will get split into two or more processes, (implying that new numbers have to be assigned) or several processes will get amalgamated into one (implying that numbers disappear) during the work of analysis. Once assigned, the process identification should not be changed, except for splitting or amalgamation, since it serves as a reference for the data flows and for the decomposition of the process to lower levels, as described in Section 3.2. For clarity, the fine lines dividing identification and description may be omitted, especially when the diagram is to be shown to users.

The *description of function* should be an imperative sentence, ideally consisting of an active verb (extract, compute, verify) followed by an object or object clause, the simpler the better, for example:

```
Extract - monthly sales
Enter - new customer details
Verify - customer is credit-worthy
```

3. DRAWING UP DATA FLOW DIAGRAMS

If in doubt, it helps to think of the function description as though it were an "order to a dumb clerk". If the description would be unambiguous to a clerk, and if you can envisage the function being carried out in a simple clerical circumstance in 5 - 30 minutes, you probably have a good function description.

Generally speaking, when you find yourself using as verbs "process," "update," or "edit," it means that you probably do not understand much about that function yet, and that it will need further analysis.

"Create," "produce," "extract," "retrieve," "store," "compute," "calculate," "determine," "verify," are all active, unambiguous verbs. "Check" can be used but may lead to confusion with the noun in accounting or financial areas. "Sort" implies that a physical solution has been chosen, since sorting is merely a physical rearrangement of the sequence of records in a file, and has no *logical* value.

Note that these imperative sentences **have** no subject; as soon as a subject is introduced, (e.g., "Sales administrator extracts monthly sales") a physical commitment has been made as to how the function will be carried out. It may be helpful when studying an existing system, to note which department, or which program, carries out a function. Similarly, when the analysis is complete, and the physical design of the new system is underway, it is convenient to be able to note how the function will physically be done. This is the purpose of the optional lower part of the process box; to carry a *physical reference,* thus:

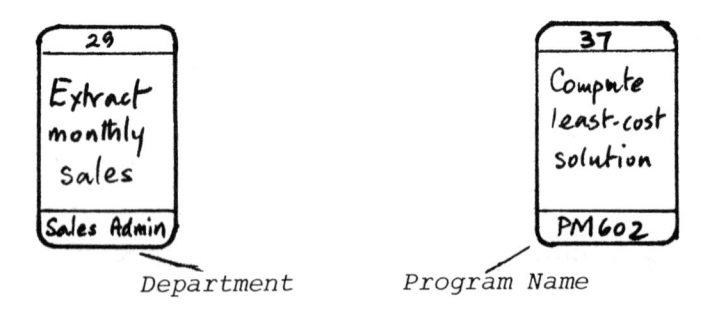

Figure 3.10

In this way, the logical function description and the physical implementation can be kept separate.

3.1.4 DATA STORE

Without making a physical commitment, we find during analysis that there are places where we need to define data as being stored between processes. Data stores can be symbolized by a pair of horizontal parallel lines, closed at one end, preferably just wide enough to hold the name, (¼" on an unreduced diagram). Each store can be identified by a "D" and an arbitrary number in a box at the left-hand end, for easy reference.

| D3 | ACCOUNTS RECEIVABLE |

The name should be chosen to be most descriptive to the user.

To avoid complicating a data flow diagram with crossed lines, the same data store can be drawn more than once on the same diagram, identifying duplicate data stores by additional vertical lines at the left.

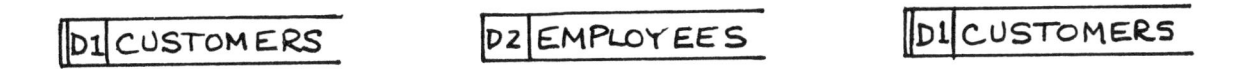

| D1 | CUSTOMERS | D2 | EMPLOYEES | D1 | CUSTOMERS |

When a process *stores* data, the data flow arrow is shown going *into* the data store, conversely, where a data store is *accessed in a read-only manner*, it is enough to show the group of data elements retrieved on the data flow coming out, thus:

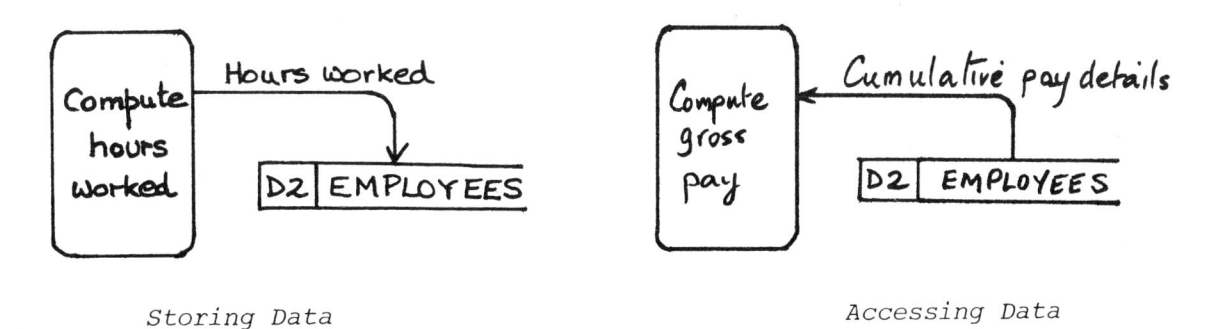

Storing Data   *Accessing Data*

Figure 3.11

3. DRAWING UP DATA FLOW DIAGRAMS

If it is necessary to specify the *search argument*, this may be shown on the opposite side of the data flow to the description; an arrowhead indicates that the search argument is passed to the data store from the process, as shown in Figure 3.12.

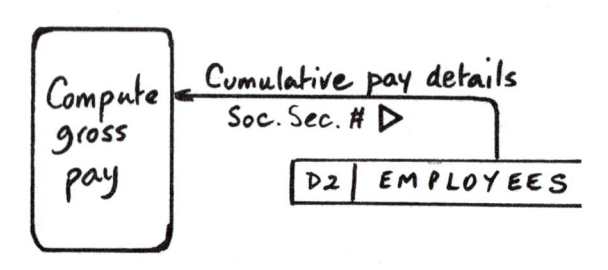

Figure 3.12

3.2 EXPLOSION CONVENTIONS

As we showed in Chapter 2, each process in the top-level data flow diagram for a system can be "exploded" to become a data flow diagram in its own right. Each process at the lower level will need to be related back to the high-level process. This can be done by giving the lower level process box an identification number which is a decimal of the high-level process box, i.e. 29 is decomposed into 29.1, 29.2, 29.3, etc., and, should it be necessary to go to a third level, 29.3 is decomposed into 29.3.1, 29.3.2, and so on.

The clearest representation of the explosion process is to draw lower level data flow diagrams within the boundary which represents the higher level process box. Obviously all data flows into and out of the higher level process box must enter and leave the boundary. Data flows which are shown for the first time at the lower level, such as error paths, can also leave the boundary. Where they do so, they can be high-lighted with an "X" at the exit point. Data stores are only shown within the boundary if they are created and accessed by this process, and by none other. For example, Figure 3.13 is an extract from Figure 2.5:

46

3.2 EXPLOSION CONVENTIONS

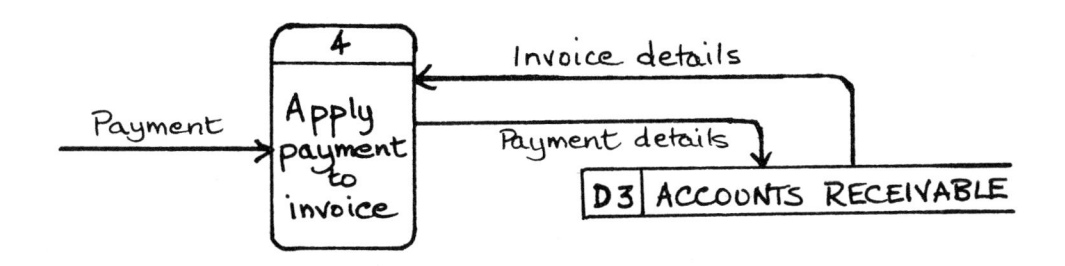

Figure 3.13

The explosion of the process to the lower level is shown in Figure 3.14. This "explosion" illustrates a number of graphical points:

1. Data stores which are "external" to the process being exploded can intrude on the boundary, if drawing them that way simplifies the diagram. D3-ACCOUNTS RECEIVABLE is really outside APPLY PAYMENT TO INVOICE, and has been drawn half-in half-out, for clarity. The three data flows, "D3-4.4 Invoice details," "D3-4.3 Invoice details," and "D3-4.9 Payment record," all cross the boundary, and "match" the flows in Figure 3.13.

2. Data store D4/1, UNTRACED PAYMENTS, exists only for the internal purposes of this process, and so is shown inside. If it were accessed by any process not part of 4 APPLY PAYMENT TO INVOICE it would have to be shown as external (like D1 CUSTOMERS). We have used a convention, D4/1, to indicate the first internal data store in process 4. It has no necessary relationship to any data store called D4 in the parent diagram.

3. External entities are not shown within the boundaries of explosions, even if, like BANK in this case, they may not be involved in any other process.

4. Where it is inevitable that one data flow has to cross another, we use the "little hoop" convention,

   as exemplified by "4.4-4.7 Invoice + payment details."

5. In the rare case when a data flow needs to cross a data store, as with "4.9-4.8 Payment details," it is also "hooped," as shown.

3. DRAWING UP DATA FLOW DIAGRAMS

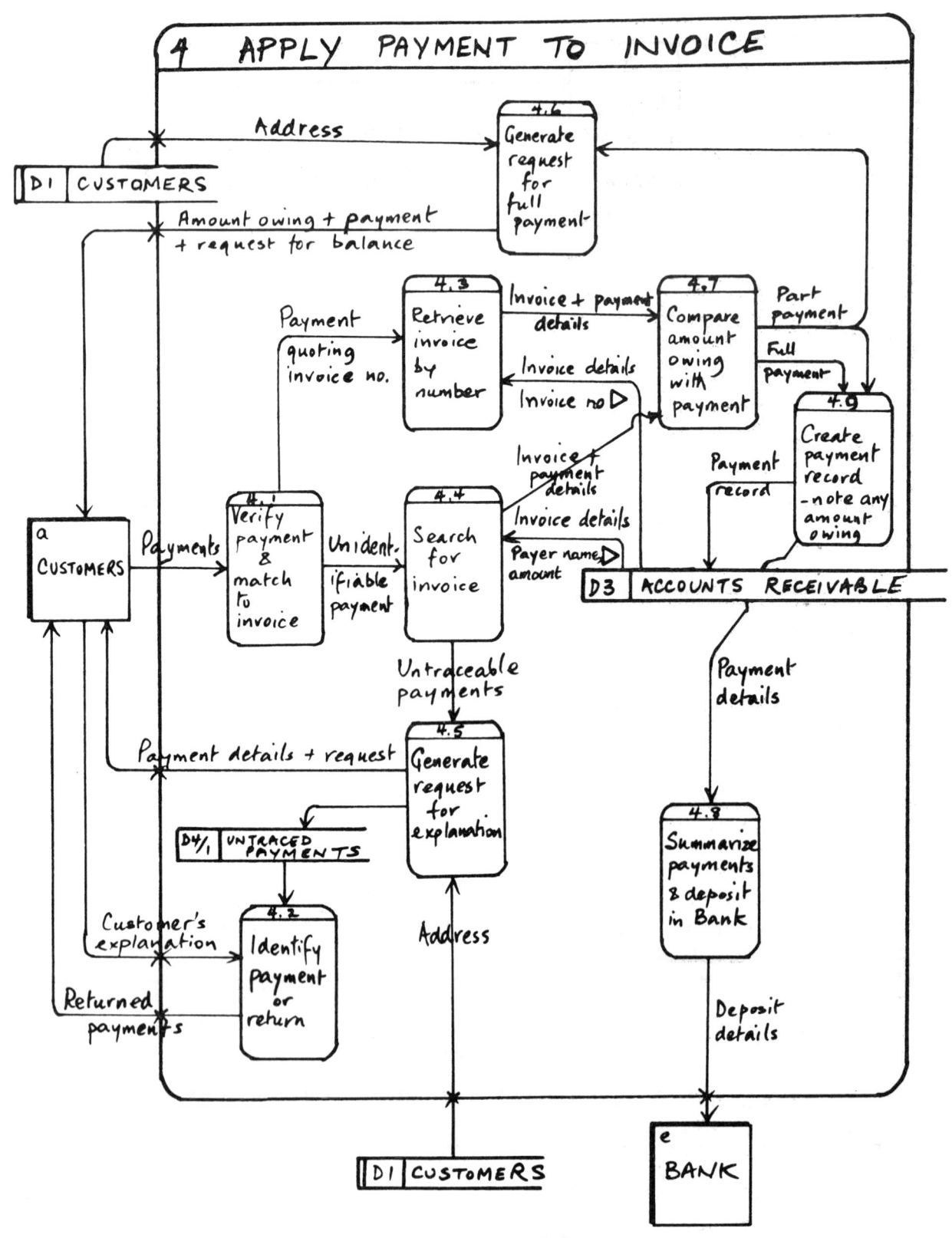

Figure 3.14

3.3 ERROR AND EXCEPTION HANDLING

As far as possible, data flows resulting from error and exception conditions should be handled within the second level diagram in which they arise. That is, bad transactions rejected from "Edit" should be handled within the explosion of "Edit." Where an external entity, such as a supervisor or adjustment clerk, must manually process the error, the entity is shown on the lower level diagram, though outside the boundary.

Where a complete process, comparable in complexity with those on the top-level diagram is involved, a low-level data flow should be worked out for it, and then a decision made as to whether the summary process should be included in the top-level diagram. This may be necessary for exceptions that have a financial consequence such as returned merchandise, dishonored checks, and so on.

This issue arises in APPLY PAYMENTS TO INVOICE over the handling of untraceable payments, that is, the case where a payment is received without any indication as to what invoice it relates to, and further, where there is no trace of any invoice being sent to this customer, or any invoice being issued for this amount. (This can happen, as when a parent company pays a bill for a subsidiary whose name is quite different, and pays the wrong amount!) Processes 4.2 and 4.5 deal with this situation, asking the customer for an explanation and holding the payment in D4/1 until either a satisfactory explanation is received, or the customer realizes the payment has been made in error and asks for its return.

Once the data flows and processes have been established, as shown in Figure 3.14, the analyst has to decide whether this function is important enough to be included on the top-level data flow diagram. Though it involves money, the need for the function will arise only on rare occasions, and the sums of money involved are coming in to the company, not going out. For these reasons, the function is probably not important enough to be shown at the top-level.

Identifying lower level functions which are additions to rather than explosions of higher level processes, can be identified by numbering the processes X1, X2, and so on.

3. DRAWING UP DATA FLOW DIAGRAMS

3.4 GUIDELINES FOR DRAWING DATA FLOW DIAGRAMS

In this section we summarize the steps involved in drawing up a data flow diagram for an existing or proposed system, before going on to develop an example in the next section.

1. Identify the external entities involved. As we said previously, this involves deciding on a preliminary system boundary. If in doubt, include within your system boundary the first "outer layer" of manual and automated systems with which you will have to interface. Remember that data flows are created when something happens in the outside world; a person decides to buy something, or an accident happens, or a truck arrives at the loading bay. If you can, get right back to the ultimate source of your data and draw the flow from there.

2. Identify the inputs and the outputs that you can expect and schedule in the normal course of business. As the list grows, try to discover logical groupings of inputs and outputs. Mark the inputs and outputs that are related solely to error and exception conditions.

3. Identify the inquiries and on-demand requests for information that could arise, involving pairs of data flows, in which one data flow specifies what is "given" to the system, and the second specifies what is "required" from the system.

4. Take a large sheet of paper (the back of used printer output is good), and starting on the left-hand side with the external entity that seems to you to be the prime source of inputs (e.g., Customers), draw the data flows that arise, processes that are logically necessary, and the data stores that you think will be required.

    Pay no attention to timing considerations except for natural logical precedence and logically necessary data stores. Draw a system that never started and will never stop. It's sometimes useful to follow a typical good input transaction all the way through the system and ask yourself, "What needs to happen to this transaction next?"

3.4 GUIDELINES FOR DRAWING DATA FLOW DIAGRAMS

Follow the notation rules set out in Section 3.1, but don't number processes until the final draft.

5. Draw the first draft freehand, and concentrate on getting everything down, except errors, exceptions, and decisions. If you find yourself drawing a diamond for a decision, slap your hand---decisions are made within low-level processes and do not appear on data flow diagrams.

6. Accept that you will need at least three drafts of the high-level data flow. Don't be concerned that the first draft looks like a hopeless tangle. It can be sorted out (see example in the next section).

7. When you have a first draft, check back with your list of inputs and outputs to ensure that you have included everything except those that deal with errors and exceptions. Note on the draft any normal inputs and outputs that you couldn't fit in. Remember that every data store which describes something outside in the real world has to be created and maintained.

8. Now produce a clearer second draft using a template for the symbols. You are aiming for a diagram with unique processes and the minimum number of crossing data flows. To minimize crossing:

    - first duplicate external entities, if needs be
    - next duplicate data stores, if needs be
    - then allow data flows to cross, if you can't see a layout that reduces crossing.

    Your second draft will look much clearer, but will probably still have some unnecessary crossings, and as you review it, you will see that the layout and relationship of process symbols could be improved.

    Check back to your list of inputs and outputs, and note on the second draft anything you still cannot fit in.

9. If you have a sympathetic user representative, or someone who knows the application, conduct a walkthrough of the second draft with them, explaining that it is only a draft, and noting any change resulting from the walkthrough.

3. DRAWING UP DATA FLOW DIAGRAMS

10. Produce a lower level explosion of each process defined on the second draft according to the conventions specified in Section 3.2. Work out the handling of errors and exceptions and incorporate changes in the top-level diagram, if necessary. The third and final version of the top-level diagram can now be completed.

11. If you think the investment worthwhile, have an artist produce a handsome copy of your finalized top-level diagram, about 36" x 48". It will serve as an invaluable aid in presentations to users, implementers, reviewers, and all concerned with the system.

3.5 EXAMPLE: DISTRIBUTION WITH INVENTORY

As an example of the construction of a full-scale data flow diagram, let us consider the CBM Computer Books by Mail Co., described at the start of Chapter 2. As an example of a DFD, we showed the present CBM system, which turned out to be a classic case of distribution-without-inventory. The system required by the new management for marketing to individuals as well as librarians, and supplying the majority of books from inventory, will be much more complex. The following narrative is taken from the old-style systems specification. We have numbered the lines for easy reference.

1.   "Orders will be received by mail, or taken over the phone by
2.   the inward WATS line. Phone orders will be taken down in a
3.   standard form, or entered directly into a CRT using a stan-
4.   dard format. Each order will be scanned to see that all im-
5.   portant information is present, that the title exists (or can
6.   be identified), that the author is correct (or can be iden-
7.   tified), and that the book is available (i.e. not out of
8.   print). If the order is defective, it is routed to a super-
9.   visor to see if, e.g., "The Programming of Management," by
10.   Does Jane, should really be "The Management of Programming,"
11.   by Jane Doe. Where payment is included, the amount is to be
12.   checked for correctness (if not correct, a request for fur-
13.   ther payment or a credit should be produced). Small disc-
14.   repancies can be ignored. Where payment is not with the order,
15.   the customer file must be checked to see if the order comes

3.5 EXAMPLE: *Distribution with Inventory*

16. from a person or organization in good credit standing; if not,
17. the person must be sent a confirmation of the order and a req-
18. uest for prepayment. If the customer is new to us, an addition
19. must be made to the customer file. For orders with payment
20. or good credit, inventory is then to be checked to see if the
21. order can be filled. If it can, a shipping note with an in-
22. voice (marked "paid" for prepaid orders) is prepared and sent
23. out with the books. If the order can only be part-filled, a
24. shipping note and invoice is prepared for the part shipment,
25. with a confirmation of the unfilled part (and paid invoice
26. where payment was sent with the order), and a back-order re-
27. cord is created. Back-orders are to be filled as soon as
28. the books are received from the publisher. Where the order
29. is for a book not held in inventory, the orders are batched
30. for purchase requisition on the publisher when a quantity
31. discount has been earned. Returned books are examined for
32. damage, and entered back into stock, with a credit or refund
33. being issued to the customer as appropriate. Where the re-
34. turned book is not an inventory item, and the publisher al-
35. lows returns, it is sent back to the publisher.
36.   When a shipment of books is received from a publisher,
37. its contents are to be checked against the original purchase
38. order, and discrepancies queried. The titles in the ship-
39. ment are checked against the back-orders for priority ship-
40. ment, and the remainder entered into inventory. Inventory
41. control policy calls for a reorder level on each title equal
42. to the (average orders over the previous four weeks) x (de-
43. livery time from publishers) plus a 50% safety factor.
44. Thus if sales of a title average 10 per week and the estimated
45. delivery time is 3 weeks, an order will be placed with the
46. publisher when the total copies in hand (and on order) have
47. fallen to 45 (3 x 10 x 150%). The safety factor may be varied
48. from time to time by management, being increased for titles
49. whose sales are rising and vice versa. The quantity for each
50. order is determined by taking the product of the average order
51. rate and delivery time, as above, multiplying by a bulk fac-
52. tor (normally 3), and rounding up to the next higher discount
53. break point, unless that increases the order by more than 25%.
54. Thus in the case above, the normal order would be 3 x 10 x 3
55. (bulk factor), or 90 copies. If the publisher offers an ad-
56. ditional discount for orders of 100 or more, 100 would be
57. ordered. If the discount is only offered for 120 or more,
58. 90 would be ordered, since to order 120 would increase the
59. order by the excessive amount of 33%. The bulk factor may
60. be varied **by** management for each title from time to time.
61. The calculation of average order rate includes not only or-
62. ders that were filled, but frustrated demand, such as back-

3. DRAWING UP DATA FLOW DIAGRAMS

63. orders, orders without payment, and inquiries that were not
64. converted to orders because the book could not be supplied
65. from stock.
66.   When payments for books supplied are received, they
67. are matched with the appropriate invoice. Where several
68. invoices are outstanding for an account, and the payment
69. does not match any one of them exactly, it is applied to
70. the oldest invoice first. Frequently a customer will send
71. one payment to cover several invoices. Where any invoice
72. is more than 30 days overdue, a statement of all invoices
73. outstanding is sent to the customer. When any invoice is
74. more than 60 days overdue, a strongly worded letter is prod-
75. uced for the Vice-President's signature.
76.   When invoices are received from publishers, they are
77. checked against the receipt-of-shipment records, and enter-
78. ed into accounts payable. If the discount for prompt pay-
79. ment given by the publisher exceeds, on an annualized basis,
80. the marginal cost of funds (as specified from time to time
81. by management), the system should produce a payment check
82. on the last day the discount is available. For example,
83. if 2½% is offered for payment in 30 days, this is equiva-
84. lent to 30% per year. The system should write a check on
85. the 29th day.
86.   Reports of invoices sent out, by day, by week, by month,
87. payments received by day, week, month, amounts overdue by
88. various periods, stock-outs, back-orders, and purchases
89. from publishers, should all be produced regularly. On-
90. demand analyses of sales by title, by subject, by publisher,
91. with trend information, should be available on an immediate
92. basis, together with information on publisher delivery times
93. and purchasing trends. Immediate access to inventory figures
94. of quantity-on-hand, quantity-on-order, and expected date
95. of delivery are all very desirable, as is the facility to
96. give a customer immediate information as to the status of
97. his particular order. If a customer calls up and says,
98. "I sent you a check for $10 five weeks ago, for Bloggs'
99. book," we would like to be able to tell him what day we
100. shipped the book to him, or on what date we will be able
101. to ship it."

    This narrative statement of requirements for the new system, like all narratives, contains information at very varying levels of detail; procedure mixed up with structure, important policy mixed up with trivia. How can we extract the information to build a top-level data flow diagram?

3.5 EXAMPLE: Distribution with Inventory

First, let us identify as many external entities as possible by scanning the narrative, before going on to identify and classify inputs and outputs.

External entities:

Name	Referred to on Line #
Customers	1-2 (by implication),15-19,70,73
Order Entry Supvsor.	8-9
Publisher	28,30,34-5,36,43,46,55,76,79
Management	48,60,81,86-9 (by implication)
Vice-President	75,89-95?

It may turn out that "management" and the "Vice-President" are the same entity, or we may find that "management" really is composed of "Sales Manager," and "Stock Manager," and "Purchasing Manager." At least we have a draft list from the narrative. Now for the inputs and outputs:

INPUT-OUTPUT LISTING

Inputs	From (EE)	Outputs	To (EE)	Ref #
Orders-mail	Customer			1
Orders-phone	Customer			1
		Defective orders	Order entry Supervisor	8
Payment	Customer			11
		Confirmation	Customer	17
		Prepayment request	Customer	18
		Shipping note	Customer	21
		Invoice	Customer	22
		Books*	Customer	23
		Purchase requisition	Publisher	30
Returned books*	Customer			31
		Credit	Customer	32
		Refund	Customer	32
		Returned books*	Publisher	34
Book shipment*	Publisher			36
		Discrepancy query	Publisher	38
		Order	Publisher	45
		(How is this different from Purchase requisition #37)		
Safety factor change	Management			47
Bulk factor change	Management			59

(continued)

* Materials movement: what data is associated with it?

3. DRAWING UP DATA FLOW DIAGRAMS

<u>INPUT-OUTPUT LISTING</u> (continued)

Inputs	From (EE)	Outputs	To (EE)	Ref #
Unfillable order	Customer			64
		Statement	Customer	72
		Strong letter	Vice-Pres. then Customer	74
Invoices	Publisher			76
		Payment check	Publisher	81
		Invoice report-daily	Management	86
		" -weekly	Management	86
		" -monthly	Management	86
		Receipt report-daily	Management	87
		" -weekly	Management	87
		" -monthly	Management	87
		Amount overdue report	Management	87
		Stock-out report	Management	88
		Back-order report	Management	88
		Publisher purchase report	Management	88

This input-output analysis identifies all the flows of data we can find in the narrative, except for on-demand queries. It is convenient to treat these separately, (see below), since they usually consist of inputs and outputs paired together.

In several places, we have noted with an asterisk an input or output that is stated as a flow of materials (shipments, returns) without a specified flow of data. Normally, of course, each materials movement is accompanied by a voucher (shipping note, claim for credit on return) which carries the relevant data. We note that these data flows must be found and analyzed.

It is worth noting, as well, that this input-output listing has been produced from a given narrative. It could equally well have been produced from the analyst's interview notes, from a pile of various memos, or a combination of the two. In any case, the reference should be noted on the input-output listing.

3.5 *EXAMPLE: Distribution with Inventory*

Queries

From/to	Given	Display	Ref #
Management	Book title	Sales analysis	90
Management	Subject	Sales analysis	90
Management	Publisher	Sales analysis	90
Management	Publisher	Delivery time analysis	92
Management	Publisher	Purchasing analysis	93
Management	Title	Quantity-on-hand	94
Management	Title	Quantity-on-order Delivery date	94
Customer	Name & order details	Order status	96-101

For the moment we are not primarily concerned with the immediacy or value of these queries; in this context, "Display" could mean "Produce a printed report next Monday morning." We will analyze queries much more thoroughly in Chapter 7. What we are concerned about is the existence of the necessary data flows (and data stores).

In keeping with the simplification rule for data flow diagrams (p 51), for our first draft, we want to eliminate those inputs and outputs which relate to error or exception conditions, and to treat flows which are logically similar as one flow. Thus we will treat "Orders-mail" and "Orders-phone" as just "Orders;" what is logically important is whether or not they are accompanied by payment. We shall omit from our top-level data flow "Defective orders," "Returned books," "Credit," "Refunds," "Discrepancy query," "Unfillable order," "Statement," and "Strong letter."

Having done this, let us produce a first draft of the DFD starting with the healthy fact that Customers are sources of "Orders."

Figure 3.15 shows a freehand sketch of the first few steps in order processing. In this first sketch, we have put together all the processes of scanning the order for completeness, checking the book file to see that the book exists and that the payment (if the customer has sent one) is the right amount, and checking the customer file to see whether we have previously dealt with him. They make up a single process "Edit order and prepayment (where present)," and we note that we will need to expand it when we come to draw the lower level data flow.

3. DRAWING UP DATA FLOW DIAGRAMS

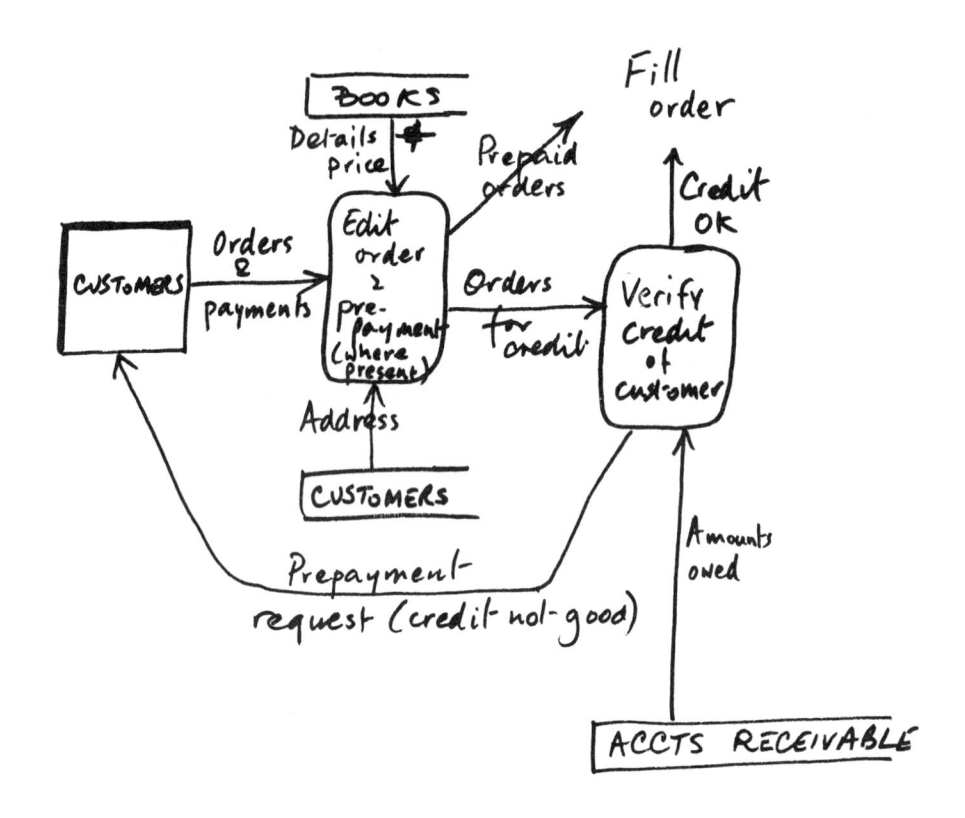

Figure 3.15
Partial First Draft

Our next task is to spell out what we mean by the note "Fill order." Let us carry our sketch a little further to the point shown in Figure 3.16. Here we take orders and access the current inventory to see whether we can fill them or not. For those that we can fill, we generate a shipping note (which may be used in the warehouse to actually take the proper books out of stock) and an invoice (which will need to be marked "Paid" in the case where we have already received the payment with the order). The sketch is becoming a little complex, but we can see how the system is growing. We continue in this way, and end up with a first draft of the whole system, shown reduced in Figure 3.17.

3.5 EXAMPLE: Distribution with inventory

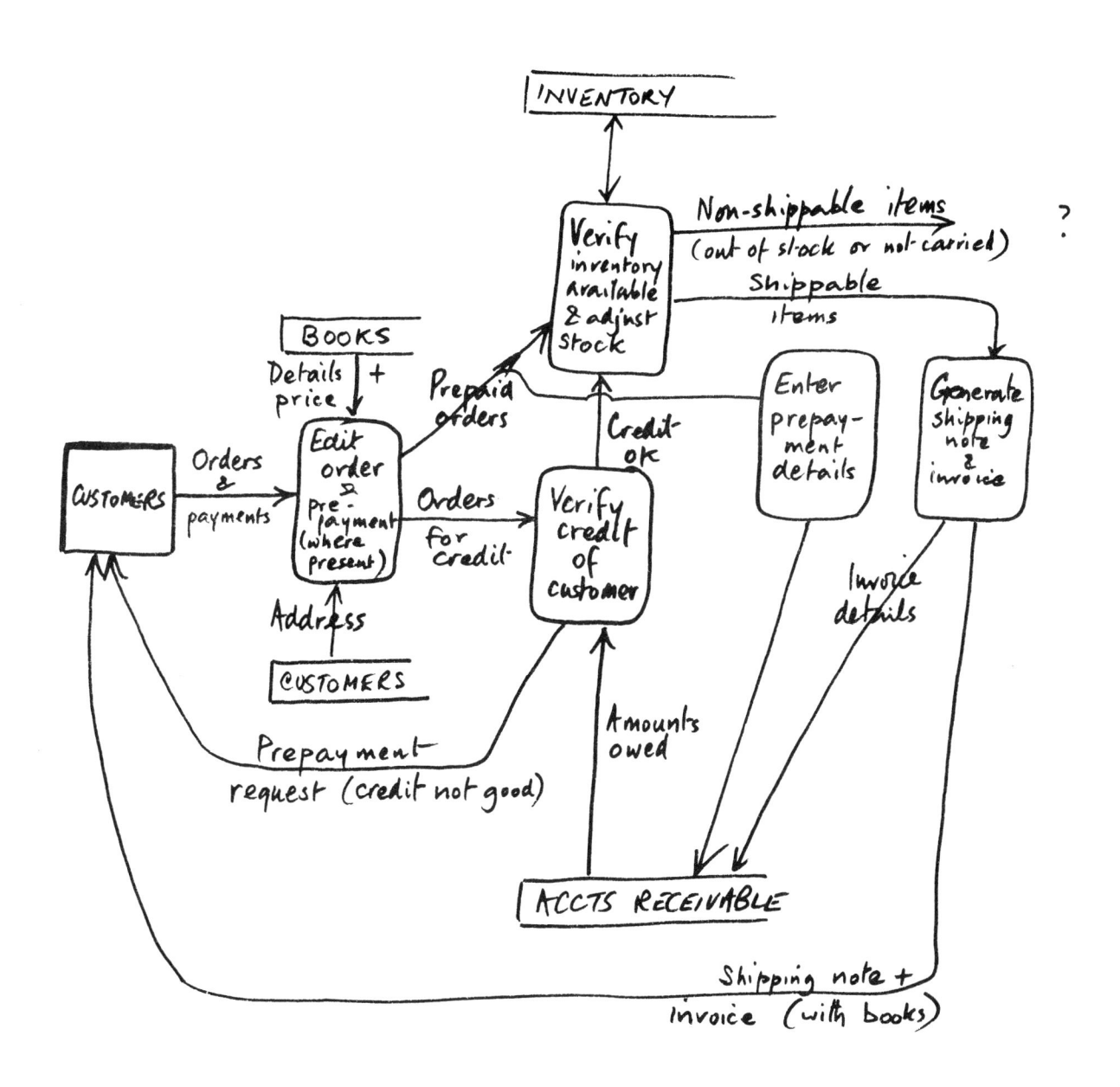

Figure 3.16
Extended First Draft

3. DRAWING UP DATA FLOW DIAGRAMS

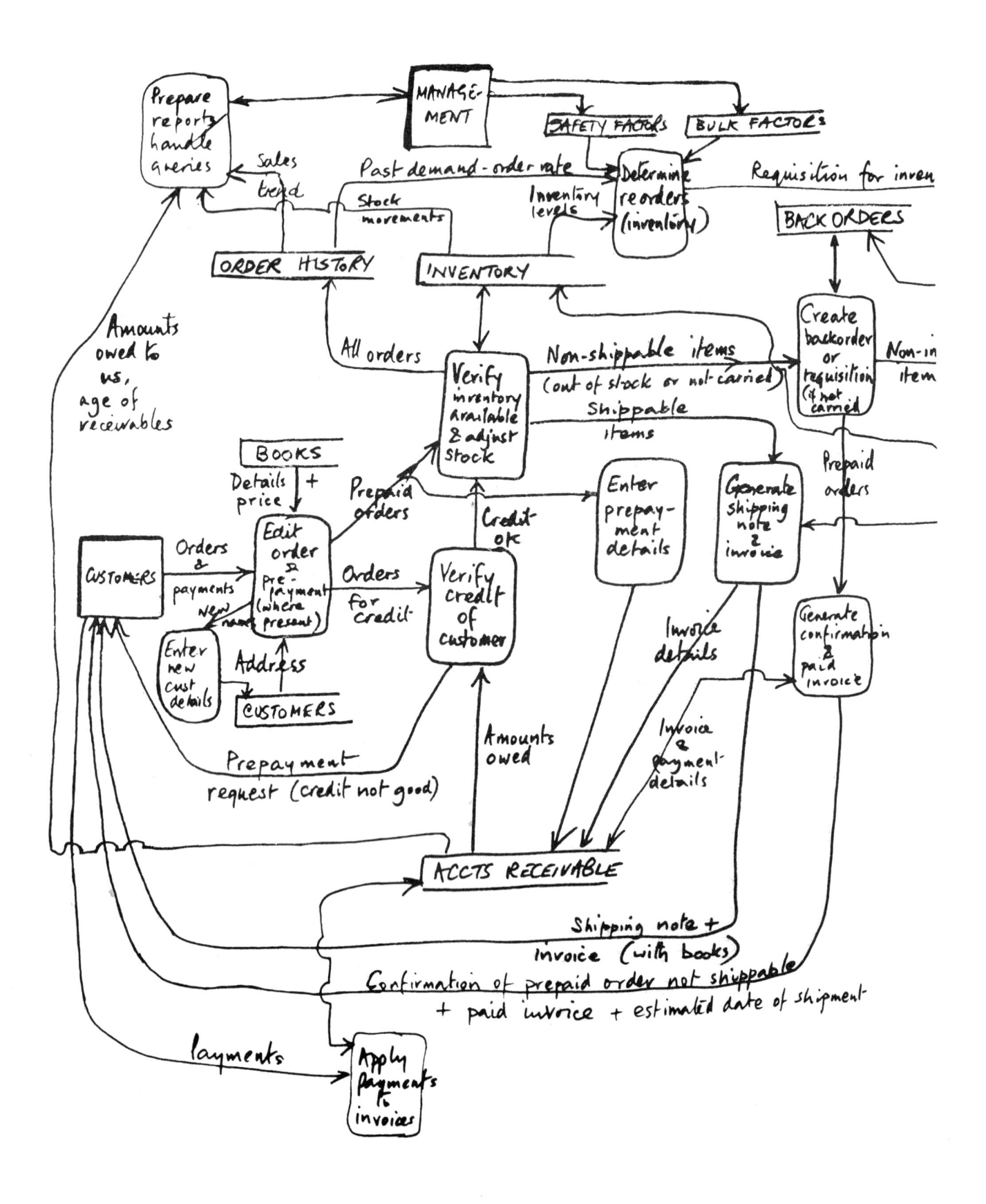

3.5 EXAMPLE: Distribution with inventory

Figure 3.17
Final First Draft

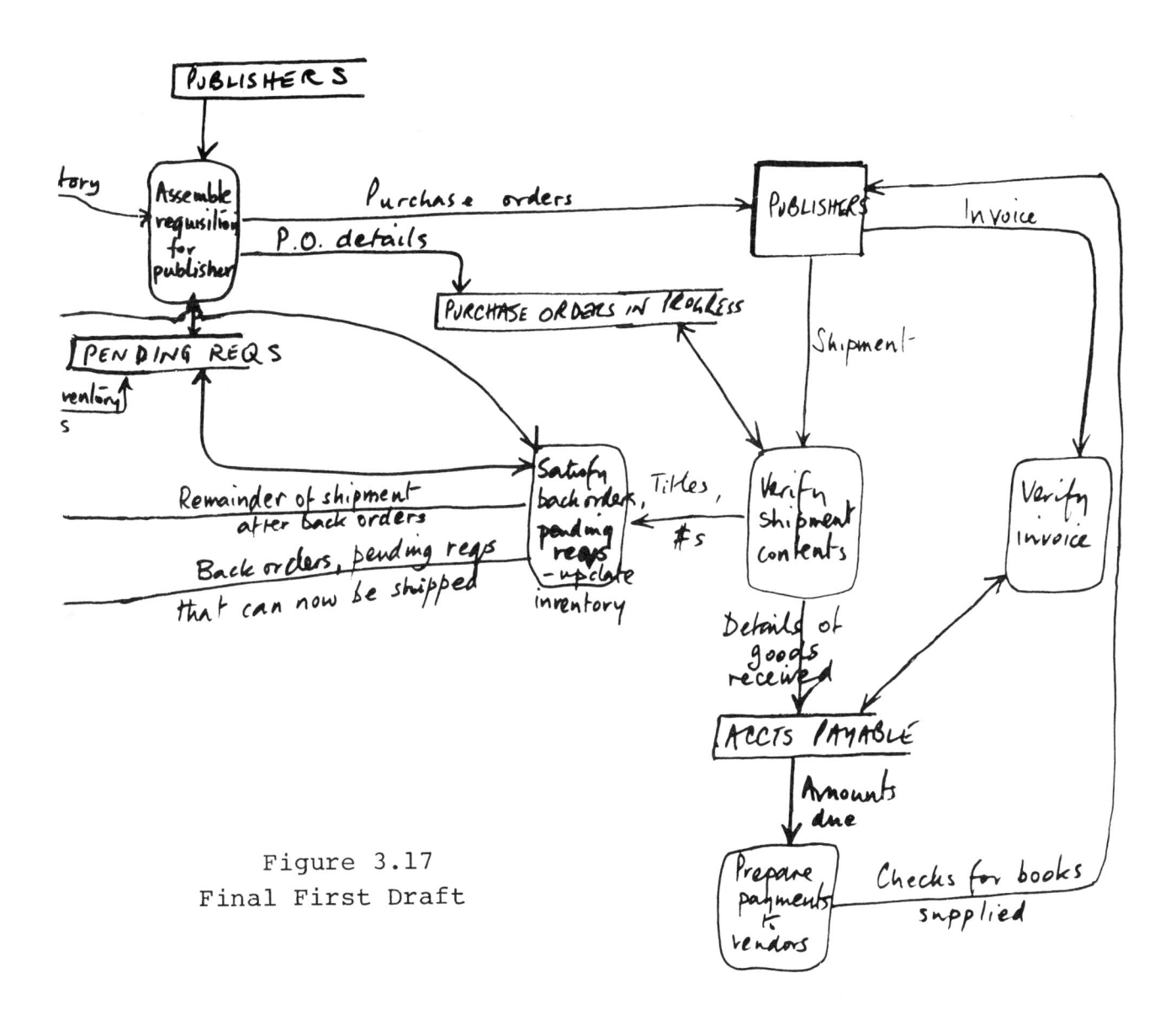

As you can see, this first draft can get quite cluttered and tangled; it's only as you identify data flows and processes that you can see how they fit together, so it is almost impossible to get a good layout first time through. At least we now have all the major features of the system on one piece of paper. The next step is to redraw the first draft using a template. Figure 3.18 shows the result.

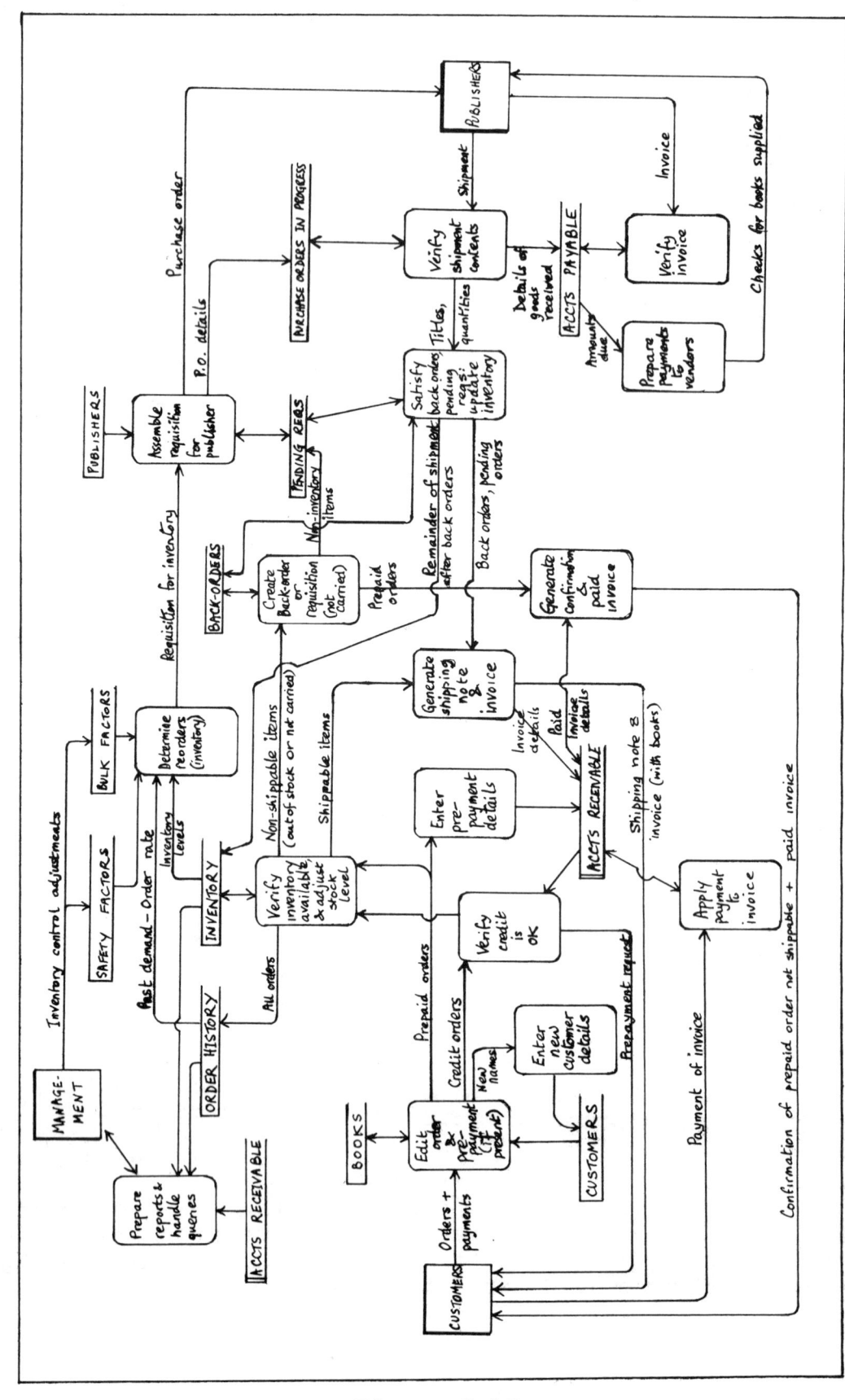

Figure 3.18

Not shown: Customer queries, management reports/queries on purchases/backorders, file creation/maintenance for books & publishers

3.5 EXAMPLE: Distribution with inventory

The second draft is much clearer and tidier, since once you see all the processes and connections, you can make a better job of positioning them on the page. Note that one very messy data flow has been cleaned up by duplicating the ACCTS RECEIVABLE data store. The chart is still very full in the upper part, around the MANAGEMENT entity, and, as is noted along the bottom, some significant (though not complex) functions have been missed out because they would destroy the clarity of the rest of the diagram. As we have noted, each data store which holds "basic" information about the outside world has to be created, and maintained as the outside world changes. The CUSTOMERS data store is maintained in our second draft to the extent that we can detect and add new names. We also need the functions of changing (e.g., address or person responsible for billing), and possibly deleting customers. The BOOKS data store will need fairly frequent updating since new books are being published all the time, and publishers occasionally change their prices. The PUBLISHERS data store may have a few changes a year once it is established, but generally is very stable.

At the cost of complicating the diagram, we can add the various data flows and functions associated with management queries about sales, inventory, purchases, and back-orders, as shown in Figure 3.19, the final data flow diagram.

The finished product was drawn by a graphics person on a full-size chart (11"x17") and is presented as a two-page spread: the layout has taken into consideration the number of data flows crossing the page break.

The reader is invited to trace through the data flows and functions and to verify that the requirements specified on pages 52-54 have been met, except for those error and exception conditions we specifically excluded. It will be noted that, even with a chart of this complexity, it has only been necessary to duplicate one external entity (m - MANAGEMENT) and three data stores (D3, D10, and D2).

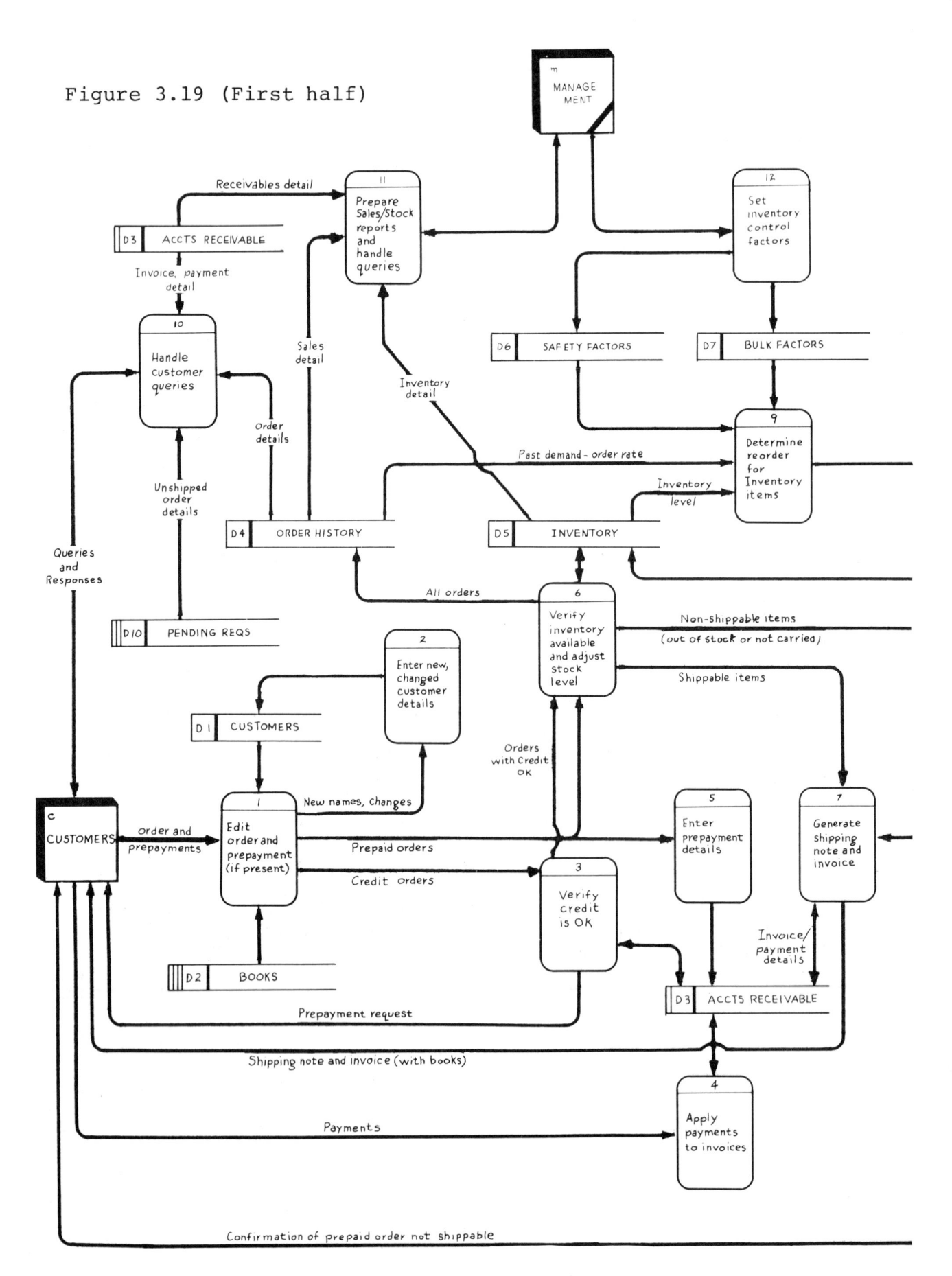

Figure 3.19 (First half)

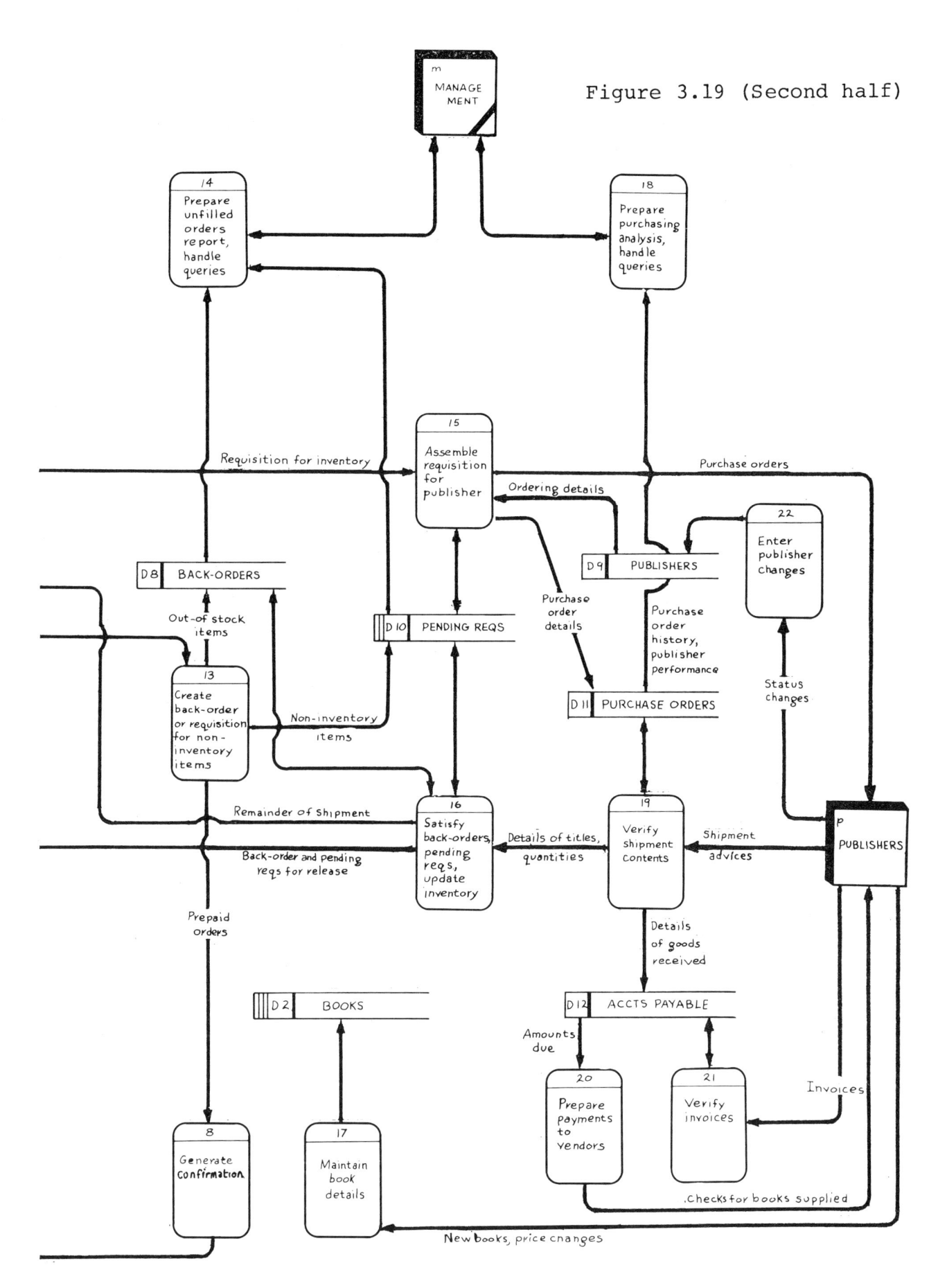

Figure 3.19 (Second half)

3.   DRAWING UP DATA FLOW DIAGRAMS

3.6   MATERIALS FLOW AND DATA FLOW

At several points when describing the system we have just charted, we commented that we were tempted to describe a materials flow instead of, as is correct, the associated data flow. It is important to keep these two concepts separate, especially in manufacturing and distribution industries, where much of the business is concerned with moving material around. Even in banking, the business of which might be thought to be as near as we can get to pure data, there are serious problems relating to the physical movement of large numbers of checks and other vouchers.

Consequently, we need a way of charting materials flow, when required, and tying the materials flow diagram to the data flow diagram. Materials flow is by its very nature physical, but we would like, as far as possible, to describe the operations performed on the materials in logical terms. Thus we will not be content with a materials flow chart of the type shown in Figure 3.20,

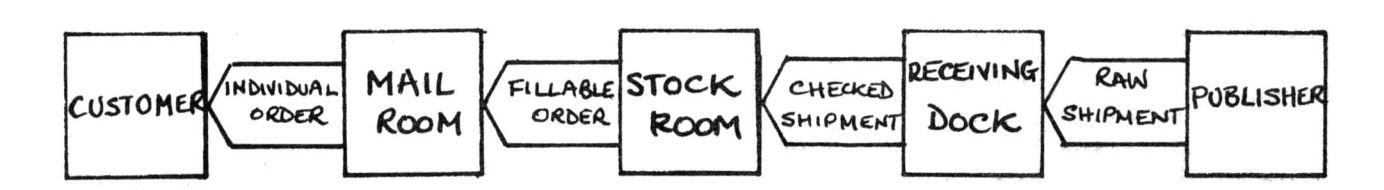

Figure 3.20

but will need some way of describing the logical function which transforms the material at each point, and indicating the associated data flows. Figure 3.21 shows both data flow and materials flow, with references to the original DFD, in Figure 3.19.

Note that only two of the material transformations correspond to processes on the data flow diagram (16 and 19), and that some materials flow can take place without a data flow (e.g., additions to stock).

3.6 MATERIALS FLOW AND DATA FLOW

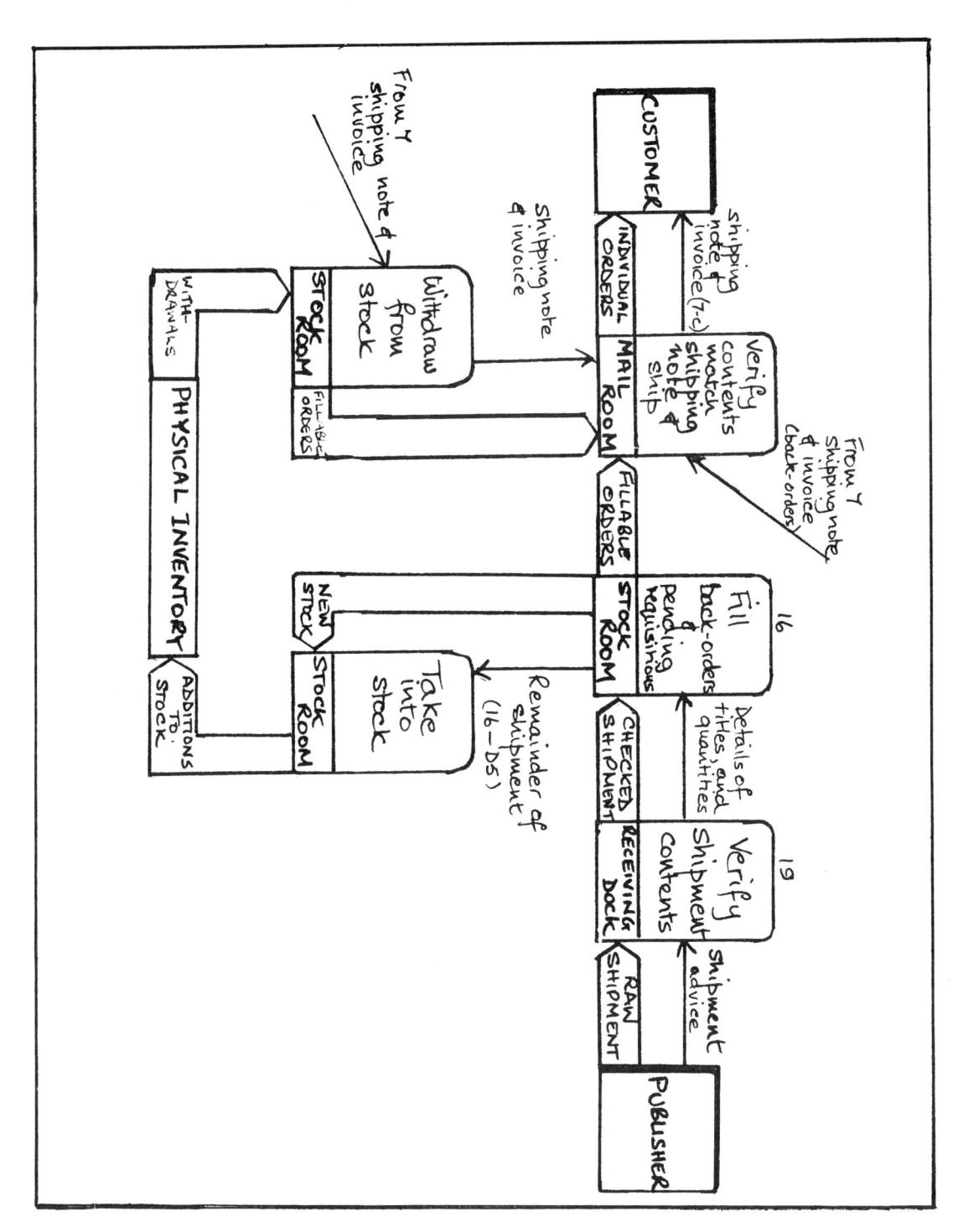

Figure 3.21

3. DRAWING UP DATA FLOW DIAGRAMS

REFERENCES

3.1 D. Martin and G. Estrin, "Models of Computations and Systems--Evaluations of Vertex Probabilities in Graph Models of Computations," *Journal of the ACM*, Vol. 14, No. 2, April 1967.

3.2 E. Yourdon and L L. Constantine, *"Structured Design,"* Yourdon inc., 1975.

3.3 G. E. Whitehouse, *"Systems Analysis and Design using Network Techniques,"* Prentice-Hall, 1973.

3.4 D. Ross, "Structured Analysis (SA): A language for Communicating Ideas," *IEEE Transactions on Software Engineering*, January 1977.

3. EXERCISES AND DISCUSSION POINTS

EXERCISES AND DISCUSSION POINTS

1. In Figure 3.19, why do "Requisitions for Inventory," (9-15) go directly to "15-Assemble Requisition for Publisher," while items that are not held in inventory get stored in D10:PENDING REQUISITIONS?

2. Why does Process 9,"Determine Reorder Level for Inventory Items" not require data from D8: BACK-ORDERS?

3. Is the reorder algorithm used by CBM (lines 44-49 in the narrative) a reasonable one? Could it be improved?

4. Assuming that the mail order supply of books is CBM's only business, what other data processing/information systems would you expect them to have, apart from the system shown in Figure 3.19?

5. Adapt Figure 3.19 to show the data flow in a business that stocks autoparts for both American and foreign cars (slogan "If We Haven't Got It, We'll Get It"), supplying them by mail order. How would the data flow be altered if they opened an over-the-counter sales department?

6. Making reasonable assumptions, explode the Processes "19-Verify Shipment Contents," "20-Prepare Payments to Vendors," and "21-Verify Invoices," (in Figure 3.19) to produce a detailed data flow diagram of the accounts-payable subsystem, dealing with likely errors and exception conditions.

7. List as many enterprises as you can think of whose business consists of maintaining inventory and distributing it, whether by mail or over the counter. Identify the similarities and differences between their data flows.

8. Draw overall data flow diagrams for the following types of enterprise:

    - manufacturing to order (a "jobbing" shop)
    - manufacturing, holding inventory, and distributing
    - making consumer loans
    - maintaining savings accounts
    - issuing and maintaining accident insurance policies and handling claims
    - selling professional time by the hour and paying salaries to professionals (e.g. a legal practice)
    - a booking agency, reserving seats for a journey.

Chapter 4

Building and using a data dictionary

In Chapter 2, we saw that as we dug further into the details of the contents of data flows, data stores and processes, we needed some structured place to keep all that detail. The data dictionary provides that place. The name "Data Dictionary" gets stretched when we start to include the details of processes, which strictly speaking are logic, rather than data. Perhaps the data dictionary should really be called a "Project Directory." However, the name data dictionary is in wide use, and we shall adopt it, with no inhibitions about using it for anything we need to describe.

4.1 THE PROBLEM OF DESCRIBING DATA

In the good old unit-record days, the terms used to describe data were quite simple. A card was divided into fields; the card itself was a record, and a number of records constituted a file. It wasn't *quite* as simple as that; a date field in the form MMDDYY could be thought of as composed of three sub-fields, so already the data description hierarchy existed at

4. BUILDING AND USING A DATA DICTIONARY

four levels:

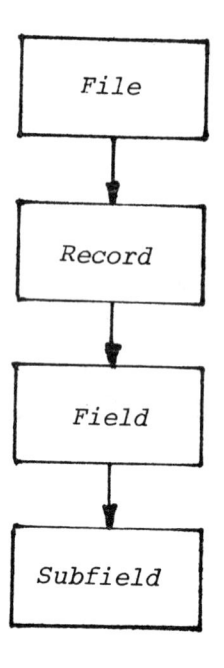

```
┌─────────┐
│  File   │
└────┬────┘
     ▼
┌─────────┐
│ Record  │
└────┬────┘
     ▼
┌─────────┐
│  Field  │
└────┬────┘
     ▼
┌─────────┐
│Subfield │
└─────────┘
```

This hierarchy was taken over to tape processing and later to disk processing. IBM called a file a "data set", and we encountered variable length records in which a header was followed by a number of trailer records or repeating groups, such as for an order:

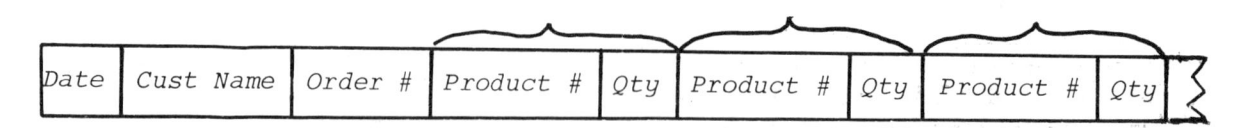

| Date | Cust Name | Order # | Product # | Qty | Product # | Qty | Product # | Qty |

Still we could use the same words to describe our data.

With the development of data base technology, the simple vocabulary no longer seemed enough. Since part of the value of data base techniques lies in taking the files from each application, and integrating them in a data base which all applications can use, the term "file" became questionable. IBM's Information Management System (IMS), [4.1], describes data as *fields*, which are combined into *segments*, which are combined into *data bases*. The Data Base Task Group (DBTG) of the Conference on Data Systems Languages (CODASYL) [4.2] describes data as *data items*, which are combined into *data aggregates*, which are combined into *records*, the relationships between which are expressed as *sets*. The COBOL Data Division provides for elementary items which are combined into *group items*, which are combined into *records*, which make up *files*. We will

4.1 THE PROBLEM OF DESCRIBING DATA

discuss other terminologies in Chapter 6; some different terms are summarized in Table 4.1.

	COBOL	*PL/I*	*IMS(DL/I)*	*CODASYL*
SMALLEST UNIT OF DATA:	Elementary Item	Element	Field	Data Item
GROUPS OF SMALLEST UNIT:	Group Item	Structure	Segment	Data Aggregate
REPEATING UNITS:	Table/ Repeating Group	Array	Segment	Vector/ Repeating Group
ENTITY WHICH IS PROCESSED AT ONE TIME:	Record	Record	Logical Record	Record
LARGEST RECOGNIZED GROUPING:	File	File	Data Base	Set/Schema

Table 4.1

Given this welter of concepts and terms, we need to choose some simple words which will not conflict unduly with these common vocabularies, and will enable us to describe both data flows and data stores at the logical level.

We find from experience that we can describe everything of interest to us as analysts at just three levels:

1. *Data elements*: these are pieces of data that it is not meaningful to decompose further for the purpose at hand. For instance, "date" is a data element for most purposes during analysis, though it might need to be regarded as a structure made up of "month, day, year" for coding a date conversion routine.

2. *Data structures* are made up of Data Elements, or of other data structures, or a mixture of both. Consider the con-

4. BUILDING AND USING A DATA DICTIONARY

      tents of the data flow "Orders" that we looked at in Chapter 2.

```
ORDER
   ORDER-IDENTIFICATION
      ORDER-DATE
      CUSTOMER-ORDER-NUM......usually present
   CUSTOMER-DETAILS
      ORGANIZATION-NAME
      PERSON-AUTHORIZING......optional
         FIRST-NAME..........may be initial only
         LAST-NAME
      PHONE
         AREA-CODE
         EXCHANGE
         NUMBER
         EXTENSION...........optional
      SHIP-TO-ADDRESS
         STREET
         CITY-COUNTY
         STATE-ZIP
      BILL-TO-ADDRESS........If not present, same as
         STREET                 SHIP-TO-ADDRESS
         CITY-COUNTY
         STATE-ZIP
      BOOK-DETAILS...............One or more iterations of this
                                 group of data elements
            AUTHOR-NAME.........One or more iterations of this
            TITLE                data element
            ISBN................Iternational Standard Book
                                 Number, (optional)
            LOCN................Library of Congress Number,
                                 (optional)
            PUBLISHER-NAME......Optional
```

      "PHONE" is a data structure made up of the four data elements, AREA-CODE, EXCHANGE, NUMBER, EXTENSION. "CUSTOMER-DETAILS" is a data structure made up of the data element "ORGANIZATION-NAME" plus the data structure "PERSON-AUTHORIZING" plus the data structure "PHONE", and so on. Indeed, "ORDER" is itself just a big data structure.

3. *Data flows and data stores*: we have already discussed this level of data description. Data flows are paths or "pipelines" along which data structures travel; data stores are places where data structures are stored until needed.
   Data flows are data structures in motion; data stores are data structures at rest.

   Our data description hierarchy is thus:

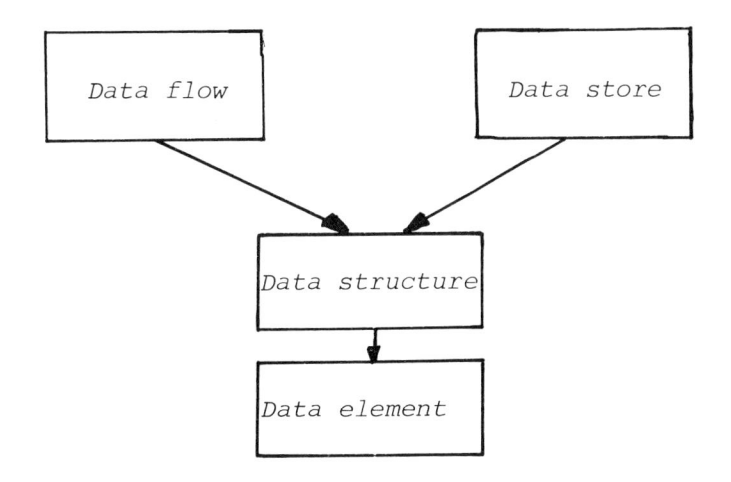

```
        Data flow              Data store
             \                  /
              \                /
               v              v
              Data structure
                    |
                    v
              Data element
```

Once we have a logical description of data at this level, it is easy to cast it into the vocabulary of the particular language or data base system that will be used for physical implementation.

4.2 WHAT WE MIGHT WANT TO HOLD IN A DATA DICTIONARY

Before discussing specific implementations of manual and automated data dictionaries, we shall examine data elements and data structures to see what we might want to record about them in various circumstances. We shall also look at the attributes of data flows, data stores, processes, external entities, and

4. BUILDING AND USING A DATA DICTIONARY

other things we encounter in analysis to note what would be worth recording about them. This will give us a background for assessing any data dictionary system, or for planning our own.

4.2.1 DESCRIBING A DATA ELEMENT

The minimum information needed to establish a data element is its *name* and a *description*. For instance,

"STREET":
- house number
- name
- street

may also include

- apartment number
- P.O. Box, or
- mailstop

The name should be chosen to be as meaningful as possible to users; it is convenient, though not necessary, to keep to the rules of programmer-defined datanames in COBOL, especially if the system is going to be implemented in COBOL. COBOL datanames can be up to 30 characters long, using only the characters "A" through "Z", 0 through 9, with embedded hyphens to separate words.

We will often need to use the same name in different contexts so it is also convenient to be able to use the qualification convention of COBOL to make the names unique. Thus we may define DATE as part of INVOICE, and also as part of ORDER and PAYCHECK. We can make DATE unique by always writing DATE OF INVOICE, DATE OF ORDER, DATE OF PAYCHECK, etc.

The description should be a thumbnail sketch of the meaning of the data element, and may include a typical example.

In addition to name and description, we want to be able to record, as we find them out:

1. Aliases for the data element
2. Related data elements
3. Range of values and meanings of values
4. Length
5. Encoding
6. Other editing information

4.2 WHAT WE MIGHT WANT TO HOLD IN A DATA DICTIONARY

1. *Aliases* may arise because different user departments have called the same thing by different names, e.g., the warehouse people call it "requisition number" and the purchasing people call it "order-number." Aliases may also arise because the same thing is defined in programs written in different languages or by different programmers. Thus REQUISITION-NUMBER may exist in the current system as REQUIS-NUM, REQNO, ORDNO, ORDNUM, POSEQNO, PUR7061, and other "internal" aliases. We want to record all of these and include them not only under the REQUISITION-NUMBER definition, but separately, with an entry of their own. So if we are going through some old documentation and find PUR7061, we can look it up in the data dictionary, find an entry saying:

   "PUR7061   Alias of REQUISITION-NUMBER"

   and look under REQUISITION-NUMBER to find the full details.

   Also, we may deliberately create aliases for our new system. For instance, in IMS, all datanames have a maximum of eight characters. Consequently we might choose to name a data element REQUISNO for IMS, and be able to pair that with its "external" alias REQUISITION-NUMBER when required.

2. *Related data elements*; sometimes we want to be able to point to data elements that have related names, though they are not aliases. For example, on a statement the date may appear as 15 JUL 1977, but the date of each invoice or payment might appear as 07/15/77. The first data element might be called DATE-IN-FULL and the second simply DATE. In the data dictionary entry for each we would want a "see also....". Of course, we have been careful to start the names of both with the same four letters, so in the alphabetic listing of data elements they will come close together.

3. *Range of values and meanings of values*; as soon as we start to examine the values that a data element can take, we realise that there are two types of data element:

   - those which for all practical purposes can take any value within a range, e.g., a dollar amount from zero to $999,999.99 to the nearest cent, or a temperature from 0° to 300°

4. BUILDING AND USING A DATA DICTIONARY

- those which only take up certain values e.g. department number which may be 36, 08, 29, or 71 but no other values. Another example of this second type would be marital status, which may be single, married, widowed, or divorced. Usually the values are a *code*. standing for some *meaning*. Thus we might have:

DEPARTMENT-NUMBER

Value	Meaning
36	Sales
08	Accounts
29	Warehouse
71	Advertising

or

MARITAL-STATUS

Value	Meaning
M	Married
S	Single
D	Divorced
W	Widowed

The first type of data element we shall call *continuous* because its values are practically continuous over its range; the second type we shall call *discrete*, because it takes on discrete values.

For *continuous* data elements, we need to note the range that they can have, a typical value, and any information about the handling of extreme values. For instance, PAYMENT-AMOUNT on a check may have a range from 1 cent to 10 million dollars. We may note that a value of zero is suspicious and a check written for amounts less than a dollar, while legal, should be flagged for attention as being unusual. Checks for amounts over $100,000 should also be flagged for verification by an officer. We might note here that demands for payment are not to be sent out for amounts under five dollars, though this information properly belongs in the logic of the process that generates the demands.

For *discrete* data elements, we need to note the values and the meaning which is given to each value. Some common discrete data elements are MARITAL-STATUS, as we saw

4.2 WHAT WE MIGHT WANT TO HOLD IN A DATA DICTIONARY

above, SEX, OWN-OR-RENT (own home), CHECKING-OR-SAVINGS (type of account), and DEPARTMENT-CODE. As an organization gets larger, the number of values for DEPARTMENT-CODE will grow, possibly into the hundreds. A familiar example of a discrete data element with many values is the three letter code used to identify airports. Figure 4.1 shows the first 25 values and meanings of AIRPORT CODE, out of more than 1,000 listed in the North American Official Airline Guide.

An example of a *DISCRETE* Data Element with many values

Code	Airport Name	Code	Airport Name
ABE	Allentown, PA	AGN	Angoon, Alaska
ABI	Abilene, TX	AGS	Augusta, GA
ABL	Ambler, Alaska	AHN	Athens, GA
ABQ	Albuquerque, NM	AIA	Alliance, NE
ABR	Aberdeen, SD	AIN	Wainwright, Alaska
ABY	Albany, GA	AIY	Atlantic City, NJ
ACA	Acapulco, Mexico	AIZ	Lake of the Ozarks, MO
ACK	Nantucket, MA	AKI	Akiak, Alaska
ACT	Waco, TX	AKK	Akhiok, Alaska
ACV	Eureka/Arcata, CA	AKN	King Salmon, Alaska
ADK	Adak Is., Alaska	AKP	Anaktuvuk Pass, Alaska
ADQ	Kodiak, Alaska	ALB	Albany, NY
AET	Allakaket, Alaska	.	
		.	
		.	

Figure 4.1

Clearly the analyst has to judge at what point it ceases to make sense to regard the data element as discrete and treat it as being a continuous element which is used as the key to a value in a data store. As a rule of thumb, if the table of values and meanings can be typed on one or two sides of a sheet, it is worthwhile tabulating the values in the data dictionary; beyond that, it is usually more sensible to define a data store which will hold the meanings.

4. BUILDING AND USING A DATA DICTIONARY

For example, PART-NUMBER can be thought of as a discrete data element. Each value has a particular meaning:

7A4601B means "Shank, slotted, 7¼ inch"

and so on. But in an engineering organization with 40,000 parts, we would not expect to hold those values and meanings in the data dictionary. The critical factor here is that we will almost certainly want to store other attributes of each part: its weight, price, who supplies it and so on. The table of values for a discrete data element should serve just to connect the value with the meaning - nothing more.

An interesting borderline case arises in specifying a telephone number. The area code is a 3 digit number, the first digit of which is never 0 or 1, and the middle digit of which is always 0 or 1, (at least in those codes available to the public). This gives a total of 144 possible area codes, 112 of which are assigned. Each value has a meaning in the sense that it specifies an area:

   212   New York City
   201   Northern New Jersey
   415   San Francisco Bay
   and so on.

We have a choice as analysts in whether we treat AREA-CODE, (and exchange and number), as a continuous data element with restrictions on its range, or whether we itemize all the values it can take and give them meanings. The meanings may be used to edit the reasonableness of the phone number by comparing the area-code with the state, or used as the basis of territories for sales analysis. The critical issue is:

"Do we care about the meanings?"

If we do, and will use the meanings in some way, then we should define the data element as discrete. If we don't, let us save some trouble by making it continuous.

Another use for discrete data elements is in defining *adjectives*. As we said, temperature is a continuous data element, with a suitable range for the purpose at hand. But we may want to describe temperatures over 85° as "HIGH", temperatures between 55° and 84° as "NORMAL" and temperatures below 55° as "COOL."

4.2 WHAT WE MIGHT WANT TO HOLD IN A DATA DICTIONARY

We would define "HIGH," "NORMAL," and "COOL" as values of a discrete data element "TEMP-RANGE," with meanings described in terms of the continuous data element "TEMPERATURE."

4. *Length*; at some point, the analyst will need to specify the length (or possible range of lengths) of each data element. He should specify the length as encountered in the real world, not the length that will be taken up e.g., in binary or a packed decimal encoding.

   The analyst should not have to specify lengths on a first pass through the creation of a data dictionary, but be free to add them at a later stage.

   In the case of a money amount, it may be convenient to specify the length implicitly by simply stating the rounded maximum amount and whether cents will be held. Thus "Max $1 million, no cents" implies a length of 6 digits, (999,999) and "Max 100 million plus cents, " implies a length of 10 digits (99,999,999.99).

5. *Encoding*; the designer and programmer need a place to record in the data dictionary the form in which the data element will physically be encoded in the system, e.g., whether a number will be stored on disk as packed decimal or whether a transmission over a communications line will be ASCII or EBCDIC. Indeed, they may need to be able to represent several encodings since the same data element may come down a communications line as ASCII characters, be processed by various programs as EBCDIC and/or binary, and be stored on disk as packed decimal. These physical decisions are not for the analyst to make and form no part of the *logical* functional specification. However, in some circumstances, there is no decision to make, since the proposed system will have to interface with another system (say by accepting an input stream of telex messages).
   In this case, we have no control over the physical format, and the analyst should note the encoding of this stream in the data dictionary. Most usually, the analyst need not get involved in this level of detail, and can confine himself to specifying the type of the external data (whether it is numeric, pure alphabetic, or alphameric).

4. BUILDING AND USING A DATA DICTIONARY

6. *Other editing information.* In this section we may hold information that assists in editing the data element, especially where it is part of an input to the system. We already have recorded range, length, and type information, so under this heading we usually note the "external" references to other data elements or data stores. We may record the fact that the user wants AREA-CODE to be checked against STATE-CODE, or ACCOUNT-NUMBER to be checked against the list of delinquent accounts, or PART-NUMBERS 7000-7999 are only supplied to authorized dealers and not the general public.

A simple form for recording data elements

As we shall see in Section 4.3, it is possible to build up a data dictionary manually, using a tub of cards, or to build an automated data dictionary using packages or custom-written software. No matter which method is chosen, it is convenient to have a form layout either as the basis for the manual file, or for data entry. The layout below is convenient for 8"x 5" index cards.

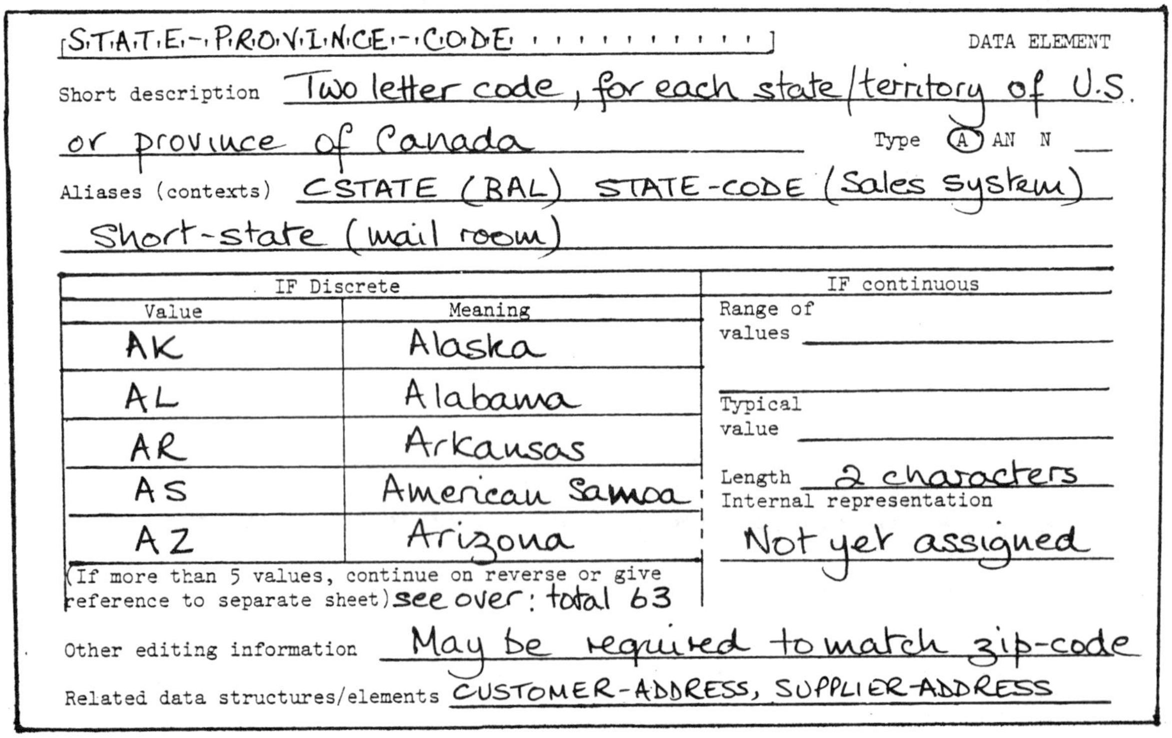

DATA ELEMENT: STATE-PROVINCE-CODE		
Short description: Two letter code, for each state/territory of U.S. or province of Canada		Type: (A) AN N
Aliases (contexts): CSTATE (BAL) STATE-CODE (Sales System) Short-state (mail room)		

IF Discrete		IF continuous
Value	Meaning	Range of values ____
AK	Alaska	
AL	Alabama	Typical value ____
AR	Arkansas	
AS	American Samoa	Length: 2 characters
AZ	Arizona	Internal representation: Not yet assigned

(If more than 5 values, continue on reverse or give reference to separate sheet) see over: total 63

Other editing information: May be required to match zip-code

Related data structures/elements: CUSTOMER-ADDRESS, SUPPLIER-ADDRESS

Figure 4.2

4.2 WHAT WE MIGHT WANT TO HOLD IN A DATA DICTIONARY

4.2.2 DESCRIBING DATA STRUCTURES

As we saw with "orders" in Chapter 2, data structures are built up out of data elements and other data structures, so in principle, we can describe any data structure we want by specifying the names of the structures and elements that make it up, provided those components are defined elsewhere in the data dictionary.

We need to know more about a structure than just its components, though. Consider a situation where a Payment Advice is sent out by a company when paying its suppliers. The Payment Advice lists the invoices covered by this payment and gives the total amount of the payment. Sometimes a check is enclosed with the Payment Advice, sometimes the payment is made directly into a previously specified bank account and the Payment Advice alone is received. For each invoice, the invoice number is cited, with the original date of issue and amount, and (if needed) some descriptive narrative. How can we represent the structure of PAYMENT-ADVICE? Let us list the components with notes:

```
        Component                              Notes

PAYMENT-ADVICE
    DATE
    NAME OF SUPPLIER
    ADDRESS OF SUPPLIER
    CHECK-NUMBER           ┐
                           ├──── either of these depending on
    BANK-PAYMENT-DETAILS ──┘     method of payment
        BANK-NAME OF SUPPLIER
        BANK-ACCOUNT-NO OF SUPPLIER
    INVOICE ─────────────────────one or more of these
        NUMBER
        DATE
        AMOUNT
        NARRATIVE ───────────────optional
    PAYMENT-TOTAL
```

We see that some components of the structure are mandatory, some are alternates, some are optional, and some are repeated (iterated) one or more times.

We need a convenient way to specify these features of a data structure; one way is to adapt the notation used in programming language manuals to show the structure of language **commands**, as follows:

4. BUILDING AND USING A DATA DICTIONARY

1. *Optional structures.* A data structure or data element within square parentheses:

    [NARRATIVE]

    means that this is an optional component of the structure.

2. *Alternate structures.* Two or more data structure or data element names, within "curly" parentheses:

    $$\begin{Bmatrix} \text{CHECK-NUMBER} \\ \text{BANK-PAYMENT-DETAILS} \end{Bmatrix}$$

    means that only one of the components will be present in any given instance of the structure.

3. *Iterations of structures.* Programming language manuals show iterations by placing three periods after the item. This is not strong enough for our purposes, so we borrow from Jackson's data structure notation, [4.3], and mark iterated structures with an asterisk:

    INVOICE*

    meaning that there may be no invoices, or one, or more. If we know the range of possibilities, then we could show the structure as:

    INVOICE* (0-10)

    meaning that there could be no invoices, or any number up to a maximum of 10.

Here is what the Payment Advice structure looks like now:

```
PAYMENT-ADVICE
   DATE
   NAME OF SUPPLIER
   ADDRESS OF SUPPLIER
  {CHECK-NUMBER          }
  {BANK-PAYMENT-DETAILS  }
      BANK-NAME OF SUPPLIER
      BANK-ACCOUNT-NO OF SUPPLIER
   INVOICE* (1-)
      NUMBER
      DATE
      AMOUNT
   [NARRATIVE]
   PAYMENT-TOTAL
```

4.2 WHAT WE MIGHT WANT TO HOLD IN A DATA DICTIONARY

A simple form for recording data structures

Since we have already defined each relevant data element, we now need to record the make-up of the data structure using the notation we have just described. Where the data structure relates to something physical, (for example, an invoice form or an existing inventory report), we may want to refer to that in the "Short Description". It is often worthwhile to give a moderately complex example of the data structure on the back of the card (if manual), or to refer to a place where this example can be found.

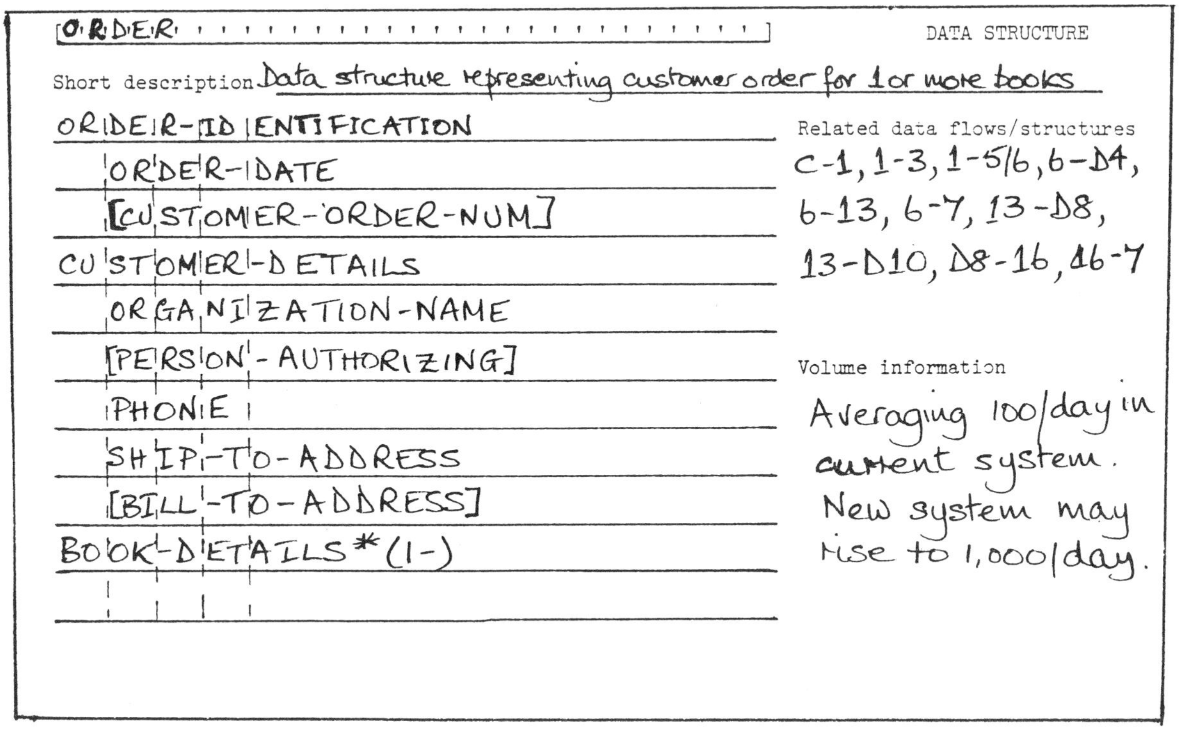

```
[ORDER                              ]        DATA STRUCTURE
Short description Data structure representing customer order for 1 or more books
ORDER-IDENTIFICATION                 Related data flows/structures
    ORDER-DATE
    [CUSTOMER-ORDER-NUM]             C-1, 1-3, 1-5/6, 6-D4,
CUSTOMER-DETAILS                     6-13, 6-7, 13-D8,
    ORGANIZATION-NAME                13-D10, D8-16, 16-7
    [PERSON-AUTHORIZING]             Volume information
    PHONE
    SHIP-TO-ADDRESS                  Averaging 100/day in
    [BILL-TO-ADDRESS]                current system.
BOOK-DETAILS * (1-)                  New system may
                                     rise to 1,000/day.
```

Figure 4.3

It is an open question whether it is better to store information on average and peak volumes for inputs and outputs at the data structure level or at the data flow level (see 4.2.3). Where a data structure goes through a number of processes more or less unchanged (such as "Orders" in the data flow diagram of Chapter 3), it makes sense to note the volumes at the data structure level. Where the data structure only takes part in one or two data flows, it is probably best to record volumes at the data flow level.

4. BUILDING AND USING A DATA DICTIONARY

4.2.3 DESCRIBING DATA FLOWS

We can now express the *contents* of a data flow by defining the names of the data structures that pass along it. (To be strict about it, the contents of a data flow just make up one large data structure, e.g. ORDERS or RETURNS or PAYMENTS, but it's a waste of an entry in the data dictionary to spell it out as such.) We also want to note:

- the source of the data flow
- the destination
- the volumes of each data structure or transaction (perhaps with their distribution throughout the day or month)
- the present physical implementation of the data flow (where we are describing an existing system)

A simple form is:

```
┌─────────────────────────────────────────────────────────────────┐
│ N.O.N.-.S.H.I.P.P.A.B.L.E.-.I.T.E.M.S. . . . . . . . ]   DATA FLOW │
│                                                                 │
│ Source ref:  6   Description: Verify inventory available        │
│ Destn. ref: 13   Description: Create back order or requisition  │
│ Expanded description  Details of each item for which an acceptable │
│ order has been received, but which cannot be shipped, either    │
│ because it is out of stock or because it is not carried in inventory │
│                                                                 │
│ Included data structures:        Volume information:            │
│    ORDER                          Out of stock — about 5 per week │
│       ORDER-IDENTIFICATION        (this is acceptable to         │
│       CUSTOMER-DETAILS                      management)          │
│       BOOK-DETAILS*              Non-inventory items             │
│       NON-SHIP-CAUSE                — about 30 per               │
│  Where the original order is for              week               │
│  multiple books, only some of them  No growth data               │
│  may appear in this data flow                                    │
└─────────────────────────────────────────────────────────────────┘
```

Figure 4.4

A short cut. As we have noted, many data flows only involve one data structure. If we take the CBM data flow of Chapter 3, examples are:

        21-p   Checks for books supplied

        6-D4   All orders

        D12-20 Amounts due

We can save entries by omitting these data flow definitions from the data dictionary and simply noting the references of the source and destination on the entry for the single data structure involved. Thus we could record data flow 21-p in the data dictionary by noting on the entry for the data structure CHECK-TO-SUPPLIER that it is the sole contents of 21-p.

4.2.4 DESCRIBING DATA STORES

Since a data store is a data structure at rest, we describe the contents of each data store in terms of the data structures which we will find in it. We also want to note the data flows which feed the data store and those which are extracted from it. If only a few defined queries are to be made, they can also be described in the data dictionary, but if the queries are of any complexity, they should be dealt with in the Immediate Access section of the functional specification (described in Chapter 7). The entry for a data store will look like Figure 4.5.

When a decision is made as to the physical organization and implementation of the store, the details of: primary key, secondary indexes, ISAM, BDAM, device residence, and so on, can be added. As can be seen from the data flow diagram, there are relatively few data stores compared with data elements and data structures; coping with the detail is not overwhelming.

4.2.5 DESCRIBING PROCESSES

As we noted in Chapter 2, the logic of processes can be documented with several tools such as decision trees, decision tables, and structured English. In a data dictionary, especially a manual one, there is sometimes not enough space to hold

4. BUILDING AND USING A DATA DICTIONARY

```
┌─────────────────────────────────────────────────────────────────┐
│  O.R.D.E.R.-.H.I.S.T.O.R.Y. . . . . . . . . . . . . ]   DATA STORE ref: D4
│  Description  All orders accepted for fulfilment – last 6 months
│  Data flows in:                    Data flows out (search arguments)
│    6 - D4   All orders              D4-10  Order details (customer name,
│                                                           order date)
│                                     D4-11  Sales detail (ISBN, Publisher-name)
│                                     D4-9   Past demand (ISBN)
│  Contents:                         Immediate access analysis is to be found in:
│    ORDER                            Functional spec, Section 8.17
│       ORDER-IDENTIFICATION         Physical organization:
│       CUSTOMER-DETAILS              Not yet specified
│       BOOK-DETAILS*(1-)
│
└─────────────────────────────────────────────────────────────────┘
```

Figure 4.5

the full logic description. What we want to do in such cases is to specify the inputs and outputs for the process, summarize the logic and enter a reference to the place in the functional specification documentation where the logic will be found.

Of course, if we can conveniently hold the entire logic of a process in the data dictionary, so much the better. A simple 8"x 5" index card form is shown in Figure 4.6.

Note that the logic is described at three levels on this card:

1. the name of the process from the data flow diagram "VERIFY CREDIT IS OK" (where the data flow diagram entry exceeds 30 characters, we should abbreviate in a meaningful way)

2. the short description of the process, which should be a "one-liner" enough to enable a person familiar with the business to understand what is meant by this process

4.2 WHAT WE MIGHT WANT TO HOLD IN A DATA DICTIONARY

```
┌─────────────────────────────────────────────────────────────────────┐
│ [V·E·R·I·F·Y·-·C·R·E·D·I·T·-·I·S·-·O·K·········]    PROCESS ref: 3  │
│ Description  Decide whether orders without prepayment can be        │
│ shipped, or whether prepayment should be demanded from customer     │
├──────────────────┬────────────────────────┬─────────────────────────┤
│      Inputs      │    Logic summary       │       Outputs           │
├──────────────────┼────────────────────────┼─────────────────────────┤
│                  │ Retrieve payment history│                        │
│ 1-3  ORDERs      │ If new customer, send   │ 3-c PREPAYMENT-REQUEST │
│                  │    prepayment request   │    [REMINDER-OF-BALANCE]│
│ D3-3 Payment history│ If regular customer, │                        │
│   DATE-ACCT-OPENED │ (average of two orders│ 3-D3 New BALANCE-ON-ORDER│
│   INVOICE*       │  per month), ok the order│                      │
│   PAYMENT*       │  unless balance overdue │ 3-6 Orders with credit OK│
│   BALANCE-ON-ORDER│ is more than two months│                        │
│                  │                    old  │                        │
│                  │ For previous customers who are                   │
│                  │ not regular, ok the order,                       │
│                  │ unless they have any balance overdue             │
├──────────────────┴────────────────────────┴─────────────────────────┤
│ Physical ref: Part of on-line order entry, OE707                    │
│ Full details of this logic can be found in: Functional specification, Section 7.2│
└─────────────────────────────────────────────────────────────────────┘
```

Figure 4.6

3. the logic summary in the center column, which should state the main functions at such a level as to enable a person familiar with the business to carry out the function (or to program it). The logic summary is not an exhaustive statement, though it should be written as unambiguously as space allows.

In Chapter 5, we discuss the relationship between these three levels, and the exhaustive and formal statement of the logic in Structured English.

4.2.6 DESCRIBING EXTERNAL ENTITIES

As can be seen from the data flow diagrams we have drawn, there are usually relatively few external entities, and it may not be worth including them in a data dictionary. However, should we wish to do so, the following information is relevant:

4. BUILDING AND USING A DATA DICTIONARY

    Name   As described on the data flow diagram

    Associated   References and descriptions
    data flows

Where the external entity is a person or group of people:

    Numbers   e.g. How many customers, growth rate, whether they are of different types

    Identification   e.g. Names of senior management, Book-keeping Supervisor

Where the external entity is another DP system:

    Language   e.g. 1401 Autocoder

    Hardware   e.g. 360/25 with emulation, System/34

    Information   Where to get more information about
    Source   the interfaces (person or document-
                          ation)

4.2.7 DESCRIBING GLOSSARY ENTRIES

In a number of applications, the users have a vocabulary of their own which can be baffling to analysts and programmers, unless they have substantial experience in the organization. When a banker speaks of "float," do you know exactly what he means? When a city official tells you about "hereditaments," do you hesitate?

    The data dictionary is a convenient place to keep these "glossary" items. Once defined they can be entered, using the data element form, with the exception that only the name, short description, and aliases sections are necessary, as shown in Figure 4.7.

4.3 MANUAL vs AUTOMATED DATA DICTIONARIES

Much of the value of a data dictionary comes from the fact that it is a *central* store of data, either for all analysts,

4.3 MANUAL vs AUTOMATED DATA DICTIONARIES

```
┌─────────────────────────────────────────────────────────────┐
│                                              GLOSSARY       │
│  [N.P.V. . . . . . . . . . . . . . . . . . . . . . ]  ~~DATA ELEMENT~~ │
│                                               ITEM          │
│  Short description  The amount of money that would have to  │
│  be invested today to produce a stated cash flow   Type A AN N __ │
│  Aliases (contexts)  Net Present Value                      │
│                                                             │
│  ┌──────────────────────────────┬──────────────────────────┐│
│  │        IF Discrete           │      IF continuous       ││
│  │  Value    │    Meaning       │  Range of                ││
│  │           │                  │  values _____        ││
│  │           │                  │                          ││
│  │           │                  │  Typical                 ││
│  │           │                  │  value _____         ││
│  │           │                  │                          ││
│  │           │                  │  Length _____        ││
│  │           │                  │  Internal representation ││
│  │           │                  │  _____        ││
│  │(If more than 5 values, continue on reverse or give      ││
│  │ reference to separate sheet) │                          ││
│  │                                                         ││
│  │ Other editing information _____       ││
│  │                                                         ││
│  │ Related data structures/elements _____       ││
└─────────────────────────────────────────────────────────────┘
```

Figure 4.7

designers and programmers working on a given project, or in a given application area, or as we shall see in Section 4.6, for the whole organization. Since the dictionary is a central store, it should be controlled by one person or one group. On a project level, such a person may be called the Data Administrator or Data Element Administrator. (At an organizational level the data dictionary is controlled by the Data Base Administration function.) The Data Administrator is responsible for keeping control over entries and changes to the data dictionary, making it as easy as possible for everyone on the project to have access to up-to-date versions, and generally helping to avoid the "reinvention of the wheel" that takes place when a team of analysts are working without coordination of data definitions. In organizations which have a Librarian as a member of the project team, the Data Administration duties fall very naturally to the Librarian.

4. BUILDING AND USING A DATA DICTIONARY

As has been shown in our examples, the data dictionary itself can be kept on the 8"x 5" cards which hold the entries, or on a larger composite card, which serves to hold all types of entry. The cards can be either grouped alphabetically within entry-type, (that is all data structure names in alphabetical order, then all data element names in alphabetical order) or in straight alpha order without regard to entry-type. A different color card may be used for each type of entry. Preferably a copying machine should be within sight, since removing a card from the tub and forgetting to put it back in the correct place has to be a crime punishable by death! The Data Administrator may need to copy all the cards regularly to give each project member an up-to-date reference copy for the uses discussed in the next section.

Depending on the resourcefulness of the Administrator and the cooperativeness of the team, quite respectable systems with upwards of 100 data elements, can be handled with file cards. As we shall see, though, there are several uses for the data dictionary which require it to be machine-readable. A first level of automation can be achieved by entering the data on the file cards using a time-sharing system with editing capability such as TSO, and allowing project members to read and print the file at will, (though only the Data Administrator should be able to update it). Simple time-sharing systems have difficulty in handling all the relationships of various entries (which can be somewhat complex), and for maximum usefulness many installations are acquiring data dictionary software packages, of which a growing number are available. A list of some of the more widely used automated packages and their vendors, appears at the end of this chapter.

In general, the automated packages provide for extensive editing of entries and build a data base with pointers and indexes to enable dictionary users to trace relationships around the dictionary, usually with on-line access. As an example, we discuss in Section 4.5 the facilities offered by DATAMANAGER, which is one such package.

4.4 WHAT WE MIGHT WANT TO GET OUT OF A DATA DICTIONARY

Once we create all the "data about data" that was described in the previous section, there are many uses for it in analysis and later in design and programming. We can identify 7 diff-

4.4 WHAT WE MIGHT WANT TO GET OUT OF A DATA DICTIONARY

erent types of output that we might want to get from a data dictionary facility; as we shall see, some of these are only practicable with software, (and the list provides a good set of criteria for evaluating data dictionary software packages). Most, however, are possible using a manual data dictionary especially on small to medium projects.

4.4.1 ORDERED LISTINGS OF ALL ENTRIES OR VARIOUS CLASSES OF ENTRY, WITH FULL OR PARTIAL DETAIL

Full listing: All details of all entries, (data flows, data structures, data elements, data stores, processes, aliases, glossary items) listed in alphabetical sequence by name. As can be imagined, this listing can become rather large when the number of data elements goes over, say, 100. Consequently, it may be convenient to provide a

Summary listing: All entries, in alphabetical sequence by name showing short description only. Figure 4.8 shows an extract from such a summary listing for a personnel system.

Where the dictionary is held on a random access device, it may be preferable to allow inquiry into the dictionary by name, or by class of entry, rather than providing extensive listings. This is because, as analysis proceeds, new elements and structures are defined and modified quite rapidly, so any printed or microfiche listing goes stale, where on-line access is not possible. The Data Administrator can combat this staleness problem by circulating an up-to-date listing to everybody on the project, say, every Monday with a list of "Additions" and "Changes" on Wednesday afternoon.

4.4.2 COMPOSITE REPORTS

When we get to the detailed analysis of the logic of a process or the structure of a data store, we need to know everything that has been defined about the relevant data structure. Not only do we want to know the contents of a data structure, we want to know the details of each component data element. Of course, we can always assemble this information from the full listing, but it is convenient to be able to call for a report that puts the information together for us in a handy form.

4. BUILDING AND USING A DATA DICTIONARY

REMOVAL-RECORD   Data structure
     Details of company paid relocation

SALARY   Data element
     Annual salary for both hourly and salaried employees

SALARY-CHANGE   Data flow
     Input from management to alter salary

SHOW-STATUS   Process
     Display current STATUS for given employee

SOC-SEC-NUM   Data element
     Social security number

SS#   Data element
     Alias for SOC-SEC-NUM

STANDARD COMPARISON   Glossary entry
     The relation between the employee's salary and the
     salaries of all those of the same grade, expressed
     as a standard deviation

STATE-ZIP   Data structure
     Combination of STATE-PROVINCE-COUNTRY-CODE and
     ZIP-POST-CODE. Covers all countries

STATUS   Data structure
     Details of all non-salary current information about
     employee

STATUS-CHANGE   Data flow
     All non-salary changes, e.g. address, dependents,
     job, training

STATUS-HISTORY   Data store
     Holds data on changes of employee STATUS and SALARY
     for entire career with company

Figure 4.8

4.4 WHAT WE MIGHT WANT TO GET OUT OF A DATA DICTIONARY

4.4.3 CROSS-REFERENCING ABILITY

After the initial work of defining the system data flow and creating the majority of entries in the data dictionary has been done, several stages of review and refinement take place during analysis and design. We often discover for example, that in the initial stage, we missed out some data elements that need to be captured, or in design, we decide to increase the length of a data element to provide for higher values in some circumstances. We thus need to *modify* some entry in the dictionary. Any change to any entry, of course can impact any number of other entries. If we add "NEXT-OF-KIN" to an employee record, we will have to modify the data structures and processes that create that record, take care of changes in it, and review all the outputs that are concerned with employee data to see if "NEXT-OF-KIN" should be included. Similarly, if we have got to the stage of defining programs and files, and decide to change the length of the "ORDER-AMOUNT" data element from 5 characters to 7 characters we need to know all the programs and files that contain or use "ORDER-AMOUNT."

This is what we call the "where used" problem; for any element or structure, we need to search the entire dictionary and find all those entries which use it. With a manual data dictionary this obviously involves a considerable amount of work, with careful study of the data flow diagram. An automated system can produce "where used" reports either by sorting, or by maintaining an index showing where each data element is used, where each data structure is mentioned, and so on.

4.4.4 FINDING A NAME FROM A DESCRIPTION

Suppose we know that each customer has a 3-month rolling average of his purchases computed. We are not sure whether this element has been defined in the data dictionary yet, and, supposing it has been defined, we are not sure of its name. How can we extract this information? We either need a super-competent Data Administrator, who knows every data element like a friend, or else we need to be able to search using keywords or strings making up parts of keywords. For example, if we could display all data elements with the strings CUST or PURCH or AV anywhere in their names, we would very likely find the data element we want.

4. BUILDING AND USING A DATA DICTIONARY

4.4.5 CONSISTENCY AND COMPLETENESS CHECKING

When we have done as much as possible to complete the data dictionary by working from the data flow diagram, it would be very convenient if we could use some software to answer these questions:

1. Are there any data flows specified without a source or without a destination?

2. Are there any data elements specified in any data stores that have no way of getting there, because they don't exist in the incoming data flows?

3. Is there any piece of process logic in which a data element is used that does not exist anywhere in the inputs to that process?

4. Are there any data elements in any data flows going into processes that either are not used in the process and/or do not appear in the output?

Especially as the system gets larger, it becomes very difficult to check the passage of every data element through every flow and process by hand. Consequently, missing elements may not be discovered until the programming stage. Test 4 above, for data elements that are redundant, may *never* be made, and possibly the redundancy may not be noticed throughout the life of the system!

4.4.6 GENERATION OF MACHINE-READABLE DATA DEFINITIONS

When the physical design of the system is complete, the programmers will want to use the data definitions in the data dictionary to create definition statements for their programs; 01's for COBOL, DSECTs for assembly language, DBD's for IMS DL/I, and so on. Rather than go through the lengthy and error-prone process of copying definitions from a data dictionary listing to a coding pad to keypunching, it makes sense to have software which transforms the data definitions in the dictionary to the format required for the target language and stores them in a copy library or punches them into cards.

4.4 WHAT WE MIGHT WANT TO GET OUT OF A DATA DICTIONARY

Thus if we have the following entries in the data dictionary:

1. Data structure: CUSTOMER-ADDRESS
   Components: ORGANIZATION
                    STREET
                    CITY
                    STATE-ZIP
                    [PHONE]

2. Data structure: PHONE
   Components: AREA-CODE
                    EXCHANGE
                    NUMBER
                    [EXTENSION]

3. Data structure: STATE-ZIP
   Components: STATE-CODE
                    ZIP

4. Data element   AREA-CODE   length   3   Numeric
5. Data element   CITY   max length   15   Alpha
6. Data element   EXCHANGE   length   3   Aphameric
7. Data element   EXTENSION   length   4   Numeric
8. Data element   NUMBER   length   4   Numeric
9. Data element   ORGANIZATION length   25   Alphameric
10. Data element   STATE-CODE   length   2   Alpha
11. Data element   STREET   length   30   Alphameric
12. Data element   ZIP   length   5   Numeric

we would like to issue the command, "Create a COBOL data division 01 entry called CUSTADDRESS using the data structure CUSTOMER-ADDRESS, and punch it into cards," and have the software, making appropriate assumptions, generate the following statements:

```
01  CUSTADDRESS
    05  ORGANIZATION         PICTURE X(25).
    05  STREET               PICTURE X(30).
    05  CITY                 PICTURE X(15).
    05  STATE-ZIP
        10  STATE-CODE       PICTURE X(2).
        10  ZIP              PICTURE 9(5).
    05  PHONE
        10  AREA-CODE        PICTURE 9(3).
        10  EXCHANGE         PICTURE X(3).
        10  NUMBER           PICTURE 9(4).
        10  EXTENSION        PICTURE 9(4).
```

97

4. BUILDING AND USING A DATA DICTIONARY

4.4.7 EXTRACTION OF DATA DICTIONARY ENTRIES FROM EXISTING PROGRAMS

This process is the reverse of the one just described. If we are concerned with maintenance of a system, or we have to develop a system which interfaces with a number of existing systems, we would like to be able to feed all the relevant existing programs into our software, and generate data dictionary entries in whatever our standard format is. This will give us a central store of information about the system we have to maintain, or interface with, which we can explore by using the other search facilities of the data dictionary. For example, we would quickly find out that one of our existing programs holds the customer's street address in a field called CUST-ST which is 22 characters long, while another holds it in a field called ST-LINE, which is only 20 characters long. These *synonyms* (different names for the same thing) and clashing definitions, may need to be resolved before we can build our new system with confidence.

4.5 AN EXAMPLE OF AN AUTOMATED DATA DICTIONARY

Several automated data dictionary software packages are listed in the appendix to this chapter. As an example of the facilities offered by such packages, we shall look at DATAMANAGER*.

DATAMANAGER is a stand-alone package, that is, it does not require the concurrent use of any data base management system (unlike, for example, UCC-TEN which requires IMS). The software consists of a single program, written in assembly language, which runs on IBM 360/370 computers under any version of OS or DOS. The size of the program varies between 50K and 80K bytes, depending on the optional features included. Each entry (called a "member") can have a name up to 32 characters long, with up to 16 aliases.

The words DATAMANAGER uses to describe processes and data are somewhat physical by comparison with the terminology we have used. Processes are described at three levels. A SYSTEM can be defined as being made up of one or more SYSTEMs, each of which is made up of PROGRAMs, each of which may be made up of MODULEs. Data can be described as a DATABASE, which may contain FILEs, which may contain GROUPs.

* *The charts and examples of DATAMANAGER output in this section are reproduced by kind permission of MSP Inc.*

4.5 AN EXAMPLE OF AN AUTOMATED DATA DICTIONARY

GROUPs may themselves be made up of numbers of GROUPs, each GROUP being composed of ITEMs. These relationships are shown in Figure 4.9.

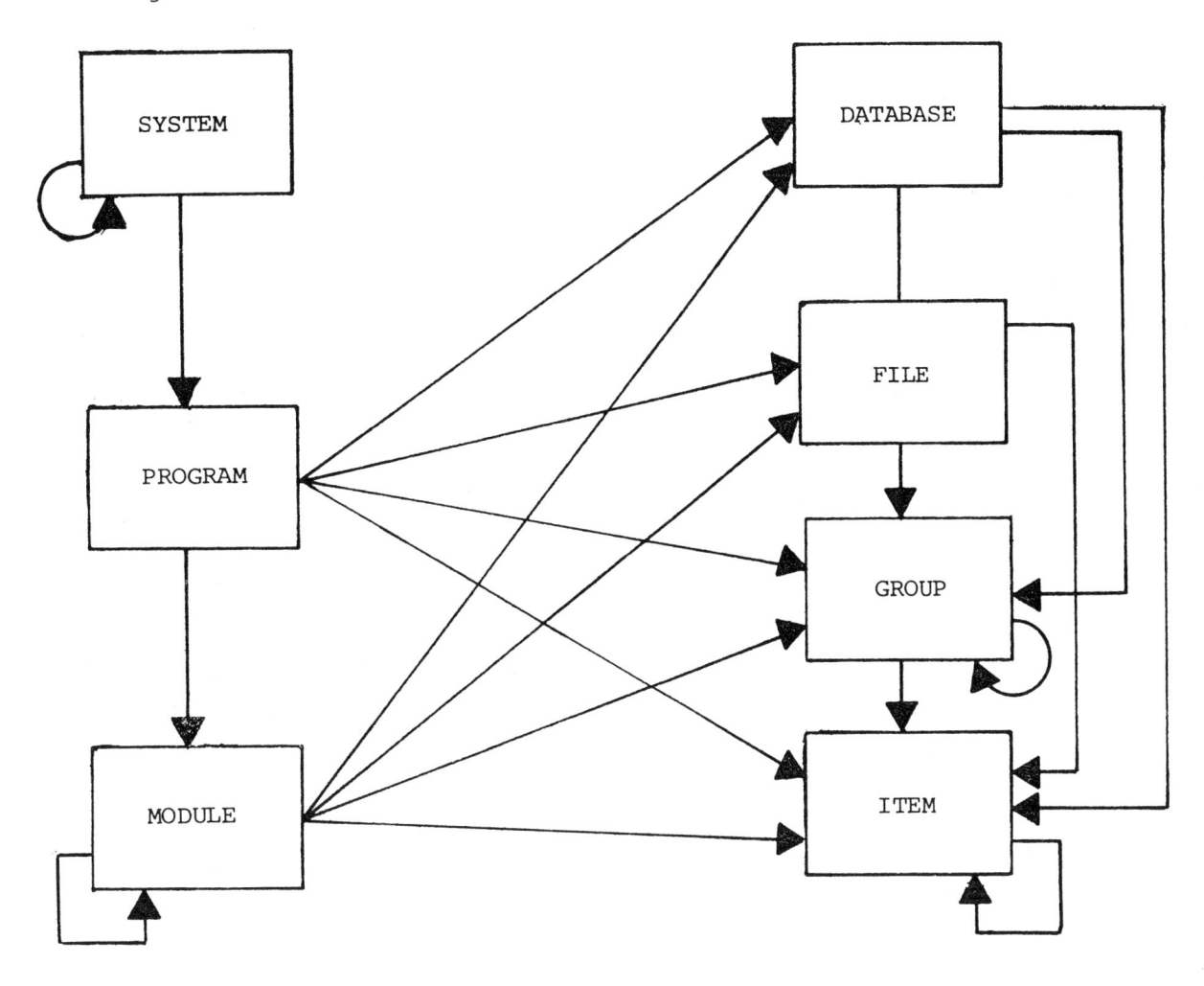

Figure 4.9

All processes are thought of as having files or databases as their inputs and outputs. Thus DATAMANAGER requires data flows to be described as files or groups, as Figure 4.10 shows.

4. BUILDING AND USING A DATA DICTIONARY

Structured Analysis Description   *DATAMANAGER description*

Data element	ITEM
Data structure	GROUP
Data flow	FILE/GROUP
Data store	FILE/DATABASE
Process	SYSTEM/PROGRAM/MODULE

Figure 4.10

DATAMANAGER provides a set of English-like commands for storing and retrieving the entries. Suppose we wanted to start defining a system which we had decided to call "MAINTAIN-EMPLOYEE-DATA"; this might correspond to one process box on the overall data flow diagram of a complex personnel system. (In fact MAINTAIN-EMPLOYEE-DATA is a subsystem, but in DATA-MANAGER, as we said, SYSTEMs can contain SYSTEMs.) We give DATAMANAGER the command "ADD MAINTAIN-EMPLOYEE-DATA" either through a terminal or on cards, and when the command is accepted, specify the characteristics of the entry.

```
28 JUL 76     21.40.08              DATAMANAGER RELEASE 1.4.1 MSP LTD (LONDON)
                                              DICTIONARY DPOOL

             00015      ADD MAINTAIN-EMPLOYEE-DATA.
  DM01131I              MAINTAIN-EMPLOYEE-DATA SUCCESSFULLY INSERTED

  DM01296I              ENCODING OF MAINTAIN-EMPLOYEE-DATA
             00100      SYSTEM CONTAINS
             00200      EMPLOYEE-VET, EMPLOYEE-MASTER-UPDATE, EMPLOYEE-REPORT
             00300      ALIAS 'MEDS'
             00400      CATALOG 'MONTHLY','COBOL'
             00500      NOTE 'THIS APPLICATION MAINTAINS THE EMPLOYEE MASTER FILE'
             00600      'AND ISSUES REPORTS ON IT'
  DM01282I              EMPLOYEE-VET ENCODED AS DUMMY
  DM01282I              EMPLOYEE-MASTER-UPDATE ENCODED AS DUMMY
  DM01282I              EMPLOYEE-REPORT ENCODED AS DUMMY
  DM01280I              MAINTAIN-EMPLOYEE-DATA SUCCESSFULLY ENCODED
```

Figure 4.11

4.5 AN EXAMPLE OF AN AUTOMATED DATA DICTIONARY

Figure 4.11 shows the beginning of this process. What we enter is shown in boxes; DATAMANAGER's responses are shown on the lines prefixed with DM. On line 100, we tell DATAMANAGER that the entry is a SYSTEM, on line 101, we specify its component processes, and enter the short name of the system "MEDS" as an alias on line 300. The CATALOG facility allows us to enter keywords which we might want to use later; in this case we foresee a need to retrieve all the systems which are run monthly, or all those which are written in COBOL.

If we enter our short description as a NOTE, when we have completed our entry, DATAMANAGER will check the statements for syntax, and create DUMMY members for those new processes we have named, but have not described yet. We proceed to fill in the details of, say, EMPLOYEE-VET, and define the files used, with their component groups and data items.

Entries can be made in any order, so we can define data elements as ITEMs before we are sure in what FILEs or data structures they will be contained.

When the system has been fully described, we may want the type of alphabetic listing of all entries that we described in Section 4.4. Figure 4.12 shows the sort of output we can get from DATAMANAGER, listing names, the type of member (ITEM, GROUP, etc.) how many other members make reference to each member and the status, (whether there is both a source record and an encoded record).

As we see from scanning this alphabetical listing, a DUMMY record has been created for TAX-CODE; we suspect this is because someone misspelled TAXCODE. We have a MODIFY command to correct this mistake.

If it were more convenient we could list members of each type (SYSTEMs, PROGRAMs, FILEs, etc). For example, if we just want to look at the data elements we would command "LIST ITEMS" and the output would look like Figure 4.13.

DATAMANAGER handles cross-referencing by setting up pointers for every possible relationship. For example, if the

4. BUILDING AND USING A DATA DICTIONARY

LIST OF ALL MEMBERS

	TYPE	MBRS USING	STATUS
ACTION-CODE	ITEM	5	SCE ENC
ADDRESS	ITEM	2	SCE ENC
ADDRESS-UPDATE	GROUP	1	SCE ENC
BASIC-UPDATE	GROUP	1	SCE ENC
DEDUCT-CODE	ITEM	2	SCE ENC
DELETE	GROUP	1	SCE ENC
DEPARTMENT	ITEM	5	SCE ENC
EMPLOYEE-HISTORY-LIST	FILE	1	SCE ENC
EMPLOYEE-HISTORY-MASTER	FILE	3	SCE ENC
EMPLOYEE-HISTORY-REPORT	PROGRAM	1	SCE ENC
EMPLOYEE-HISTORY-UPDATE	PROGRAM	1	SCE ENC
EMPLOYEE-LIST	FILE	1	SCE ENC
EMPLOYEE-MASTER	FILE	4	SCE ENC
EMPLOYEE-MASTER-UPDATE	PROGRAM	1	SCE ENC
EMPLOYEE-NUMBER	ITEM	5	SCE ENC
EMPLOYEE-RECORD	GROUP	1	SCE ENC
EMPLOYEE-REPORT	PROGRAM	1	SCE ENC
EMPLOYEE-TRANSACTIONS	FILE	1	SCE ENC
EMPLOYEE-TRANSACTIONS-SORTED	FILE	3	SCE ENC
EMPLOYEE-VET	PROGRAM	1	SCE ENC
FILLER00002	ITEM	13	SCE ENC
HISTORY-RECORD	GROUP	1	SCE ENC
HISTORY-REPORT-RECORD	GROUP	1	SCE ENC
JOB-COUNT	ITEM	2	SCE ENC
JOB-ENTRY	GROUP	1	SCE ENC
JOB-STATUS	ITEM	5	SCE ENC
JOB-TITLE	ITEM	5	SCE ENC
MAINTAIN-EMPLOYEE-DATA	SYSTEM	0	SCE ENC
MAINTAIN-EMPLOYEE-HISTORY	SYSTEM	0	SCE ENC
NAME	ITEM	4	SCE ENC
REPORT-COUNT	GROUP	1	SCE ENC
REPORT-RECORD	GROUP	1	SCE ENC
SALARY	ITEM	5	SCE ENC
SOCIAL-SECURITY-NUMBER	ITEM	3	SCE ENC
TAX-CODE	*ITEM	1	DUM
TAXCODE	ITEM	1	SCE ENC
TRANSACTION-RECORD	GROUP	2	SCE ENC
TYPE	ITEM	6	SCE ENC

```
DM01500I        LIST CONTAINED
DM01501I         15 ITEMS
DM01502I         10 GROUPS
DM01503I          6 FILES
DM01505I          5 PROGRAMS
DM01506I          2 SYSTEMS
DM01508I          1 DUMMYS
```

Figure 4.12

4.5 AN EXAMPLE OF AN AUTOMATED DATA DICTIONARY

```
00051        LIST ITEMS.

LIST OF ITEMS

                                  TYPE     MBRS USING  STATUS

   ACTION-CODE                    ITEM         5       SCE ENC
   ADDRESS                        ITEM         2       SCE ENC
   DEDUCT-CODE                    ITEM         2       SCE ENC
   DEPARTMENT                     ITEM         5       SCE ENC
   EMPLOYEE-NUMBER                ITEM         5       SCE ENC
   FILLER00002                    ITEM        13       SCE ENC
   JOB-COUNT                      ITEM         2       SCE ENC
   JOB-STATUS                     ITEM         5       SCE ENC
   JOB-TITLE                      ITEM         5       SCE ENC
   NAME                           ITEM         4       SCE ENC
   SALARY                         ITEM         5       SCE ENC
   SOCIAL-SECURITY-NUMBER         ITEM         3       SCE ENC
   TAX-CODE                      *ITEM         1       DUM
   TAXCODE                        ITEM         1       SCE ENC
   TYPE                           ITEM         6       SCE ENC


   DM01500I              LIST CONTAINED
   DM01501I                 15 ITEMS
   DM01508I                  1 DUMMYS
```

Figure 4.13

group TRANSACTION-RECORD (shown in Figure 4.12) contains AD-DRESS-UPDATE, which contains ADDRESS, we know there are two direct relationships, which may be shown as in Figure 4.14. But those two direct relationships imply another four *indirect* relationships, making a total of six between the three members, as shown by the dotted lines in Figure 4.15.

103

4. BUILDING AND USING A DATA DICTIONARY

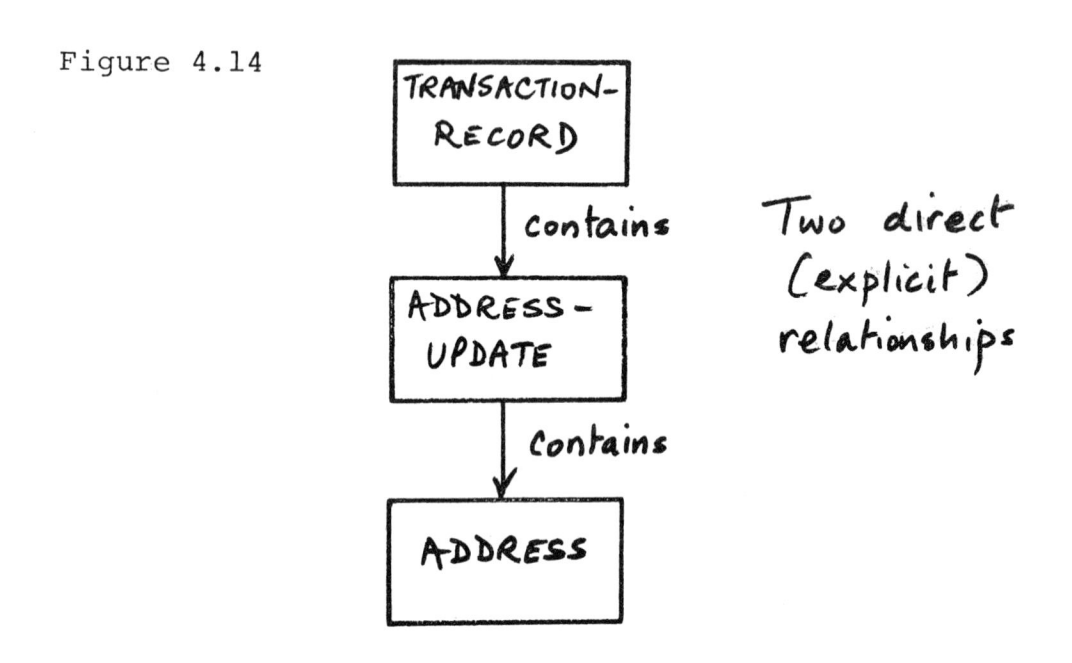

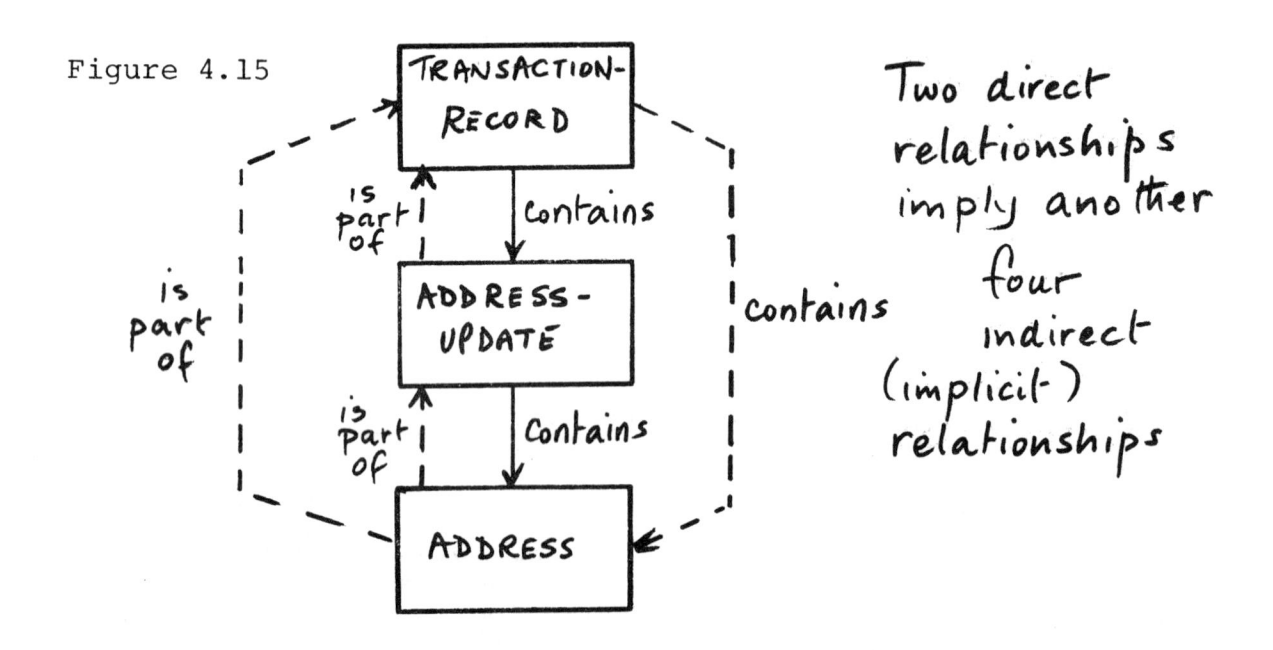

Figure 4.14

TRANSACTION-RECORD
↓ contains
ADDRESS-UPDATE
↓ contains
ADDRESS

Two direct (explicit) relationships

Figure 4.15

TRANSACTION-RECORD
↕ is part of / contains
ADDRESS-UPDATE
↕ is part of / contains
ADDRESS

(with dashed "is part of" and "contains" relationships connecting TRANSACTION-RECORD and ADDRESS directly)

Two direct relationships imply another four indirect (implicit) relationships

104

4.5 AN EXAMPLE OF AN AUTOMATED DATA DICTIONARY

DATAMANAGER sets up and maintains complete bi-directional cross-references, based on the direct relationships entered by the user. This makes it relatively simple to get "where used" information. For example, suppose we are concerned with the item DEPARTMENT and want to trace through the system to see what data flows, processes, and data stores it is involved in. We can issue the "WHAT USES...." command, and get a tracing of all the relationships, as shown in Figure 4.16. (For brevity the PROGRAM and SYSTEM relationships have been omitted.

The "WHICH...USE" command enables more specific questions to be asked:

```
00058         WHICH FILES USE DEPARTMENT.

THE FOLLOWING USE ITEM DEPARTMENT
 FILES EMPLOYEE-MASTER
       EMPLOYEE-HISTORY-MASTER
       EMPLOYEE-TRANSACTIONS-SORTED
       EMPLOYEE-TRANSACTIONS
       EMPLOYEE-HISTORY-LIST
       EMPLOYEE-LIST

00059         WHICH PROGRAMS USE DEPARTMENT.

THE FOLLOWING USE ITEM DEPARTMENT
 PROGRAMS EMPLOYEE-HISTORY-REPORT
          EMPLOYEE-MASTER-UPDATE
          EMPLOYEE-REPORT
          EMPLOYEE-HISTORY-UPDATE
          EMPLOYEE-VET
```

Other DATAMANAGER commands enable cross-references between members and keywords to be explored as required.

At the time of writing, DATAMANAGER can read data definitions from existing programs in COBOL and PL/I. It can automatically generate data definitions for COBOL, PL/I, and IBM 370 Assembler and put them in named libraries ready for use by the programmer. It can generate data definitions for TOTAL, ADABAS, and IMS.

4. BUILDING AND USING A DATA DICTIONARY

```
    00061        WHAT USES DEPARTMENT.

ITEM DEPARTMENT IS USED BY
   GROUP    EMPLOYEE-RECORD
   GROUP    HISTORY-RECORD
   GROUP    TRANSACTION-RECORD
   GROUP    HISTORY-REPORT-RECORD
   GROUP    REPORT-RECORD

GROUP EMPLOYEE-RECORD IS USED BY
   FILE     EMPLOYEE-MASTER

GROUP HISTORY-RECORD IS USED BY
   FILE     EMPLOYEE-HISTORY-MASTER

GROUP TRANSACTION-RECORD IS USED BY
   FILE     EMPLOYEE-TRANSACTIONS-SORTED
   FILE     EMPLOYEE-TRANSACTIONS

GROUP HISTORY-REPORT-RECORD IS USED BY
   FILE     EMPLOYEE-HISTORY-LIST

GROUP REPORT-RECORD IS USED BY
   FILE     EMPLOYEE-LIST

FILE EMPLOYEE-MASTER IS USED BY
   PROGRAM EMPLOYEE-HISTORY-REPORT
   PROGRAM EMPLOYEE-MASTER-UPDATE
   PROGRAM EMPLOYEE-MASTER-UPDATE
   PROGRAM EMPLOYEE-REPORT

FILE EMPLOYEE-HISTORY-MASTER IS USED BY
   PROGRAM EMPLOYEE-HISTORY-REPORT
   PROGRAM EMPLOYEE-HISTORY-UPDATE
   PROGRAM EMPLOYEE-HISTORY-UPDATE

FILE EMPLOYEE-TRANSACTIONS-SORTED IS USED BY
   PROGRAM EMPLOYEE-HISTORY-UPDATE
   PROGRAM EMPLOYEE-MASTER-UPDATE
   PROGRAM EMPLOYEE-VET

FILE EMPLOYEE-TRANSACTIONS IS USED BY
   PROGRAM EMPLOYEE-VET

FILE EMPLOYEE-HISTORY-LIST IS USED BY
   PROGRAM EMPLOYEE-HISTORY-REPORT

FILE EMPLOYEE-LIST IS USED BY
   PROGRAM EMPLOYEE-REPORT
```

Figure 4.16

4.5 AN EXAMPLE OF AN AUTOMATED DATA DICTIONARY

Multiple versions of entries can be held and retrieved, so it is possible to carry out testing with one version of a data structure, and retain the previous version for production work in the meanwhile.

There are several levels of security protection. A password system is provided and any given user may be limited to accessing certain entries and not others (e.g., SALARY is inaccessible to everyone except the Personnel Manager) or limited to certain commands (e.g., MODIFY can only be issued by the Data Administrator).

It is possible to have DATAMANAGER and the data dictionary available to other programs on-line in "active-mode". Thus, since the allowable range of a data element is held in the data dictionary, an edit program can retrieve that allowable range during the editing of a transaction. Should we want to change the allowable range, we change the value held by DATAMANAGER and do *not* need to recompile the edit program (or programs).

DATAMANAGER enables us to do a very limited amount of consistency checking. Since data flows have to be represented as files, there is no way of ensuring that each data flow has both a source and a destination; we have to supply that ourselves. While it is possible to hold the logic of processes as a NOTE in the program entry or module entry, it is not possible to manipulate this logic in any way, or to compare the data elements in the input data flows.

External entities, though not represented as a separate member type, can be entered as SYSTEMs (or as FILEs or GROUPs).

This description does not exhaust the facilities of DATAMANAGER, which, like other data dictionary packages, is an evolving product. Nonetheless, we hope it gives a good idea as to the power, flexibility, and usefulness of automated packages as a support for systems analysis, and later for design and programming.

4. BUILDING AND USING A DATA DICTIONARY

4.6 CROSS-PROJECT OR ORGANIZATION-WIDE DATA DICTIONARIES

Much of the discussion in this Chapter has tacitly assumed that each project is building a data dictionary for its own use. Once we have an automated data dictionary capability, though, it becomes attractive to consider holding *all* the data elements, data structures and other entries for an entire applications area (say order entry and inventory control) or for the whole organization.

Building an organization-wide data dictionary involves a major commitment; since the hundreds of files and applications that exist in an organization were created separately, there will be many aliases, incompatible definitions of the same element, conflicting codes, redundant items, and cases where the same name is used for two different things. Producing an integrated data dictionary needs a great deal of work to resolve data incompatibilities: though the data dictionary software can handle aliases, the existence of incompatible definitions and duplicates means that programs may have to be changed before data can be shared. People need to change habits they may have practised for years; suppose the Sales Department has divided the country into regions, for reporting purposes, with say, New England being defined as Maine, Vermont, and New Hampshire, whereas the Shipping Department has always defined New England as Maine, Vermont, New Hampshire, Massachusetts, and Rhode Island. If they must now move in the direction of an integrated data base, somebody's definition of REGION must change, or we will always appear to have shipped more to the Yankee traders than we sold them! Who wants to change their description? Neither department benefits directly, though corporate management benefits because reports comparing sales and shipments can now be produced. The Data Base Administrator will need to use tact and persuasion to slowly get compatible data definitions.

Once an organization-wide data dictionary has been built, though, the benefits are great:

4.7 DATA DICTIONARIES AND DISTRIBUTED PROCESSING

- users benefit because they can have a uniform list of what data is available to them. This is particularly valuable where users have on-line access to data bases and can use a query language, which enables them to ask questions of the data, such as "What PRODUCT had SALES in any REGION last week that were more than 10% greater than the previous week's SALES in that REGION?" For the query language software (explained in more detail in Chapter 7) to be able to handle this query there must be a data dictionary definition of PRODUCT, SALES, and REGION. If the data dictionary covers the whole organization, users can ask questions about all aspects. If, as with many query systems installed at present, the data dictionary just covers one application area, the possibilities are much more limited.

- analysts benefit because they start a new project with readily available definitions of the data that will be provided to the new system. Much of the spadework of building a data dictionary will have been done for them.

- designers and programmers benefit because the physical characteristics of much of the data they will need in their programs are already available in machine-readable form.

4.7 DATA DICTIONARIES AND DISTRIBUTED PROCESSING

In the 1960's there were important economies of scale in computing; it was cheaper to have one large computer than two computers half the size. For various reasons, this is no longer true; in the late 1970's we are going through a phase where small computers have become much cheaper, size for size, than large ones. Consequently, it is very attractive to give each department, branch office, factory and warehouse its own small computer, and only have a computer at headquarters for the purposes of pulling together management information from all these distributed smaller computers.

For example, the New York branch of a Chicago-based firm might have a small computer, say a System/34 or a PDP-11 with a few CRT displays, and use it for entering orders and generating invoices. Each night the New York computer automatically dials up the main Chicago computer and transmits details of the orders and amounts to Chicago, which receives all payments for the corporation; this information is used to update the

4. BUILDING AND USING A DATA DICTIONARY

main corporate files. In turn, the main Chicago computer transmits details to New York of payments received from New York's customers during the previous day, and information on say, current delivery times for all products. By the start of the next day, New York's computer has updated its local files so that inquiries can be made as to which customers are behind in their payments, which products are out of stock in which warehouses, and so on. In fact New York now has enough information to do the day's business, only communicating with Chicago to handle exception conditions.

If, as has been the case in many corporations during the mid 70's, all processing is done on a large computer in Chicago, the only way to give New York comparable service has been to have an expensive (and fault-prone) leased line between each branch and headquarters.

The advantages of distributed processing over this centralized processing are:

1. *More reliability*. The branches are not paralyzed by failures of the central computer, or the telephone line; since each has its own processor, they can continue to do business, even if they can't exchange information with headquarters

2. *Less cost*. The total equipment cost for the distributed system is often cheaper than the centralized system plus the telephone lines

3. *More flexibility*. Each branch manager has the possibility of programming his own computer to take account of local variations in business or different reports that he requires. Indeed, it is this feature of distributed processing, as much as any, that is accounting for its rapid spread. Many user managers are dissatisfied with the level of service they have been getting from centralized computer facilities, and the quality of response to their needs they have been getting from centralized computing services, so they have jumped at the opportunity to get computing service from cheap, easy-to-instal machines that appear easy to program. A minicomputer has been defined as "any computer so cheap that you can buy it without corporate headquarters knowing what you're doing!"

The dangers in this trend, however, are that unless each organization is very careful, the distributed systems will

4.7 DATA DICTIONARIES AND DISTRIBUTED PROCESSING

each evolve in their own way, with incompatible data definitions and programs so that it will be impossible for the various parts of the organization to share data. This is where the data dictionary comes in. Provided each local system keeps rigidly to the data elements and structures defined in the data dictionary it will always be possible for the central site to gather and use the local processors' data, and vice versa, no matter what programs may be written at each local site. No-one can conform to definitions in a dictionary unless they have the dictionary, so it follows that we are likely to see an upsurge of activity in the creation and diffusion of organization-wide data dictionaries.

4. BUILDING AND USING A DATA DICTIONARY

APPENDIX

COMMERCIALLY AVAILABLE DATA DICTIONARY SOFTWARE PACKAGES

DATA CATALOGUE

    Synergetics Corporation
    1 De Angelo Drive
    Bedford, MA 01730
    (617) 275 0250

DATAMANAGER

    MSP Inc.
    594 Marrett Road
    Lexington, MA 02173
    (617) 861 6130

DATA DICTIONARY

    Cincom Systems Inc.
    307 Maples Ave. West
    Vienna, VA 22180
    (703) 281 2121

IBM DB/DC DATA DICTIONARY SYSTEMS

    see IBM SB21 1256

LEXICON

    Arthur Anderson and Co.
    69 West Washington Street
    Chicago, IL 60602
    (312) 346 6262

UCC TEN

    University Computing Company
    8303 Elmbrook Drive
    Dallas, TX 75247
    (214) 688 7100

REFERENCES

4.1 *IMS/VS General Information Manual, GH20-1260,* IBM Corporation, White Plains, N.Y., 1974.

4.2 National Bureau of Standards Handbook 113, *"CODASYL Data Description Language Journal of Development,"* Government Printing Office, Washington D.C. 20402, 1974.

4.3 M. Jackson, *"Principles of Program Design,"* Academic Press, 1975.

4.4 H. Lefkovitz, *"Data Dictionary Systems,"* QED Information Sciences, Wellesley, MA 02181, 1977.

4.5 B. Leong-Hong and B. Marron, *"Technical Profile of Seven Data Element Dictionary/Directory Systems,"* National Bureau of Standards Special Pub. 500-3, Government Printing Office, Washington D.C. 20402, 1977.

4. BUILDING AND USING A DATA DICTIONARY

EXERCISES AND DISCUSSION POINTS

1. Make an exhaustive list of all the data elements (name and concise description only) in a simple system with which you are familiar (or use the non-inventory CBM system described in Chapter 2). How many data elements did you end up with? How long did it take per element? If you have access to a time-sharing system, use it to sort the data elements by name.

2. List some instances of data incompatibilities in your organization.

3. What data element of your knowledge has the most aliases?

4. How many different formats of calendar date are in use in your organization?

5. Describe some complex data structures using the notation for option, alternation, and iteration.

6. Design a multipurpose form which could be used for making any entry into a data dictionary.

7. Become familiar with the facilities and commands of the data dictionary software in your installation (or in the nearest installation to have such software).

8. Build a table relating the 2-letter standard abbreviations for the United States to the valid zip-code ranges for each state. Under what circumstances would it be worth editing an address using this table?

9. Where should the data dictionary be in a distributed processing system?

10. Do you think the creation of a project data dictionary is worth the effort? If not, why not?

Chapter 5

Analyzing and presenting process logic

5.1 THE PROBLEMS OF EXPRESSING LOGIC

In defining a process in the data dictionary, we specified the inputs and outputs, and wrote a summary of the logic as clearly as we could in English. As we noted in Chapter 2, there are a number of pitfalls in expressing logic in narrative English; in this chapter we examine these problems, and then examine the tools which enable us to express process logic clearly and unambiguously.

5.1.1 NOT ONLY BUT NOTWITHSTANDING, AND/OR UNLESS....

What is the difference between the following five statements?

1. "Add A to B unless A is less than B, in which case subtract A from B."

2. "Add A to B. However, if A is less than B, the answer is the difference of A and B."

3. "Add A to B, but subtract A from B where A is less than B."

5. ANALYZING AND PRESENTING PROCESS LOGIC

4. "The total is found by adding B to A. Notwithstanding the previous sentence, in the circumstance where B is greater than A the result is the difference between B and A."

5. "The total is the sum of A and B: only when A is less than B should the difference be used as the total."

The answer, of course, is that there is *no* logical difference. The forms of the five English narrative sentences clearly obscure that similarity, since each can be reduced to one statement in the standard IF-THEN-ELSE-SO form:

```
IF    A is less than B
  THEN subtract A from B
ELSE  (A is not less than B)
  SO add A to B.
```

In trying to understand narrative in policy documents, memos, and specifications, we continually run up against the variety of possible forms that English allows. As analysts, we need to be able to reduce these varieties to simple logical statements for use as program and system specs. Our aim is to reduce all the documents into imperative phrases and conditions. We shall use the term *action* to refer to some imperative phrase such as "Add A to B" or "Send out prepayment request," and *condition* to refer to some fact which determines which action is taken (e.g., "A is less than B," and "Customer is regular").

Statement 1: "Add A to B unless A is less than B, in which case subtract A from B,"

reduces to:

"Action-1, *unless* Condition-1, *in which case* Action-2

which, in our standard IF-THEN-ELSE-SO form, is:

```
IF    Condition-1
  THEN Action-2
ELSE  (not Condition-1)
  SO  Action-1
```

5.1 THE PROBLEMS OF EXPRESSING LOGIC

Note that the order of actions in the IF-THEN-ELSE-SO structure is the reverse of the order in the English sentence, unless we rephrase the condition to ask a negative question first:

```
   IF     not Condition-1            IF     A is not less than B
     THEN  Action-1                    THEN  add A to B
     ELSE  (Condition-1)               ELSE  (A is less than B)
       SO  Action-2                      SO  subtract A from B
```

Statement 2: "Add A to B. However, if A is less than B, the answer is the difference of A and B,"

reduces to:

      Action-1. *However, if* Condition-1, Action-2

(The phrase "the answer is the difference of A and B," is an implied imperative and so counts as an action.)

Statement 3: "Add A to B, but subtract A from B where A is less than B,"

reduces to:

      Action-1, *but* Action-2 *where* Condition-1

Statements 4 and 5 can be reduced in similar ways; they are shown in Figure 5.1 together with some other common sentence structures. They indicate the variety of English sentence structures that the poor analyst must deal with to get his information on the policies and procedures which are to be built into the system.

The guiding principle is:

> *"Identify the actions and the conditions and use the common sentence structure set out in Figure 5.1 to cast the sentence into the IF-THEN-ELSE-SO model."*

5. ANALYZING AND PRESENTING PROCESS LOGIC

Some English conditional sentence structures expressing the logic:

```
        IF    Condition-1
        THEN  Action-A
        ELSE  (not Condition-1)
        SO    Action-B
```

1. Action-B, *unless* Condition-1, *in which case* Action-A.

2. Action-B, { *but* / *however* } *if* Condition-1, Action-A.

3. Action-B, { *but* / *however* } Action-A *where* Condition-1.

4. Action-B. *Notwithstanding the former,* { *where* / *if* } Condition-1, Action-A.

5. Action-B, *only if* Condition-1, Action-A.

6. Action-A, *subject to* Condition-1. *Failing that,* Action-B.

Figure 5.1

5.1.2 GREATER THAN, LESS THAN

As if the variety of structures were not a confusing enough situation, the English language often gives us trouble when it comes to expressing a range of values as part of a condition. Suppose we say, "Up to 20 units, no discount. More than 20 units, 5% discount." What could be clearer than that? But what discount is given for exactly 20? The problem is that English needs the awkward phrase "inclusive," or "up to and including." We could say "1-19 units," "Quantity less than or equal to 20," and we are back in the problem of too many ways to say the same thing, as shown in Figure 5.2.

We tend to be very sloppy about the way we specify ranges, because *we* know which of the possibilities we mean, but users are often surprised when the strict interpretation of their words turns out to be different from what they had in mind. To verify that all the possible ranges are dealt with we can go through the narratives, looking for the words "less," "up to," "greater," "more," and substitute for them the programming terms GT, GE, LE, LT, as appropriate, where:

5.1 THE PROBLEMS OF EXPRESSING LOGIC

```
.
.
.
18
     1-19  Up to but not including 20
19         Quantity less than 20        1-20  Up to and including 20
------------------------------                Quantity less than or
20                                                  equal to 20
     20 or Quantity greater than or   ----------------------
21    more      equal to 20
                                      More
22                                    than  Quantity greater than 20
                                       20
.
.
.
```

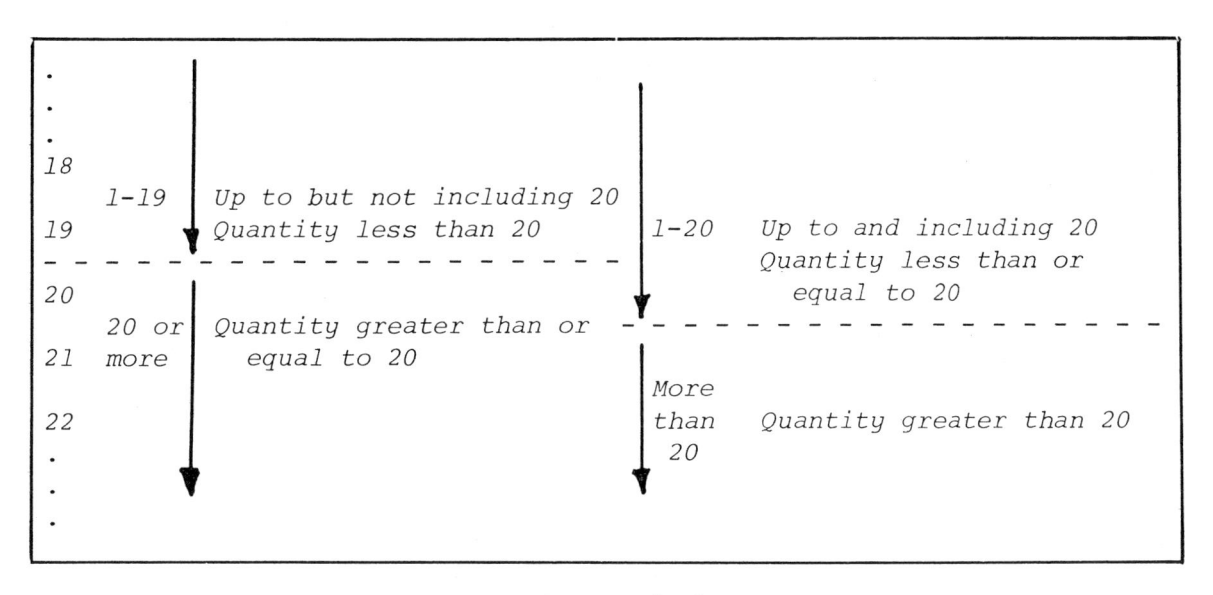

Figure 5.2

    GT   means "greater than"
    GE   means "greater than or equal to"
    LE   means "less than or equal to"
    LT   means "less than"

Clearly, if condition-1 is LT 20 and condition-2 is GT 20, we still have a problem knowing what to do when the quantity is exactly 20. We are looking for the matching pairs of either GT and LE, or GE and LT. We should ensure that all users are familiar with these distinctions, even if the abbreviations are too much like jargon to be easily accepted.

5.1.3 AND/OR AMBIGUITY

Consider this policy statement:

> *"Customers who place more than $10,000 business per year and have a good payment history or have been with us more than 20 years are to receive priority treatment."*

Is it enough to place more than $10,000 per year to get priority treatment? Clearly not, because of the phrase "*and* have a good payment history." But is it enough to have been a customer more than 20 years? Well, it all depends on the tone

5. ANALYZING AND PRESENTING PROCESS LOGIC

of voice with which one reads the policy. If you read it as:

("more than $10,000 a year")

and

("good payment history" or "more than 20 years")

you will say "No; you must also place more than $10,000 business per year." But if you read it as:

("more than $10,000 a year" and "good payment history")

or

("more than 20 years")

you will say "Yes, being a customer more than 20 years gets me priority treatment, no matter how little business I do, or how poorly I pay."

The person who wrote (or more likely dictated) the original policy statement knows what he meant; unfortunately, any English sentence with both "and" and "or" joining conditions in it is clear when spoken, because of the emphasis given, but not clear when written down. Worse still, if the analyst does not question the and/or ambiguity, and writes the policy into the specification, the programmer may not question it and will code is, say in COBOL, as it is written. The COBOL compiler has built-in rules for deciding on the precedence of "and" and "or" unless given specific instructions; in this case, it will cause the program to behave as though the policy were:

("more than $10,000 a year" and "good payment history")

or

("more than 20 years")

and give priority treatment to old Joe, who has ordered $100 worth of goods every Christmas for the last 35 years, and paid us when the next harvest is in. This may or may not be what the management of the business want to do, but if the decision is left to the COBOL compiler, we have only a 50:50 chance of being right!

5.1 THE PROBLEMS OF EXPRESSING LOGIC

The moral is that if we cannot avoid writing policy statements which mix "and" with "or" in the same sentence, we should rephrase the combinations to express our true meaning, so that the programmer will know unambiguously what we mean. For example, we could write:

"Customers who place more than $10,000 business per year, and in addition, either have a good payment history or have been with us more than 20 years, are to receive priority treatment."

5.1.4 UNDEFINED ADJECTIVES

What is a "good" payment history?
What is a "regular" customer?

People in organizations develop experience in their jobs, and as a result can assign meanings to adjectives like these. As analysts, we have to be sure that we can define the adjectives and be able to tell a good history from a bad one. Often the defining rule is quite complex, and unless we scan each policy statement, identify the adjectives, and be sure we have each one defined, we will never find out.

As we noted in Section 4.2, the data dictionary provides us with a good format for defining the adjectives that are used in logic. The very phrase "good payment history" implies the possibility of a *bad* payment history, and that implies the existence of a data-element "PAYMENT-HISTORY-TYPE," which can have the value "GOOD" or "BAD." Going back to the data dictionary format, we know that we have to give a meaning to each value of a discrete data element, as shown in Figure 5.3.

5.1.5 HANDLING COMBINATIONS OF CONDITIONS

So far, we have looked at the logical problems that can happen in fairly simple statements, and seen how to make them unambiguous. A different class of problem arises when, as frequently happens, the policy statement involves *combinations* of conditions.

The customer treatment policy specifies two sets of combined conditions:

121

5. ANALYZING AND PRESENTING PROCESS LOGIC

```
┌─────────────────────────────────────────────────────────────────┐
│ [P.A.Y.M.E.N.T.-.H.I.S.T.O.R.Y.-.T.Y.P.E. . . . . . . . . .]      DATA ELEMENT │
│ Short description  Categorizes whether customer is regarded      │
│ as a good payer or not                         Type  (A)  AN  N __│
│ Aliases (contexts) _____│
│ _____ │
│ ┌──────────────────────────────────┬────────────────────────────┐│
│ │         IF Discrete              │      IF continuous         ││
│ │  Value  │      Meaning           │  Range of values _____ ││
│ │  GOOD   │ No invoice more than   │  _____  ││
│ │         │ 30 days overdue in     │                            ││
│ │         │ last 6 months          │  Typical value _____  ││
│ │  BAD    │ One or more invoices   │                            ││
│ │         │ more than 30 days      │  Length _____  ││
│ │         │ overdue in last        │  Internal representation   ││
│ │         │       6 months         │  _____  ││
│ │(If more than 5 values, continue on reverse or give             ││
│ │ reference to separate sheet)     │                            ││
│ └──────────────────────────────────┴────────────────────────────┘│
│ Other editing information _____│
│ Related data structures/elements _____│
└─────────────────────────────────────────────────────────────────┘
```

Figure 5.3

"more than $10,000 business" *and* "good payment history"

or

"more than $10,000 business" *and* "with us more than 20 years"

The policy specified that these combinations led to a certain action (priority treatment) and implied that all other combinations led to normal treatment. We could draw a picture of this policy as a *decision tree*:

5.1 THE PROBLEMS OF EXPRESSING LOGIC

```
                    good payment ─────────────── priority treatment
                   /history
        more than /
        $10,000  <
       /business  \  bad payment   with us more
      /            \ history      /than 20 years ── priority treatment
     /                           <
    <                             \with us 20 years
     \                             or less      ──── normal treatment
      \
       \$10,000 ────────────────────────────────── normal treatment
        or less
```

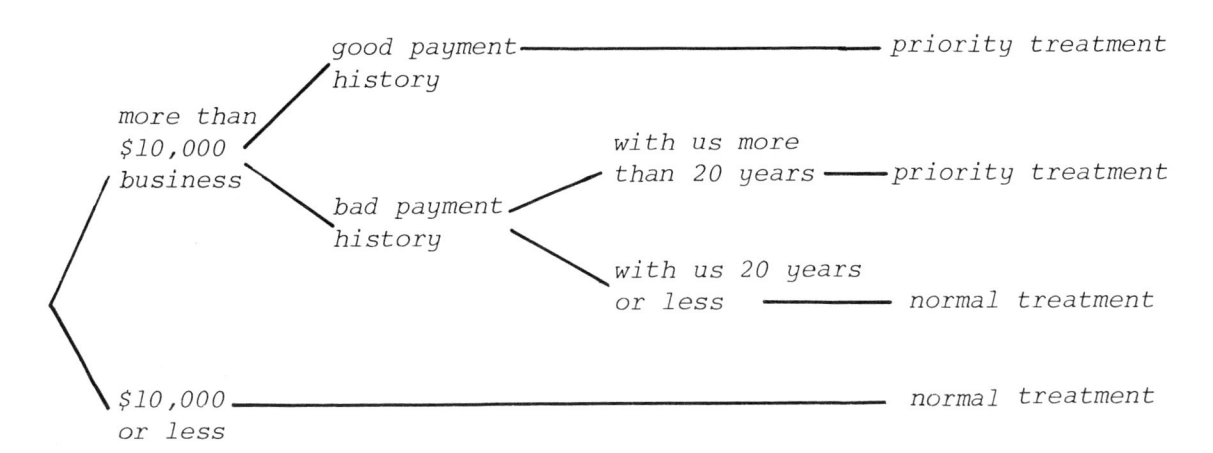

Figure 5.4

or cast it into the IF-THEN-ELSE-SO format:

```
IF       customer does more than $10,000 business
  and-IF    customer has good payment history
      THEN  priority treatment
      ELSE  (bad payment history)
         so-IF     customer has been with us for more than 20 years
            THEN  priority treatment
            ELSE  (20 years or less)
               SO normal treatment
ELSE    (customer does $10,000 or less)
   SO   normal treatment
```

Figure 5.5

Notice that combinations of conditions cause us to "nest" IFs within IFs and ELSEs. We have indented the IFs and ELSEs that are nested within other IFs and ELSEs, and aligned each ELSE with its corresponding IF.

The IF-THEN-ELSE-SO format or so-called "Structured English" is easier to type for good presentation than the decision tree, and allows for conditions and actions to be written out in full, but it does not show the structure of the policy as vividly as the decision tree. We will discuss the pros and cons of these techniques later; for a decision of this

5. ANALYZING AND PRESENTING PROCESS LOGIC

simple sort *both* are less handy than the unambiguous sentence we worked out:

> *"Customers who place more than $10,000 business per year, and in addition, either have a good payment history or have been with us for more than 20 years are to receive priority treatment."*

But what if the decision is really more complex than this? Suppose customers who do less than $10,000 business per year, and have a good payment history, can also get priority treatment?

The decision tree would then become:

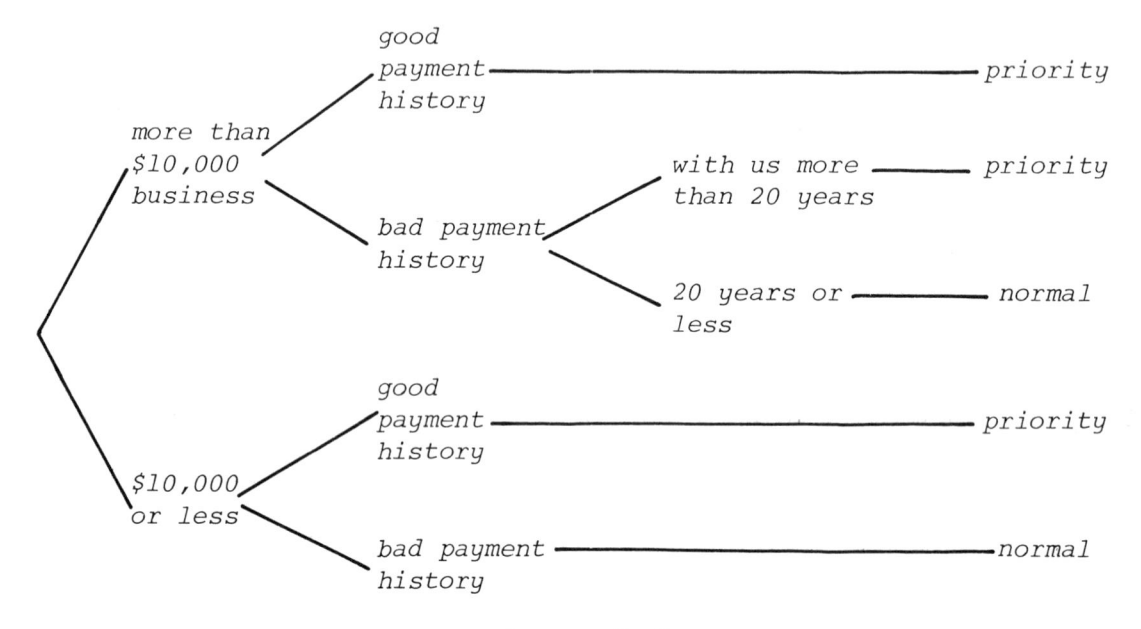

```
                       good
                       payment ─────────────────── priority
                       history
         more than   /
         $10,000    /
         business  /                with us more ── priority
                   \  bad payment  /than 20 years
                    \ history   ──<
                                   \ 20 years or ── normal
                                     less

                       good
                       payment ─────────────────── priority
                       history
         $10,000    /
         or less   <
                    \ bad payment ─────────────── normal
                      history
```

Figure 5.6

Nothing in the format of the decision tree encourages us to ask whether any other combinations of conditions should be tested. This is where the *decision table* technique is useful. Figure 5.7 shows a decision table representing our new policy in three stages of completion:

5.1 THE PROBLEMS OF EXPRESSING LOGIC

Decision table for customer priority

a) With conditions and actions filled in

Conditions
- c1: more than $10,000 a yr?
- c2: good payment history?
- c3: with us more than 20 yrs?

Actions
- a1: priority treatment
- a2: normal treatment

8 rows:
3 conditions each of which has 2 possibilities
$2^3 = 8$

b) With all the possible combinations of conditions filled in

c1: more than $10,000 a yr?	Y	Y	Y	Y	N	N	N	N
c2: good payment history?	Y	Y	N	N	Y	Y	N	N
c3: with us more than 20 yrs?	Y	N	Y	N	Y	N	Y	N
a1: priority treatment								
a2: normal treatment								

Y = Yes
N = No

c) With the combinations of conditions connected to the actions

c1: more than $10,000 a yr?	Y	Y	Y	Y	N	N	N	N
c2: good payment history?	Y	Y	N	N	Y	Y	N	N
c3: with us more than 20 yrs?	Y	N	Y	N	Y	N	Y	N
a1: priority treatment	X	X	X		X	X		
a2: normal treatment				X			X	X

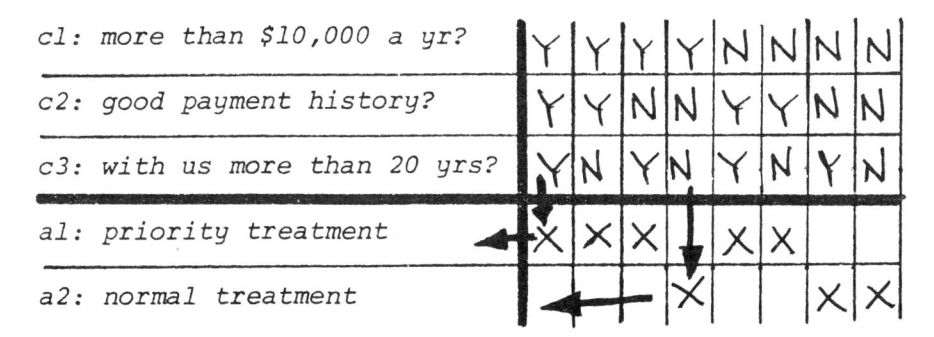

Figure 5.7

5. ANALYZING AND PRESENTING PROCESS LOGIC

Since we identify three conditions, each of which has only two possibilities, we know that there will be 2^3 or 8 possible combinations. In stage (b), we exhaustively list all those combinations, as groups of "Y"s and "N"s. Then we work out the action that corresponds to each combination of conditions and mark it with an "X".

As you can see, the decision *table* is more compact than the decision *tree*; also we are certain, when we have completed an exhaustive decision table like this, that we have considered all the possible combinations of conditions that might ever occur. The drawback is that it doesn't give us a vivid picture of the structure. Decision tables can also be baffling if you haven't seen them before.

In the next three sections, we shall work out a case-study of a somewhat complex process, looking in turn at decision trees, decision tables, and forms of Structured English, discussing the advantages and disadvantages of each, and see the role that each technique has to play in the analysis and presentation of logic.

5.2 DECISION TREES

To see the scope of the decision tree technique, let us analyze a moderately complex policy statement.

The CBM company, whose data flow we analyzed in Chapter 3, ships parcels of books to customers, charging them for the shipping costs as part of the invoice (unless the invoice was prepaid). The shipping and handling costs are expressed in units whose dollar value can be varied from time to time to adjust for actual rates, wage inflation, and so on. The current dollar value of a unit is 50 cents. The following is extracted from the company's existing explanatory leaflet on rates:

5.2 DECISION TREES

"Air shipping charges are set depending on the weight of the parcel. The basic rate is 3 units per pound, reducing to 2 units per pound for excess over 20 pounds, with a minimum of 6 units. Surface freight (including handling) is 2 units per pound for express delivery; however, this rate applies only in the local delivery area. If the shipping address is outside the local area and the parcel weighs over 20 pounds or express delivery is not required, the surface rate is the same as for local delivery express. Normal delivery of packages up to 20 pounds is 3 units per pound with a 1 unit express surcharge (per pound).

Notwithstanding the provisions of the previous paragraph, air freight to destinations West of the Mississippi is charged at double rates."

Our first step in analyzing this policy document should be to go through it identifying:

> conditions
> actions
> "unless," "however," "but"...structures
> greater than/less than ambiguities
> and/or ambiguities
> undefined adjectives

as we discussed in Section 5.1.

Figure 5.8 shows the text of the shipping leaflet marked up in this way. We find one "and/or" confusion, two undefined adjectives ("Basic," "Local"), and a range confusion possibility (up to 20 pounds, over 20 pounds). Based on this brief examination we will ask some questions of the Shipping Manager, and incorporate the answers into a revised text. The questions asked, with the Shipping Manager's answers, appear on the page following the marked up text.

5. ANALYZING AND PRESENTING PROCESS LOGIC

Air shipping charges are set depending on the weight of the parcel. The basic rate is 3 units per pound, reducing to 2 units per pound for excess over 20 pounds, with a minimum of 6 units. Surface freight, including handling, is 2 units per pound for express delivery; however, this rate applies only in the local delivery area. If the shipping address is outside the local area and the parcel weighs over 20 pounds or express delivery is not required the surface rate is the same as for local delivery (express). Normal delivery of packages up to 20 pounds is 2 units per pound with a 1 unit express surcharge (per pound).

Notwithstanding the provisions of the previous paragraph air freight to destinations West of the Mississippi is charged at double rates.

— defined?
— GT/GE?
— defined?
— and/or
LE/LT?

Figure 5.8

☐ actions
__underscore__ conditions

5.2 DECISION TREES

> *Q. The "basic rate" referred to in line 2 of the leaflet; is that for air or for surface as well?*
>
> *A. That's the air shipping rate. Surface is dealt with in the next sentence.*
>
> *Q. The leaflet sometimes refers to freight, sometimes to shipping and handling. Is there a difference?*
>
> *A. No; all the rates include freight and handling.*
>
> *Q. What exactly is meant by the local area?*
>
> *A. It's the area served by our own trucks; in practice, it's anywhere within the city limits.*
>
> *Q. The leaflet mentions "up to 20 pounds" and "over 20 pounds." Which would apply for a package that weighed exactly 20 pounds?*
>
> *A. It's generally understood that "up to 20 pounds" means "up to and including" we can't spell out every little thing, you know.*
>
> *Q. The third sentence of the leaflet could in theory be taken two ways. It could read as "both outside the local area and also over 20 pounds, or alternatively, express not required," or it could be "outside the local area, and in addition, either over 20 pounds or express not required." Which is correct?*
>
> *A. The second one. The first meaning couldn't be right because you would end up charging the local express rate when express delivery wasn't required. I see your point though, it is a bit confusing.*

Dialogue between the analyst and the Shipping Manager

5. *ANALYZING AND PRESENTING PROCESS LOGIC*

Now we can rewrite the original narrative, line by line. We shall use the abbreviations:

        GT - greater than
        LE - less than or equal to
        u/p - units per pound

ORIGINAL	REVISED
1. Air shipping charges are set depending on the weight of the parcel.	Useful background but no logic, omit.
2. The basic rate is 3 units per pound reducing to 2 units per pound for excess over 20 lbs. with a minimum of 6 units.	Air rate is: If weight LE 2 lb - flat 6 units If weight GT 2, LE 20 - 3 u/p If weight GT 20 - flat 60 units plus 2 units for each lb. over 20.
3. Surface freight, including handling is 2 units per pound for express delivery; however this rate applies only in the local delivery area.	Surface rate, if in the local area, and if express delivery: 2 u/p
4. If the shipping address is outside the local area and the parcel weighs over 20 lbs or express delivery is not required, the surface rate is the same as for local delivery (express)	Surface rate, if outside the local area, *and, in addition, either* if weight GT 20 *or* if normal delivery, is 2 u/p
5. Normal delivery of packages up to 20 lbs is 2 units per pound with a 1 unit express surcharge (per pound)	Surface rate is: If weight LE 20 and *normal* delivery, then rate is 2 u/p If weight LE 20 and *express*, then rate is 3 u/p.
6. Notwithstanding the provisions of the previous paragraph, air freight to destinations West of the Mississippi is charged at double rates	If shipment is by air, and if destination is West of Mississippi, rate is doubled.

5.2 DECISION TREES

Already, we are beginning to make considerably more sense out of the tangle. Let us draw a piece of a decision tree corresponding to each revised sentence.

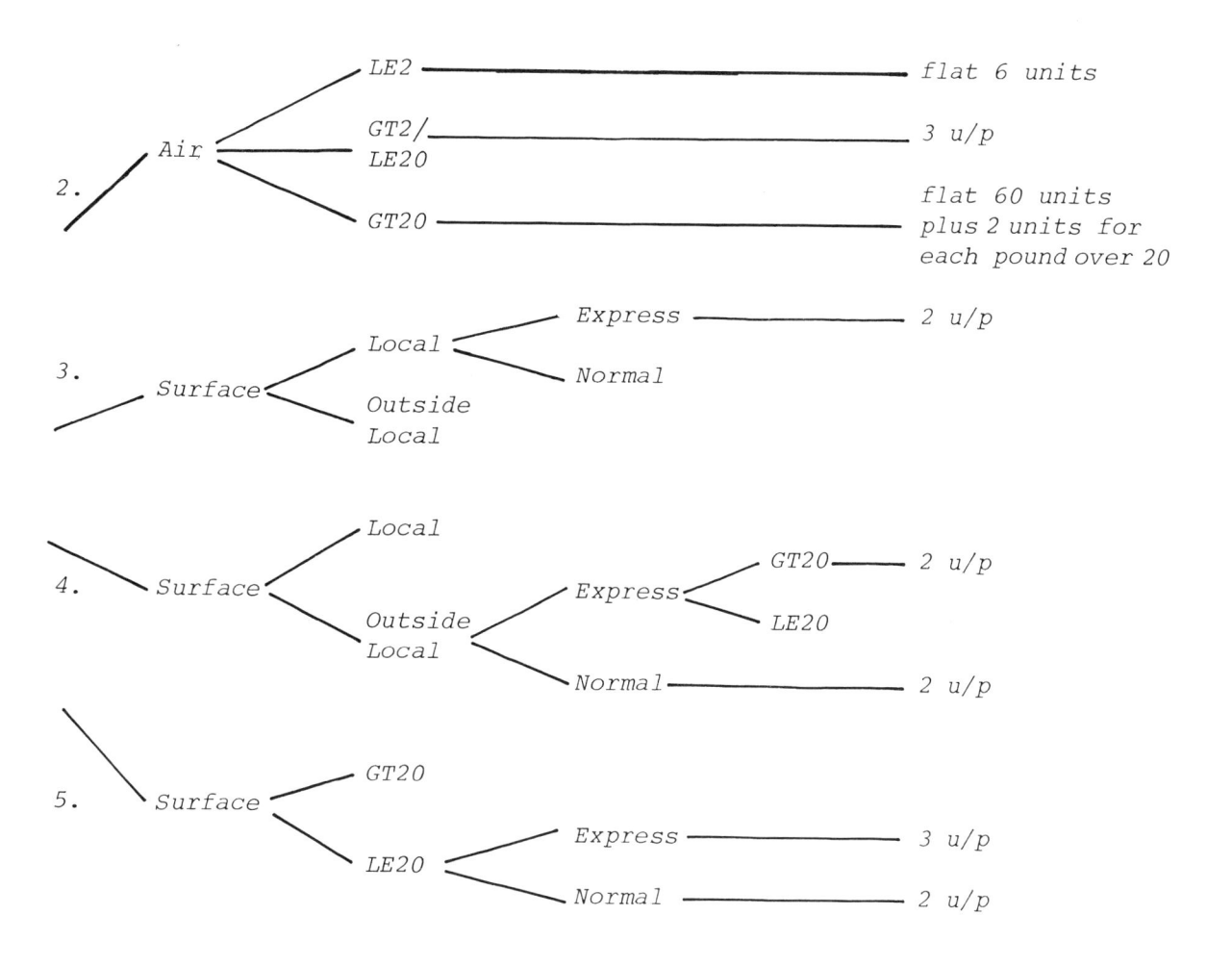

2.
```
         ┌── LE2 ─────────────────── flat 6 units
    Air ─┼── GT2/LE20 ─────────────── 3 u/p
         └── GT20 ────────────────── flat 60 units
                                     plus 2 units for
                                     each pound over 20
```

3.
```
             ┌── Local ──┬── Express ── 2 u/p
    Surface ─┤           └── Normal
             └── Outside
                 Local
```

4.
```
             ┌── Local
    Surface ─┤              ┌── Express ─┬── GT20 ── 2 u/p
             └── Outside ───┤            └── LE20
                 Local      └── Normal ──────────── 2 u/p
```

5.
```
             ┌── GT20
    Surface ─┤         ┌── Express ── 3 u/p
             └── LE20 ─┤
                       └── Normal ──── 2 u/p
```

Sentence 6, concerning air shipments to the West, we will leave for separate consideration.

Looking at the pieces of the decision tree, we see that 2, 3, and 4 fit together very neatly. What about 5? It appears to ask questions in a different order from 3 and 4, and it is not clear whether 5 applies to the local area or outside it. However, since in 4 we have an unspecified rate for Outside-Express-LE20, we can assume that 5 relates to this possibility. In that case, the lower branch of 5 is redundant; we know that Outside-Normal rate is 2. So (bearing in mind the need to check our assumptions) we can put the whole tree together, each branch reflecting the possible values of a condition.

131

5. ANALYZING AND PRESENTING PROCESS LOGIC

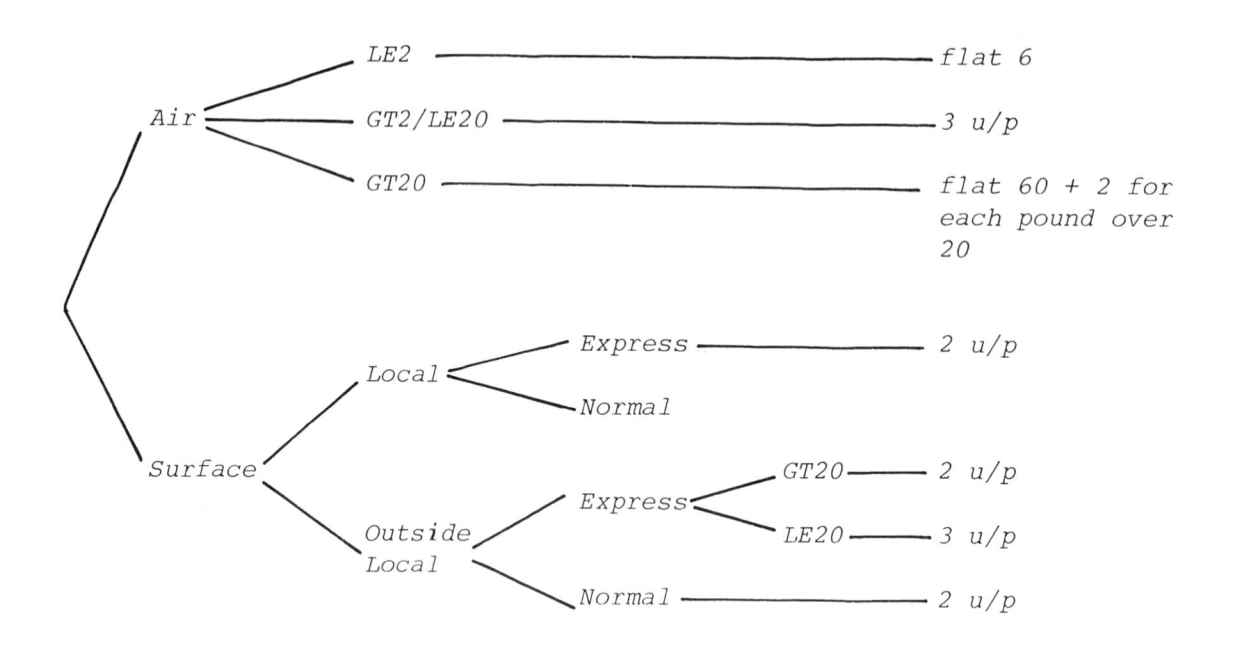

```
            ┌── LE2 ─────────────────────── flat 6
       Air ─┼── GT2/LE20 ─────────────────── 3 u/p
       │    └── GT20 ────────────────────── flat 60 + 2 for
       │                                     each pound over
       │                                     20
       │
       │           ┌── Express ──────────── 2 u/p
       │     Local ┤
       │           └── Normal
       Surface ─┤
                   ┌── Express ─┬── GT20 ── 2 u/p
             Outside┤           └── LE20 ── 3 u/p
              Local │
                   └── Normal ────────────── 2 u/p
```

Figure 5.9

We see immediately that one of the possibilities shown up by the decision tree (Surface-Local-Normal) has no rate specified for it. We will obviously have to go back to the Shipping Manager and check what the rate is in this case. Maybe the 1 unit express surcharge mentioned in Sentence 5 applies, so that the normal rate for local deliveries is 1 unit. Maybe all local delivery is express, so normal delivery is irrelevant. We just can't tell on the information we have in the leaflet.

While not being a strict logical inconsistency, the tree highlights something which seems contrary to common sense: what are the rates for two packages, one weighing 19 pounds, and one weighing 21 pounds, both sent surface, outside the local area, and express? From the tree as it stands, the 10 pound package costs 3 units x 19 pounds = 57 units, and the 21 pound package costs 2 units x 21 = 42 units. There seems to be something funny about the rate break at 20 pounds. Maybe the leaflet should have been worded to have the same sense as the air rate sentence; the 2 units per pound applies to the whole amount. Once again, we note another question for the Shipping Manager.

Sentence 6, relating to air freight to the West, can be handled in one of two ways:

132

5.2 DECISION TREES

1. splitting each of the three prices for air freight into two

2. creating a second step in the decision for air freight only.

It's better to have all the factors in the decision incorporated in one chart, because there's less possibility of missing one out, but the tree begins to "sprout" and get unwieldy. Figure 5.10 shows the tree written out in full.

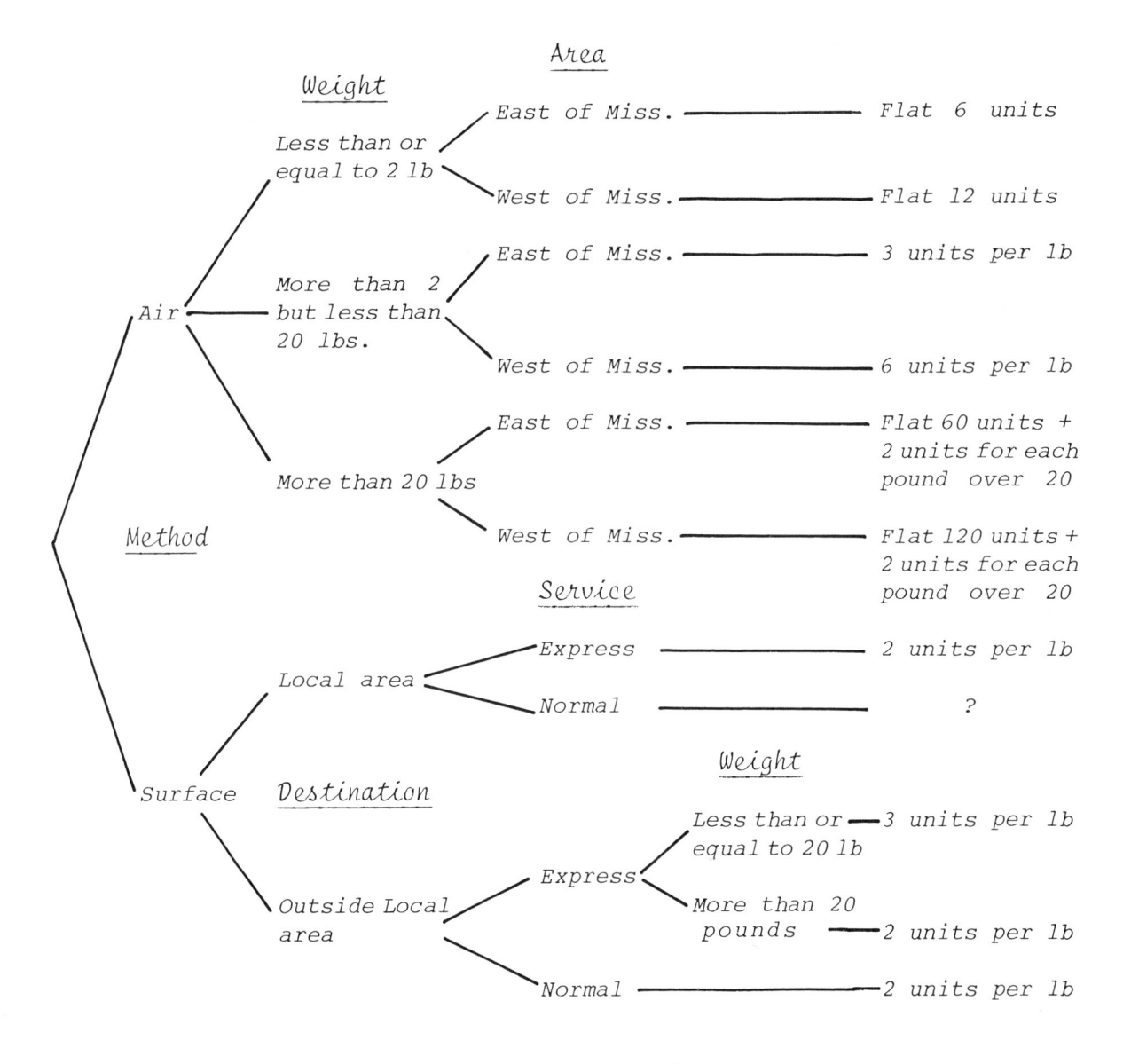

Method
- *Air*
  - *Weight*
    - Less than or equal to 2 lb
      - *Area*
        - East of Miss. —— Flat 6 units
        - West of Miss. —— Flat 12 units
    - More than 2 but less than 20 lbs.
      - East of Miss. —— 3 units per lb
      - West of Miss. —— 6 units per lb
    - More than 20 lbs
      - East of Miss. —— Flat 60 units + 2 units for each pound over 20
      - West of Miss. —— Flat 120 units + 2 units for each pound over 20
- *Surface*
  - *Destination*
    - Local area
      - *Service*
        - Express —— 2 units per lb
        - Normal —— ?
    - Outside Local area
      - Express
        - *Weight*
          - Less than or equal to 20 lb —— 3 units per lb
          - More than 20 pounds —— 2 units per lb
      - Normal —— 2 units per lb

Figure 5.10

133

5. ANALYZING AND PRESENTING PROCESS LOGIC

If the tree appears confusing or unfamiliar to users, we can express it as a conventional table. Note that we have changed the order of questions slightly to make the columns of the table more uniform. (N/A = not applicable)

METHOD OF SHIPMENT	DESTI-NATION	WEIGHT OF PACKAGE	SERVICE CLASS	SHIPPING AND HANDLING RATES
AIR	EAST OF MISSIS-SIPPI	LESS THAN OR EQUAL TO 2 LBs.	N/A	FLAT 6 UNITS
		MORE THAN 2 LB. BUT LESS THAN 20 LBs.	N/A	3 UNITS PER POUND
		MORE THAN 20 LBs.	N/A	FLAT 60 UNITS PLUS 2 FOR EACH POUND OVER 20.
	WEST OF MISSIS-SIPPI	LESS THAN OR EQUAL TO 2 LBs.	N/A	FLAT 12 UNITS
		MORE THAN 2 LB, BUT LESS THAN 20 LBs.	N/A	6 UNITS PER POUND
		MORE THAN 20 LBs.	N/A	FLAT 120 UNITS PLUS 4 FOR EACH POUND OVER 20.
SURFACE	LOCAL AREA	N/A	EXPRESS	2 UNITS PER POUND
			NORMAL	?
	OUTSIDE LOCAL AREA	LESS THAN OR EQUAL TO 20 LBs.	EXPRESS	3 UNITS PER POUND
			NORMAL	2 UNITS PER POUND
		MORE THAN 20 LBs.	EXPRESS	2 UNITS PER POUND
			NORMAL	2 UNITS PER POUND

Figure 5.11

The table throws up another anomaly that is not obvious from the decision tree. For a package outside the local area, more than 20 pounds, there is no difference between express and normal rates! This strengthens our suspicion that the "Over 20 pounds" rates are really excess surcharges, but as we noted before, we will have to check with the management to be sure.

Another aspect of the table is the number of times we have to insert "N/A." As far as we could tell from the decision tree, service class (Express or Normal) has no bearing on air freight. But is this really true? Is it really true that the weight of a package has no bearing on the local rates? Or is that just another omission from the leaflet? Drawing up a table forces us to consider more combinations of possibilities.

This is the virtue of the *standard decision table*; it provides us with a straightforward way of identifying all the possible combinations of conditions that might arise, and checking them out systematically so that we can be sure we have got to the bottom of all the complications. In the next section we will describe the conventions for drawing up decision tables, and then apply the technique to our freight rate problem.

5.3 DECISION TABLES

To see the conventions for decision tables, let us go back to Figure 5.7c, which is reproduced here:

c1: more than $10,000 a yr?	Y	Y	Y	Y	N	N	N	N
c2: good payment history?	Y	Y	N	N	Y	Y	N	N
c3: with us more than 20 yrs?	Y	N	Y	N	Y	N	Y	N
a1: priority treatment	X	X	X		X	X		
a2: normal treatment				X			X	X

135

5. ANALYZING AND PRESENTING PROCESS LOGIC

5.3.1 CONDITIONS, ACTIONS AND RULES

In this form of table, the various actions that are to be taken as a result of the decision are listed in the lower left-hand section; known as the *action stub*. Similarly the various conditions that affect the decision are listed in the upper left-hand section, the *condition stub*. We phrase the conditions as questions in such a way that they can be answered "Yes" or "No," so that we can put down all the possible combinations by listing a pattern of "Y"s and "N"s in the upper right-hand section, with no repetitions or omissions. Each combination is called a *rule* in decision table work; for example, rule 3 is:

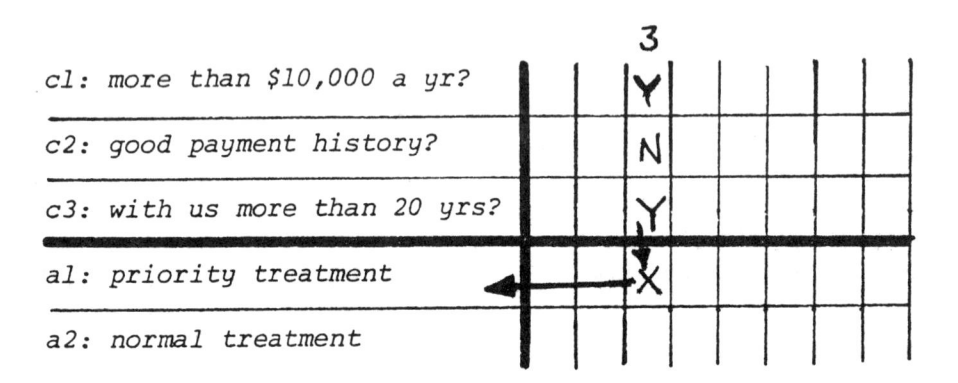

		3					
c1: more than $10,000 a yr?		Y					
c2: good payment history?		N					
c3: with us more than 20 yrs?		Y					
a1: priority treatment		X					
a2: normal treatment							

which is equivalent to saying:

"If the customer does more than $10,000 business

and

if he does *not* have a good payment history

and

if he *has* been with us more than 20 years..."

The "X" is equivalent to completing the sentence with:

"then he gets priority treatment"

136

5.3.2 BUILDING UP THE RULE MATRIX

There are several ways of being sure you have covered every possibility yet not repeated yourself. A simple way is:

1. Work out the total number of rules by multiplying together the number of possibilities for each condition; e.g.,

C1: more than $10,000	2 possibilities
C2: good payment history	x2 possibilities
C3: with us more than 20 yrs	x2 possibilities
TOTAL	8

2. Create the condition and action stubs, and provide enough columns for all the rules:

c1: more than $10,000 a yr?								
c2: good payment history?								
c3: with us more than 20 yrs?								
a1: priority treatment								
a2: normal treatment								

3. Take the last condition and alternate its possibilities all along the row:

c1: more than $10,000 a yr?								
c2: good payment history?								
c3: with us more than 20 yrs?	Y	N	Y	N	Y	N	Y	N
a1: priority treatment								
a2: normal treatment								

5. ANALYZING AND PRESENTING PROCESS LOGIC

4. Note how often the pattern repeats itself. C3 has only two possibilities, so the "YN" pattern repeats itself every two columns. If the condition has 3 possibilities, it would repeat itself every 3 columns and so on. Take the condition immediately above the one you just filled in, and cover each pattern group with a value for this next condition, repeating as needed:

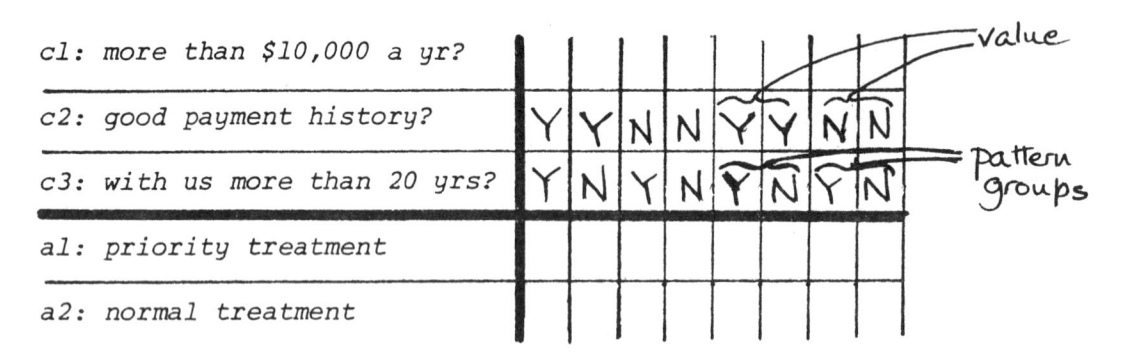

c1: more than $10,000 a yr?								
c2: good payment history?	Y	Y	N	N	Y	Y	N	N
c3: with us more than 20 yrs?	Y	N	Y	N	Y	N	Y	N
a1: priority treatment								
a2: normal treatment								

(value; pattern groups)

5. Note how often the new pattern repeats itself (every four columns in this case), and cover each pattern group with a value for the condition above:

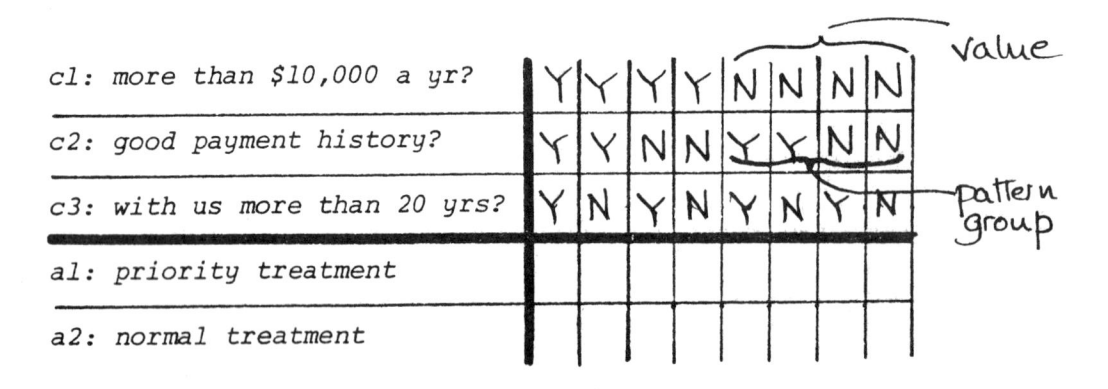

c1: more than $10,000 a yr?	Y	Y	Y	Y	N	N	N	N
c2: good payment history?	Y	Y	N	N	Y	Y	N	N
c3: with us more than 20 yrs?	Y	N	Y	N	Y	N	Y	N
a1: priority treatment								
a2: normal treatment								

(value; pattern group)

Provided you calculated the number of columns correctly in the first place, the values for the top condition should fit exactly, and you should find that you have covered every possible combination of values.

5.3 DECISION TABLES

This technique for setting up the matrix works equally well when there are more than two values per condition, as we shall see when we come to deal with the freight rate example.

5.3.3 INDIFFERENCE

Once we have set up the matrix of all possible combinations, we can enter the "X"s in the lower right-hand portion of the table, to show what actions are to be taken for each combination. In cases where we are not sure that we have all the information about the policy in question, we can sit down with the user and consider each rule in turn, to make completely sure that our table is accurate. When we have done this, we can scan the table to see whether any of the columns can, in fact be eliminated.

Take the completed table for deciding on priority treatment, reproduced again here:

	1	2	3	4	5	6	7	8
c1: more than $10,000 a yr?	Y	Y	Y	Y	N	N	N	N
c2: good payment history?	Y	Y	N	N	Y	Y	N	N
c3: with us more than 20 yrs?	Y	N	Y	N	Y	N	Y	N
a1: priority treatment	X	X	X		X	X		
a2: normal treatment				X			X	X

Consider rules 7 and 8. What they say, in effect, is that if a customer does less than $10,000 business a year, and does *not* have a good payment history, then he will get normal treatment if he has been with us more than 20 years, (rule 7), and he will get normal treatment if he has been with us *less* than 20 years. In other words, if you're a small customer and a bad payer, it *makes no difference* how long you've been with us; you still don't get priority treatment. In decision table jargon, rules 7 and 8 are *indifferent* to the value of C3. We can replace them with a single column, using a "-" (the indifference symbol) for C3:

5. ANALYZING AND PRESENTING PROCESS LOGIC

	1	2	3	4	5	6	7/8
c1: more than $10,000 a yr?	Y	Y	Y	Y	N	N	N
c2: good payment history?	Y	Y	N	N	Y	Y	N
c3: with us more than 20 yrs?	Y	N	Y	N	Y	N	—
a1: priority treatment	X	X	X		X	X	
a2: normal treatment				X			X

The technique for consolidating the decision table in this way is:

1. Find a pair of rules for which

    - the action is the same
    - the condition values are the same except for one *and only one* condition where they're different

2. Replace that pair with a single rule using the indifference symbol for the only condition which was different.

3. Repeat for any other pair which meets the criteria.

In fact, the customer priority table reduces to just five rules:

	1/2	3	4	5/6	7/8
c1: more than $10,000 a yr?	Y	Y	Y	N	N
c2: good payment history?	Y	N	N	Y	N
c3: with us more than 20 yrs?	—	Y	N	—	—
a1: priority treatment	X	X		X	
a2: normal treatment			X		X

Why can't rules 3 and 4 be consolidated, since they have the same entries except for C3? Because they lead to different actions, so C3, far from making no difference, makes all the difference in the decision.

The reader might find it useful to draw up a decision tree which expresses the table above, and then compare it with Figure 5.6. Though the table and the tree are equivalent, the table has been arrived at by considering every possibility; the tree most likely, has not.

5.3.4 EXTENDED ENTRY; THE FREIGHT RATE PROBLEM

Tables of the sort we have been considering, where the values a condition may have are limited to two ("Yes" and "No" in our case) are known as *limited-entry* tables. Where a condition can have more than two values, the table is known as an *extended-entry* table. We can see that this will be required for the freight rate problem since the weight of a package can take three values (LE2, GT2 but LE20, GT20), and since the company is located in New Jersey, we could consider the destination as taking three values (local, outside the local area but still East of the Mississippi, outside the local area and West of the Mississippi). Let us go through the steps in developing a decision table for this more complex case.

1. First, we should list the conditions and actions we have identified giving suitable abbreviations for use in the table.

    Conditions:

    C1: Method of delivery: A - Air
                              S - Surface

    C2: Destination: L - Local
                              E - Outside the local area but
                                    East of the Mississippi
                              W - Outside the local area but
                                    West of the Mississippi

    C3: Weight: L - Light, LE 2 lbs
                              M - Medium, GT 2 but LE 20 lbs
                              H - Heavy, GT 20 lbs

    C4: Service: E - Express
                              N - Normal

    (Notice that we have just defined four discrete data elements, and named their values.)

5. ANALYZING AND PRESENTING PROCESS LOGIC

>   Actions:
>
>   A1: Flat 6 units
>   A2: Flat 12 units
>   A3: Flat 60 units plus 2 units per pound over 20
>   A4: Flat 120 units plus 4 units per pound over 20
>   A5: 2 units per pound
>   A6: 3 units per pound
>   A7: 6 units per pound
>
> There is nothing sacred about the order of conditions and actions; the order of conditions is approximately what seems to us to be in order of importance. We should bear in mind that as the development of the table proceeds, the Shipping Manager may tell us about other conditions and actions we haven't heard of yet.

2. Next, we establish how many columns there will be in the decision table:

   > C1 has 2 possibilities
   > C2 has 3 possibilities
   > C3 has 3 possibilities
   > C4 has 2 possibilities

   giving a total of 2 x 3 x 3 x 2 or 36 combinations in all.

3. Figure 5.12a shows the condition and action stubs of the first 12 columns, and the alternating possibilities filled in for the first condition, C4.

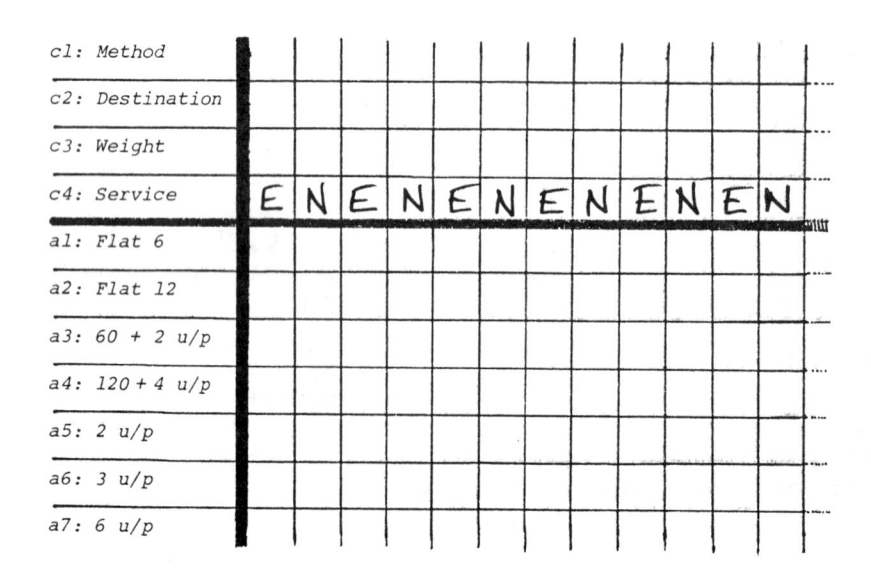

c1: Method												
c2: Destination												
c3: Weight												
c4: Service	E	N	E	N	E	N	E	N	E	N	E	N
a1: Flat 6												
a2: Flat 12												
a3: 60 + 2 u/p												
a4: 120 + 4 u/p												
a5: 2 u/p												
a6: 3 u/p												
a7: 6 u/p												

Figure 5.12a

5.3 DECISION TABLES

The pattern repeats itself every two columns. As we said in Section 5.3.2, the next step is to cover each pattern group with a value of the next condition, as shown in Figure 5.12b.

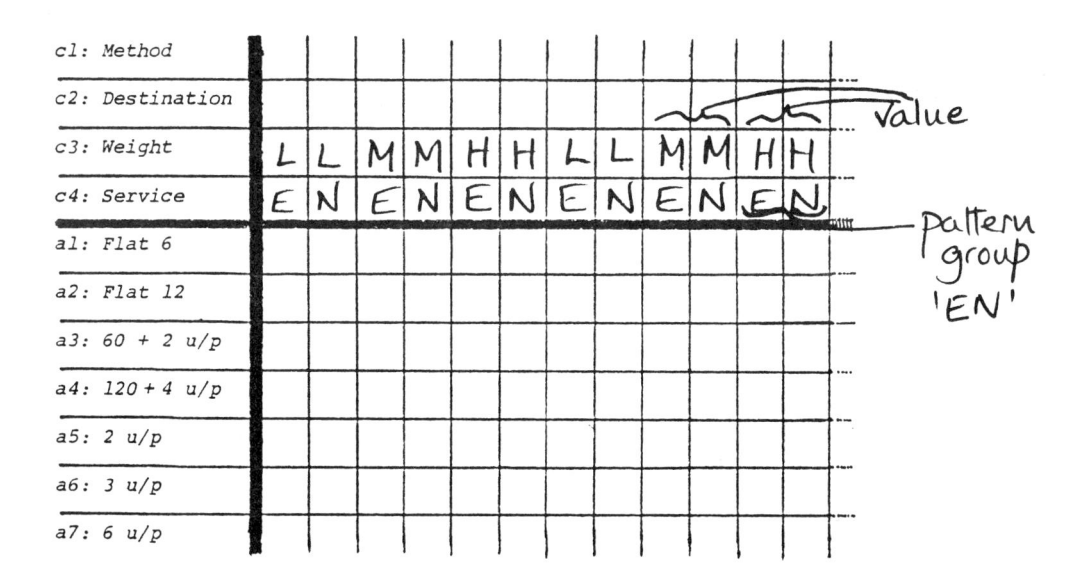

c1: Method												
c2: Destination												
c3: Weight	L	L	M	M	H	H	L	L	M	M	H	H
c4: Service	E	N	E	N	E	N	E	N	E	N	E	N
a1: Flat 6												
a2: Flat 12												
a3: 60 + 2 u/p												
a4: 120 + 4 u/p												
a5: 2 u/p												
a6: 3 u/p												
a7: 6 u/p												

(annotations: "value" pointing to c3 row; "pattern group 'EN'" pointing to c4 row)

Figure 5.12b

Now we cover each instance of the 6 column pattern in C3 with a value of C2, and so on. The result is shown in Figure 5.13.

4. With the matrix set-up in the table, we have to consider the actions that go with each combination of conditions. Before we can do that, we must ask ourselves whether there are any combinations that are inherently impossible i.e. any of our rules that cannot exist. We find an example right away; rules 1-6 of Figure 5.13 all deal with air delivery to a local destination. This is a manifest impracticality, and as is shown, the six rules have been deleted. This leaves only 30 rules to be worked out; we have completed the section that related to air shipment, and the reader is invited to complete the others before seeing which can be consolidated.

5. ANALYZING AND PRESENTING PROCESS LOGIC

Figure 5.13

5.3.5 DECISION TABLES VS DECISION TREES

Suppose instead of the seven separate actions we identified for the freight rate problem, there had been 30 separate actions, one for each possible combination of conditions? The action stub of the decision table would have been larger, but not inherently any more complex. Can you imagine what the decision tree would have looked like? This gives us a clue to:

Guideline 1   *Use a decision tree when the number of actions is small, and not every combination of conditions is possible; use a decision table when the number of actions is large, and many combinations of conditions occur.*

As we dug further into the freight rate leaflet we kept on turning up new ambiguities or omissions. The judgement as to whether it is worth the effort of developing a decision table depends in part on how confident you are that your tree is exhaustive; that you have completely covered everything. So we get:

Guideline 2   *Use a decision table if you doubt that your decision tree shows the full complexity of the problem.*

Perhaps the most important factor, though, is the vividness given by a decision tree. Even though abbreviations may be used, it is possible to read off the combinations of conditions for each branch of the tree and "see how it all fits together." Contrast this with a decision table where the meaning of each rule must be assembled in the mind, condition by condition.

Guideline 3   *Even if you need a decision table to get to the bottom of the logic, end up presenting it as a tree if you can do so without violating Guideline 1.*

5.   ANALYZING AND PRESENTING PROCESS LOGIC

5.4   STRUCTURED ENGLISH, PSEUDOCODE, AND "TIGHT ENGLISH"

Decision trees and decision tables are the tools of choice for dealing with complex branching processes, such as we typically find in discount calculation, rate calculation, sales commissions and productivity bonus calculation, inventory control procedures, and so on. Many of the processes we have to document are not that complex, however. We find that they involve operations ("Do this, then that..."), some decisions, and a few loops ("Repeat this procedure until..."). This is not surprising since we know it can be proved [5.4] that any program can be made up out of suitable combinations of step-by-step instructions (like MOVE or ADD), binary decisions (IF-THEN-ELSE-SO), and loops; the logical processes we are defining in Structured Analysis are just that, programs for execution by either a clerk or a computer. These few structures provide the basis for Structured Programming, which gets part of its effectiveness from the simplification and standardization that comes from using just a few structures, rather than the great variety allowed by the particular programming language being used. If we could write our specifications using the same approach in English, we will achieve some of the same benefits. Let us look at these structures in a little more detail, to see how to incorporate them into English.

5.4.1   THE "STRUCTURES" OF STRUCTURED PROGRAMMING

1.   *Sequential instructions*. This structure covers any instruction or group of instructions that has no repetition or branching built into it.

    For example:

    "Multiply hours-worked by pay-rate to get gross-pay"
    "Ship books to ship-to-address"
    "Add freight charge to invoice"
    "Calculate freight charge. (Freight charge is 60 units for each pound over 20.)"
    "Terminate the employee"
    "Give a 25% raise"

5.4 STRUCTURED ENGLISH, PSEUDOCODE, AND "TIGHT ENGLISH"

It is sometimes necessary to include descriptions or definitions of terms used, as in the case of "Freight-Charge" above. What we must not do is sneak in any hidden branches or repetitions, such as:

"Ship books to ship-to-address or bill-to-address, whichever is relevant."
(hidden branch)

"Continue to allocate space, a unit at a time, stopping when all requests are satisfied"
(a hidden loop, with a clear test to tell you when to stop looping)

If we get a group of truly sequential statements and put them all together, the resultant "Block" of statements can be looked at as a single sequential statement. If we create the following group of statements and call them

"Compute-deductions"

- get gross pay
- get pay-to-date details
- compute Federal withholdings
- compute State withholdings
- compute FICA

and we also create a group of statements called

"Issue-pay-check"

- enter gross and withholdings on check stub
- enter net pay on check
- sign check
- mail check and stub

we can legitimately write as sequential instructions:

DO compute-deductions
DO issue-pay-check

without implying any hidden decisions or loops. The "DO" means "Carry out all the instructions listed under the following name...".

5. ANALYZING AND PRESENTING PROCESS LOGIC

2. *Decision instructions*. As we have seen, the standard format for a decision is the structure:

```
IF      Condition-1
   THEN    action-A
   ELSE    (not Condition-1)
      SO   action-B
```

Each action can be a set of sequential instructions, or a loop, *or another decision*. Suppose we replace "action-A" with another IF-THEN-ELSE-SO structure; we get:

```
IF      Condition-1
   THEN    IF       Condition-2
              THEN   action-C
              ELSE   (not Condition-2)
                 SO  action-D
   ELSE    (not Condition-1)
      SO   action-B
```

An example of this case would be:

```
IF       you need a vacation
   THEN     IF       you can afford a vacation
               THEN  take a vacation
               ELSE  (you can't afford a vacation)
                  SO paint the house
   ELSE    you don't need a vacation
      SO   keep working.
```

The phrase "IF..THEN IF..," while correct English, reads a little oddly. As in our examples earlier in the chapter, we prefer to write "IF... and-IF...," which means the same, reads a little more naturally, and saves the THEN to denote an action.

If you can substitute an IF-THEN-ELSE-SO for an action in one place, you can do it anywhere. We could substitute another IF for "take a vacation," thus:

```
IF you need a vacation
   and-IF you can afford a vacation
      and-IF you have someone to go with
         and-IF you have somewhere you want to go
            THEN.......
            ELSE
```

148

5.4 STRUCTURED ENGLISH, PSEUDOCODE, AND "TIGHT ENGLISH"

This is how "nested-IFs" arise, as the logic of the decision we are representing gets more complex. This nested logic can get difficult to follow, so it is important to use the conventions of aligning each ELSE with the IF to which it belongs. As a rule of thumb, once you have more than 3 IFs nested inside one another, it is clearer to represent the logic with a decision tree, at the price of not being able to write conditions out in full.

2.1 *"Case" decision instructions.* A special kind of decision structure arises when there are several possibilities of a condition, which *never occur in combination* i.e. they represent different cases which are mutually exclusive.

This sort of structure can always be represented in a simple table, for example:

Condition	Action
Transaction is order	Add to sales-to-date
Transaction is return	Subtract from sales-to-date
Transaction is payment	Add to cash received
Transaction is complaint	Pass to Complaints Department
Transaction is cancellation	Subtract from Sales-to-date

The important facet of this situation is that the transaction must be one of the five types, and *only one*. It can't be *both* an order and a cancellation, for instance.

Only one case applies at any one time. This special structure is known as the "case" structure and is conventionally represented in Structured English like this:

```
IF        case-1
          action-1
ELSE IF   case-2
          action-2
ELSE IF   case-3
          action-3
ELSE.....
```

To apply this convention to our transaction structure would give:

5. ANALYZING AND PRESENTING PROCESS LOGIC

```
IF        transaction is order
          add to sales-to-date
ELSE-IF   transaction is return
          subtract from sales-to-date
ELSE-IF   transaction is payment
          add to cash received
ELSE-IF   transaction is complaint
          pass to Complaints Department
ELSE-IF   transaction is cancellation
          subtract from sales-to-date
ELSE      (none of the above)
    SO    refer to Supervisor
```

Note the last ELSE; this is required so that there are an equal number of IFs and ELSEs. If this "none-of-the-above" or "default" case is omitted, there is a possibility of a case arising in practice which is not covered by the specified logic. The analyst should weigh that case and provide accordingly. The possibility may seem remote when the cases represent arbitrary splits along a complete continuum, for instance:

```
IF        person is under 18
          add to total of "minors"
ELSE IF   person is 19 or over but under 35
          add to total of "young adults"
ELSE IF   person is 35 or over but under 65
          add to total of "mature adults"
ELSE IF   person is 65 or over
          add to total of "senior citizens"
ELSE      (none of the above)
```

What could possibly represent (none of the above)? The analyst may allow the last ELSE to "dangle" in this situation; it is for the programmer to consider how to deal with a case which fails all the age tests (perhaps because age has been incorrectly encoded).

3. *Repetition (loop) instructions*. This structure applies to any situation in which an instruction or group of instructions is repeated until some desired result is obtained.

To take a very simple example, where an invoice covers the sale of a number of items, each will be listed on a

5.4 STRUCTURED ENGLISH, PSEUDOCODE, AND "TIGHT ENGLISH"

separate line with the quantity sold and the unit price, thus:

```
                           INVOICE

TO:   Customer Name
      Customer Address


QUANTITY         DESCRIPTION           UNIT PRICE         AMOUNT

    7            Saddles               $200.00              ..
   28            Horseshoes            $  1.00              ..
    7            Bridles               $ 10.00              ..
    1            Shotgun               $ 50.00              ..
                                                          _____

                                  TOTAL OF INVOICE          ..
```

Figure 5.14

We could specify the completion of the invoice by writing:

> "For each item-line, multiply the quantity by the unit price, to get the line totals. When all the item-lines have been dealt with, add all line-totals to get the invoice total."

More formally, we could specify:

"REPEAT EXTEND-ITEM-LINE UNTIL all lines have been extended"

and define EXTEND-ITEM-LINE as:

"Multiply quantity by unit-price to get line total"

5. ANALYZING AND PRESENTING PROCESS LOGIC

With only one instruction making up the body of the loop, this is taking a sledge-hammer to crush a nut, but it can be a valuable structure to use where the situation calls for a number of instructions to be repeated over and over as a group. More importantly, it highlights the two aspects that we must specify to define a repetition structure:

1. Exactly what it is that is to be repeated

2. How you know when to stop.

While the analyst should be prepared to specify repetitions when they are clearly part of the logical function, it is more usual for the programmer and designer to get involved in setting up loops for execution by the machine. This is because the analyst is primarily concerned with the logical processing of each data structure, and the designer/programmer is concerned with the physical processing of a stream of such structures. The definition of loops thus becomes more important, the closer we get to physical implementation, as we shall see in Section 5.4.3 on.

5.4.2 CONVENTIONS FOR STRUCTURED ENGLISH

As we noted earlier, it is possible to write any logical specification using the four basic structures we have just discussed. When the logic is written out in English sentences, using the capitalization and indentation conventions, it is known as "Structured English." We can summarize the conventions for Structured English as follows:

1. The logic of all processes in a system is expressed as a combination of sequential, decision, case and repetition structures.

2. The rules of unambiguous English should be observed, as set out in Section 5.1.1.

3. Keywords IF, THEN, ELSE, SO, REPEAT, UNTIL, should be capitalized, and structures indented to show their logical hierarchy.

5.4 STRUCTURED ENGLISH, PSEUDOCODE, AND "TIGHT ENGLISH"

4. Blocks of instructions may be grouped together and given a meaningful name describing their function, which should also be capitalized.

5. Where a word or phrase is used, which is defined in a data dictionary, the word or phrase should be capitalized.

As an example to illustrate the strengths and weaknesses of Structured English, we will take part of function 7, "Generate Shipping Note and Invoice" from the data flow diagram of Chapter 3. The generation of the invoice involves three steps:

- the calculation of the total for each item and for the invoice

- the calculation of the discount (if any)

and

- the calculation of the shipping and handling charge (using a simplified version of the policy we examined with the decision tree and the decision table).

Figure 5.15 shows a Structured English representation of most of the logic.

Several points should be made about this sample of Structured English:

1. It has much of the precision of a computer program, but it is *not* a computer program. There is no specification of the reading and writing of physical files, no setting of counters or switches, or any physical design. The procedure as written could quite well be carried out by a clerk, though we might choose to present the instructions differently.

2. The procedure has been written as a hierarchy of "blocks" of instructions, as shown in Figure 5.16.

5. ANALYZING AND PRESENTING PROCESS LOGIC

```
GENERATE INVOICE
    DO COMPUTE-INVOICE-TOTAL
    DO COMPUTE-DISCOUNT
    DO COMPUTE-SHIPPING-HANDLING
    Subtract DISCOUNT from INVOICE-TOTAL to get INVOICE-NET
    Add SHIPPING-HANDLING-FEE to INVOICE-NET to get TOTAL-PAYABLE
    Write INVOICE.

COMPUTE-INVOICE-TOTAL
    REPEAT  EXTEND-ITEM-LINE UNTIL all ITEM-LINEs have been extended
    Add all ITEM-LINE-TOTALs to get INVOICE-TOTAL

EXTEND-ITEM-LINE
    Multiply QUANTITY by UNIT-COST to get ITEM-LINE-TOTAL.

COMPUTE-DISCOUNT
    IF        INVOICE-TOTAL is GE $1000
              DISCOUNT is 5% of INVOICE-TOTAL
    ELSE IF   INVOICE-TOTAL is GE $250 but LE $1000
              DISCOUNT is 2½% of INVOICE-TOTAL
    ELSE IF   INVOICE-TOTAL is GE $100 but LE $250
              DISCOUNT is 1% of INVOICE-TOTAL
    ELSE      (INVOICE-TOTAL is LT $100)
        SO    DISCOUNT is nil

COMPUTE-SHIPPING-HANDLING
    IF     order specifies air shipment
      THEN DO COMPUTE-AIR-FREIGHT
    ELSE   (order specifies surface shipment or method is open)
        SO DO COMPUTE-SURFACE-FREIGHT
    Multiply RATE by CURRENT-UNIT-VALUE to get SHIPPING-HANDLING-FEE

COMPUTE-AIR-FREIGHT
    IF        WEIGHT is LE 2
              RATE is 6 units
    ELSE IF   WEIGHT is GT 2 but LE 20
              Multiply each pound of WEIGHT by 3 units to get RATE
    ELSE      (WEIGHT is GT 20)
        SO    Subtract 20 from WEIGHT to get EXCESS
              Multiply EXCESS by 2 units per pound and add 60
                 (20 pounds at 3 units per pound) to get RATE

COMPUTE-SURFACE-FREIGHT
    IF     destination is local
      and-IF   SERVICE-CODE is EXPRESS
              THEN Multiply each pound of WEIGHT by 2 units to get RATE
    .....
    .....
    and so on
```

Figure 5.15

5.4 STRUCTURED ENGLISH, PSEUDOCODE, AND "TIGHT ENGLISH"

```
                    ┌─────────────┐
                    │  GENERATE   │
                    │   INVOICE   │
                    └──┬───┬───┬──┘
Causes the            ╱    │    ╲
performance          ╱     │     ╲
of                  ╱      │      ╲
          ┌────────┐  ┌──────────┐  ┌──────────┐
          │COMPUTE-│  │ COMPUTE- │  │ COMPUTE- │
          │INVOICE │  │ DISCOUNT │  │ SHIPPING │
          │ TOTAL  │  │          │  │ HANDLING │
          └────┬───┘  └──────────┘  └──┬────┬──┘
which in       │                       ╱      ╲
turn cause     │                      ╱        ╲
the per-       │                     ╱          ╲
formance   ┌───┴───┐           ┌─────────┐  ┌─────────┐
of         │EXTEND-│           │COMPUTE- │  │COMPUTE- │
           │ LINE- │           │  AIR-   │  │SURFACE- │
           │ ITEM  │           │ FREIGHT │  │ FREIGHT │
           └───────┘           └─────────┘  └─────────┘
```

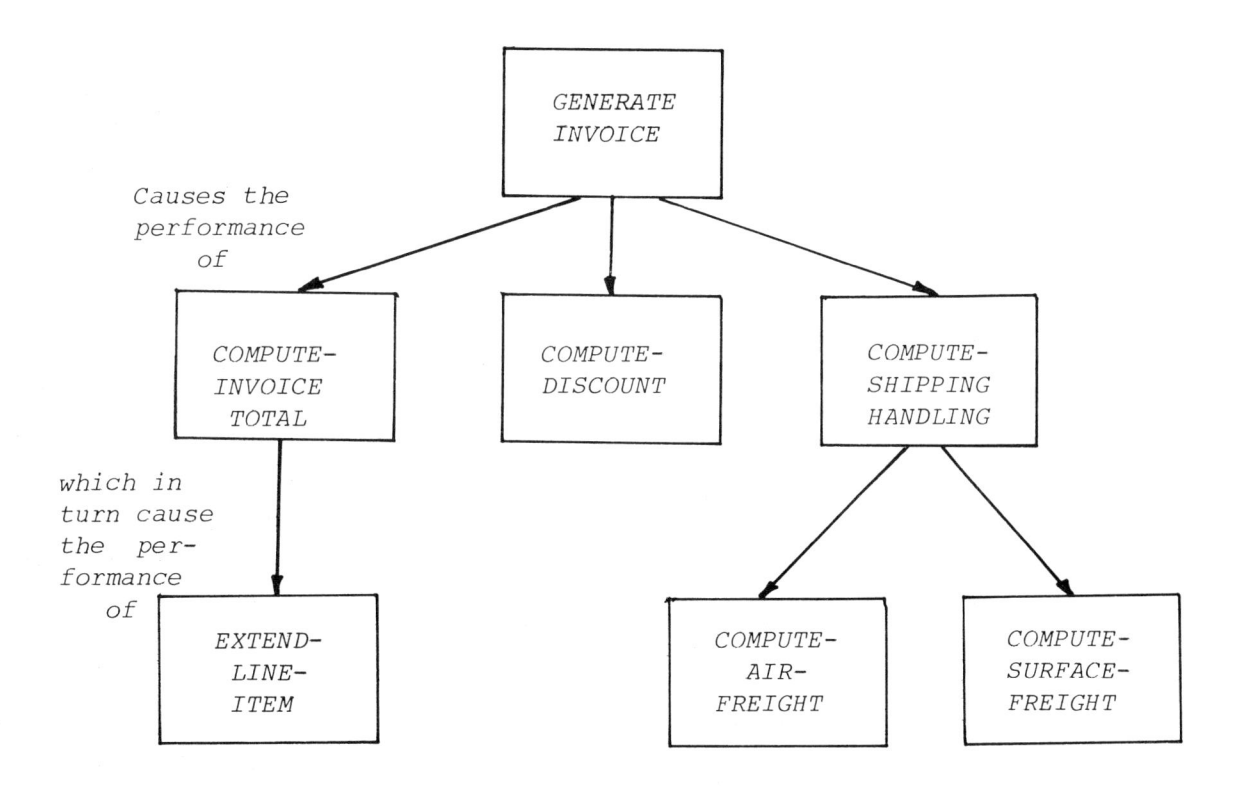

Figure 5.16

If you want to get a summary of the procedure, you just have to read the top block. If you want the details of any specific part, you just read that part.

3. The capitalization of items defined in the data dictionary highlights the use of terms which still need definition. In Figure 5.15 the condition "destination is local" written in small letters implies that no strict definition of "local" had yet been done. If it had been, we might find the condition expressed as "DESTINATION - TYPE is LOCAL."

4. Subject to the need for more definitions, the instructions are complete. They include step-by-step accounts of what to do, unlike the decision tree and decision table, which merely show the underlying branching logic.

5. The precision and completeness of the Structured English is bought at the price of an unfamiliar and verbose presentation, which spells out every little detail for the person who is not familiar with the procedure, but quickly becomes irritatingly nitty-gritty for a person who merely wants to be reminded of the bare facts.

5. ANALYZING AND PRESENTING PROCESS LOGIC

"Tight English" is a derivation of Structured English which addresses this problem; we discuss it in Section 5.4.4.

5.4.3 PSEUDOCODE

We noted that a process defined in Structured English is not a computer program. However, once the main work of physical design has been done, and physical files defined, it becomes extremely convenient to be able to specify physical program logic using the conventions of Structured English, but *without getting into the detailed syntax of any particular programming language*. This "almost-code" notation is known as pseudocode.[5.5]

To specify a program in pseudocode, we need to be able to handle initialization and termination, read and write to files, handle end-of-file, and specify counters and flags. Thus if the physical designer had decided that "Generate Invoices" would be a separate program, reading a file called, say "TODAYS-SHIPMENTS," the pseudocode for the top-level of the program hierarchy might be as in Figure 5.17.

```
Initialize the program (open files, set counters)
Read the first order-record
DO-WHILE there are more order-records
     DO-WHILE there are more items on the order

          Compute ITEM-TOTAL
          Add ITEM-TOTAL to INVOICE-TOTAL

     END-DO

     Compute discount
     Compute shipping and handling fee
     Compute invoice-net, total-payable
     Print invoice
     Write invoice to accounts-receivable file
     Add invoice-detail to summary counters
     Read next order record

END-DO

Print summary of day's invoices
Terminate  program
```

Figure 5.17

5.4 STRUCTURED ENGLISH, PSEUDOCODE, AND "TIGHT ENGLISH"

In pseudocode we distinguish the DO-WHILE loop structure from the REPEAT-UNTIL loop structure, which takes on a special meaning. The DO-WHILE loop structure implies that the termination test is made before the body of the loop is executed; the REPEAT-UNTIL structure implies that the body of the loop is executed before the test is made. In flowchart terms, these structures are:

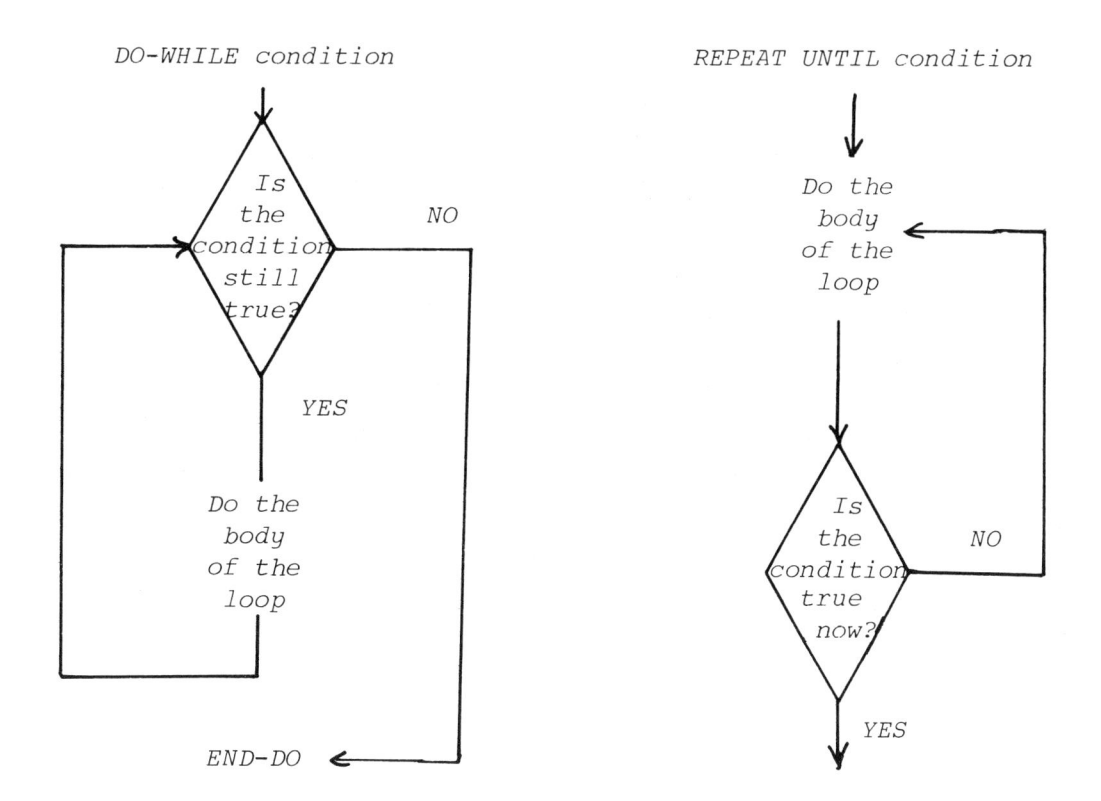

DO-WHILE condition   *REPEAT UNTIL condition*

[Flowchart: DO-WHILE loop — test "Is the condition still true?" first; if YES, "Do the body of the loop" and repeat; if NO, END-DO.]

[Flowchart: REPEAT UNTIL loop — "Do the body of the loop" first, then test "Is the condition true now?"; if NO, repeat; if YES, exit.]

The important difference is that in a situation where the condition is other than what would be expected, on the first time round, the DO-WHILE loop will not execute the body, jumping straight to the END-DO, whereas the REPEAT UNTIL will plow through the body before making the test and finding, for example, that there are no orders to be processed. In Structured English where we are specifying non-physical logic, we don't care about this distinction. In pseudocode, where we are laying down the program logic, we care very much.

The program consists of a DO-WHILE loop within a DO-WHILE loop, the outer loop processing each order record. The designer has specified that after initialization, the program will read the first record of the orders file, so that, in the case where

5. ANALYZING AND PRESENTING PROCESS LOGIC

there are no orders in the file, the test of the DO-WHILE will fail on the first time, control will jump to the outer END-DO, a nil summary will be printed, and the program will terminate normally.

We see therefore, that pseudocode represents a very detailed program design. Especially in an environment where all data structure and data elements are defined in a machine-readable data dictionary, the task of translating pseudocode into COBOL or BASIC-PLUS, say, is a relatively simple one. Given Structured English, on the other hand, several important physical design decisions have to be taken before a program can be written, since the Structured English represents the "external" logic only.

5.4.4 LOGICALLY "TIGHT ENGLISH"

We commented that the conventions of Structured English, while producing a precise and complete statement of logic, involved too many words and too much unfamiliar notation to be ideal for presentation to users. This leads us to ask whether it is possible to get the benefits of the rigor of Structured English without the disadvantages. Once we have a firm grasp of the reasons why Structured English works to eliminate ambiguity and incompleteness, we can throw away the clumsy parts of the notation and write an equivalent style of English that is "logically tight" in the sense of reflecting precise and complete analysis. Figure 5.18 shows the "Generate Invoice" process written in "Tight English."

It is allowable to include decision trees in a "Tight English" write-up, if you are sure they will be comprehensible to everybody using the document. "Tight English" is familiar in appearance, and structurally equivalent to Structured English. The irony is that it is hard to write until the logic of a process is worked out in Structured English and understood in detail in terms of the four structures (sequence, decision, case, repetition).

We can summarize the conventions for "Tight English" as follows:

1. Sequential operations are presented as imperative instructions for doing them clerically

2. IF-THEN-ELSE-SO structures are presented with decimal notation and indentation to show nesting, thus:

5.4 STRUCTURED ENGLISH, PSEUDOCODE, AND "TIGHT ENGLISH"

To generate an invoice:

Step 1: Work out the invoice total as follows:

    1.1 Take each line of the invoice and multiply the quantity of the item by the unit price to get the item total

    1.2 Add the item totals to get the invoice total

Step 2: Work out the discount (if any):

The quantity discount is based on the invoice total according to the following table

INVOICE TOTAL	% DISCOUNT
Less than $100	none
More than $100 but less than $250	1%
More than $250 but less than $1,000	2½%
More than $1,000	5%

Subtract the discount from the invoice total giving the invoice net amount.

Step 3: Work out the shipping and handling charge (S&H).
S&H is based on units, currently each costing 50 cents.
Orders specify either air freight or surface freight, when no freight specified, the default is surface freight.

    3.1 For orders specifying air freight:

- *up to and including 2 pounds - flat rate of 6 units*
- *more than 2 but less than 20 - 3 units per pound*
- *20 pounds or more - 60 units + 2 units for each pound over 20*

    3.2 For orders specifying surface freight (or not specifying any method)

        3.2.1 For local destinations
            a. For express service - 2 units per pound
            b. For normal service - 1 unit per pound
        3.2.2 For destinations outside the local area
            a. For express service
- *weight greater than 20 lb. - 2 units per pound*
- *weight less than or equal to 20 pounds - 3 units per pound*

            b. For normal service - 2 units per pound

Figure 5.18

5. ANALYZING AND PRESENTING PROCESS LOGIC

    5.
        5.1
            5.1.1
                5.1.1a

They may also be presented as decision trees

3. ELSE conditions are presented as: "For (explanation of condition)"

4. Case structures are presented as tables

5. Where genuine exception conditions are involved, the "Action-1 *unless* condition *in which case* Action-2" structure (see Section 5.1.1), may be used for clarity.

Thus when we wrote the logic summary for the process defined in the data dictionary (repeated below) it was enough to use Tight English with "unless" clauses to specify exception handling. The Tight English shown in "Verify Credit is OK" was written as a summary of the Structured English which expresses full details of the logic. Figure 5.19 shows the full Structured English for "Verify Credit is OK"

VERIFY-CREDIT-IS-OK   PROCESS ref: 3

Description: <u>Decide whether orders without prepayment can be shipped, or whether prepayment should be demanded from customer</u>

Inputs	Logic summary	Outputs
1-3 ORDERs	Retrieve payment history	3-c PREPAYMENT-REQUEST [REMINDER-OF-BALANCE]

D3-3 Payment history
  DATE-ACCT-OPENED
  INVOICE*
  PAYMENT*
  BALANCE-ON-ORDER
	If new customer, send prepayment request
	If regular customer, (average of two orders per month), ok the order unless balance overdue is more than two months old
	For previous customers who are not regular, ok the order, unless they have any balance overdue
		3-D3 New BALANCE-ON-ORDER
		3-6 Orders with credit OK

Physical ref: <u>Part of on-line order entry, OE707</u>

Full details of this logic can be found in: <u>Functional specification, Section 7.2</u>

5.4 STRUCTURED ENGLISH, PSEUDOCODE, AND "TIGHT ENGLISH"

```
Process name:      VERIFY CREDIT IS OK                        Ref: 3


For each ORDER,
    Retrieve PAYMENT-HISTORY
     IF       PAYMENT-HISTORY is not found (new customer)
      THEN    generate PREPAYMENT-REQUEST
     ELSE     (existing customer)
       SO     compute AVERAGE-ORDER-FREQUENCY
                     (average number of orders per
                      month over last 3 months)
              compute total of balance overdue
              IF       AVERAGE-ORDER-FREQUENCY is GT 2.0
                       (customer is regular)
                and-IF     age of oldest balance overdue is LT 60 days
                   THEN    mark order as OK for credit
                           send down data flow 3-6
                           (orders with credit OK)
                   ELSE    (balance overdue GE 60 days)
                     SO    generate PREPAYMENT-REQUEST plus
                           REMINDER-OF-OVERDUE-BALANCE
              ELSE    (customer is not regular)
                so-IF      balance overdue is GT zero
                   THEN    generate PREPAYMENT-REQUEST
                   ELSE    (no balance owing)
                     SO    mark order as OK for credit
                           send down data flow 3-6
                           (orders with credit OK)
```

Figure 5.19

5.4.5 PROS AND CONS OF THE FOUR TOOLS

Table 5.1 summarizes the relative strengths and weaknesses of the four tools for process logic that we have discussed in this chapter.

5. ANALYZING AND PRESENTING PROCESS LOGIC

COMPARISON OF TOOLS FOR PROCESS STRUCTURES

USE	*DECISION TREES*	*DECISION TABLES*	*STRUCTURED ENGLISH*	*"TIGHT ENGLISH"*
LOGIC VERIFICATION	Moderate	Very good	Good	Moderate
DISPLAY LOGIC STRUCTURE	Very good (but decision only)	Moderate (Decision only)	Good (All)	Moderate (All but dependent on author)
SIMPLICITY (easy to use)	Very good	Very poor to Poor	Moderate	Good
USER VERIFICATION	Good	Poor (unless user trained)	Poor to Moderate (slightly jargon)	Good
PROGRAM SPECIFICATION	Moderate	Very good	Very good	Moderate
MACHINE "READABLE"	Poor	Very good	Very good	Very good
MACHINE "EDITABLE"	Poor	Very good	Moderate (needs syntax)	Poor
CHANGEABILITY	Moderate	Poor (unless simple rule change)	Good	Good

Table 5.1

The table compares the four logic tools we have discussed, in terms of eight considerations. These eight considerations can be categorized into three basic areas:

1. How easy are the tools for the analyst to use?

    Three considerations are tabulated: "Logic verification," "Display (of) logic structure," and "Simplicity." "Logic

5.4 STRUCTURED ENGLISH, PSEUDOCODE, AND "TIGHT ENGLISH"

verification" describes the ease with which each tool can be used to ensure that all the logical possibilities have been covered. "Display logic structure" describes the extent to which each tool provides a graphic representation of the four basic types of structure (sequence, decision, loop, and case). "Simplicity" rates each tool on how easy or difficult it is to learn to use it.

2. How easy are the tools for the user community to use?

   The ratings for "user verification" deal with this.

3. How easy are the tools for the designer/programmer to use?

   The heading "program specification" describes the suitability of each tool for that purpose. The considerations of "machine readable" and "machine editable" describe the ease with which each tool could be processed by a text-editor, and the ease with which a software package could be used to verify the consistency and syntax of the logic.

   "Changeability" describes the ease with which the tool allows logic to be changed, because, for example, the user's requirements have changed.

We can draw on the table to summarize the most convenient situation for use of each of the tools, as follows:

Decision trees are best used for logic verification or moderately complex decisions which result in up to 10-15 actions. Decision trees are also useful for presenting the logic of a decision table to users.

Decision tables are best used for problems involving complex combinations of up to 5-6 conditions. Decision tables can handle any number of actions; large numbers of combinations of conditions can make decision tables unwieldy.

Structured English is best used wherever the problem involves combining sequences of actions with decisions or loops.

"Tight English" is best used for presenting moderately complex logic once the analyst is clear that no ambiguities can arise.

5. ANALYZING AND PRESENTING PROCESS LOGIC

5.4.6 WHO DOES WHAT?

Throughout this book we have distinguished the three roles of analyst, designer, and programmer:

1. the analyst who establishes the logical functional specifications

2. the designer who specifies how the system will be put together to meet the logical requirements

3. the programmer who implements the designer's specification.

We have said that these roles may be played by one person (as on a small project), or by two or three people or teams. The project management style of the organization, and the abilities of the people available, dictate who on the project discharges which roles. One of the most-debated points in the division of labor is "who writes the detailed logic?" We are clear that the analyst-role does not include the writing of pseudocode; that is physical and should be done by the designer or programmer. But should the analyst-role write structured English, or should that person confine himself to decision trees and hopefully "Tight English," leaving detailed logic analysis to the designer/programmer? On the one hand the analyst is the person most closely in touch with the organization's business and with the details of the logic; on the other hand many programmers find a good part of their job satisfaction in working out the logic detail from a clear but concise logic summary such as that in the data dictionary entry and when presented with data structure definitions and Structured English cry, "You're making me into a mere coder!"

We take the compromise position, namely that the analyst should be trained in and comfortable with decision tables and Structured English, but should use them only as tools of last resort to get to the bottom of a complex process. The analyst normally should present process logic to the designer/programmer as decision trees or "Tight English" (which, happily, are also the most useful tools for communicating with the users). *If the designer or programmer require it*, the analyst must be prepared to provide decision tables or Structured English. The test should always be "is this logic statement unambiguous enough to be able to write a program from it?"

REFERENCES

5.1 G. E. Whitehouse, *Systems Analysis and Design using Network Techniques*, Prentice-Hall, 1973.

5.2 S. L. Pollack, H. T. Hicks, and W. J. Harrison, *Decision Tables: Theory and Practice*, Wiley, 1971.

5.3 K. R. London, *Decision Tables*, Auerbach, 1972.

5.4 C. Böhm and G. Jacopini, "Flow diagrams, Turing Machines, and Languages with only two formulation rules," *Communications of the ACM*, May 1966.

5.5 P. Van Leer, "Top-down development using a Program Design Language," *IBM Systems Journal*, Vol. 15, No. 2, 1976.

5. ANALYZING AND PRESENTING PROCESS LOGIC

EXERCISES AND DISCUSSION POINTS

1. Apply the indifference test to the extended entry decision table in Figure 5.13.

2. Convert the decision table produced in the previous exercise into Structured English

3. The following narrative describes the procedure carried out by the clerical staff in the warehouse of an electrical parts distributor:

    "When an order is received from the Sales Department, each item on the order is checked to see if it can be met from current inventory. If sufficient inventory is held, the warehouse clerk adjusts the stock records, and passes the item for picking and dispatch. Each time the stock records are changed, the new inventory-level is compared to the safe-reorder level which is marked on the stock-item card. If the new inventory-level is below the safe-reorder level, the warehouse clerk writes out a purchase order form, notes the quantity ordered on the stock-item card, and passes the purchase order form to the Chief Buyer for approval and dispatch to the supplier.

    If there is some inventory, but not enough to fill the order, (part-fulfilment), the warehouse clerk dispatches the items available, adjusts the stock records, and creates a back-order for the required amount. The back-order is filed in part-number order awaiting receipt of a shipment. If the item has been previously ordered, the warehouse clerk sends an expedite-delivery notice to the Chief Buyer. If the item is not on order, a purchase order form is prepared.

    If the item is completely out-of-stock, a back-order is created and filed as above, and the clerk sends an expedite-delivery notice to the Chief Buyer whether or not the item is already on order."

    Draw up a decision table to show the logic of this process. When you have the decision table, present the logic in a decision tree, and also write it out in the form of Structured English. Produce a set of instructions for a newly-hired warehouse clerk, using "Tight English," and test them for clarity by walking through them with a clerical person.

5. EXERCISES AND DISCUSSION POINTS

4. Take a program that has moderately complex logic, and write the logic section out in pseudocode. Does this aid your understanding of the program?

5. A number of software packages exist which convert decision tables directly into code. Why do you think that these packages are not in wider use today?

6. If you have access to decision table processing software, use it to convert the freight-rate decision table developed in Exercise 1 into COBOL code. Compare the output with the Structured English produced in Exercise 2.

7. The next time you have to write a memo containing the work "if," write it first in Structured English, then convert it to "tight" English.

8. If your organization has a pension plan, analyze the policy statement which specifies the amount of the pension, using the logic tools of this chapter. Which representation is the easiest to understand?

Chapter 6

Defining the contents of data stores

6.1 WHAT COMES OUT MUST GO IN

As we noted in Chapter 2, once we have specified the data structures in the data flows coming out of a data store, we know what the minimum contents of that data store should be. For a data element to be extracted, it must be put into the store in the first place. We can therefore examine the details of the data flows going into the data store, to check that all the necessary elements are being stored, and to see whether anything is being stored that never gets used.

Figure 6.1 shows part of the data flow for a personnel system.

6. DEFINING THE CONTENTS OF DATA STORES

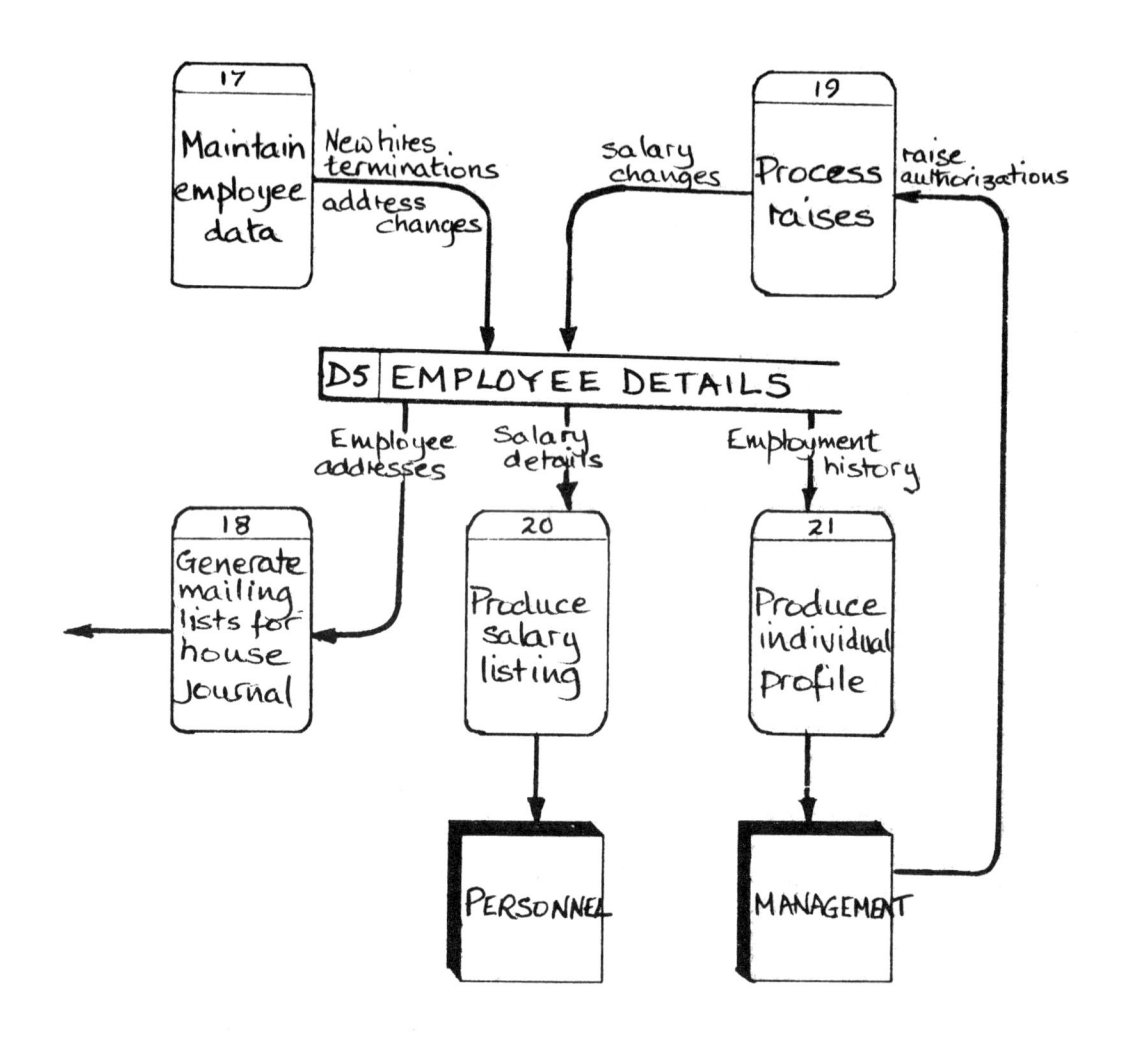

Figure 6.1

We can see from the nature of the data flows and related processes that the data store D5-EMPLOYEE-DETAILS must hold information on all employees, their names, addresses and salaries. But what does that mean in detail? We can look up each of the five data flows in the data dictionary to find their component data structures, and look up the data structures to find their contents. Then we can compare the contents of the flows into the data store with the flows out of the data store, which might be as shown in Figure 6.2:

6.1 WHAT COMES OUT MUST GO IN

```
                DATA STRUCTURES CONTAINED IN:


    FLOWS INTO D5                      FLOWS OUT OF D5


    NEW-EMPLOYEE    (17-D5)            EMPLOYEE-ADDRESS    (D5-18)
        DATE-HIRED                         NAME
        NAME                               ADDRESS
        PERSONNEL-NO
        ADDRESS
        JOB-TITLE                      SALARY-DETAILS     (D5-20)
        SALARY-START                       NAME
                                           PERSONNEL-NO
                                           CURRENT-SALARY
    TERMINATIONS    (17-D5)
        NAME
        PERSONNEL-NO                   EMPLOYMENT-HISTORY  (D5-21)
                                           NAME
                                           PERSONNEL-NO
    ADDRESS-CHANGE  (17-D5)                DATE-HIRED
        NAME                               JOB-HISTORY*
        PERSONNEL-NO                           JOB-TITLE
        ADDRESS-OLD                            DATE-EFFECTIVE   ⎫ for whole
        ADDRESS-NEW                        SALARY-HISTORY*      ⎬ employment
                                               SALARY           ⎭ with orga-
                                               DATE-EFFECTIVE     nization
    SALARY-CHANGE   (19-D5)                [REVIEW-SUMMARY]
        NAME
        PERSONNEL-NO
        SALARY-OLD
        SALARY-NEW
        DATE-EFFECTIVE
```

Figure 6.2

The next step is to produce a draft of the contents of D5, based on our analysis of inflow and outflow. A first attempt might be as shown in Figure 6.3.

6. *DEFINING THE CONTENTS OF DATA STORES*

```
          STRUCTURE OF D5 (First Draft)

         Data structure/element

                NAME
                PERSONNEL-NO
                ADDRESS
                CURRENT-SALARY
                DATE-HIRED
                JOB-HISTORY*
                   JOB-TITLE
                   DATE-EFFECTIVE
                SALARY-HISTORY*
                   SALARY
                   DATE-EFFECTIVE
                [REVIEW-SUMMARY]
```

Figure 6.3

Comparing the flows out with the flows in, we note the following points:

1. EMPLOYEE-ADDRESS clearly requires details of the current address, yet ADDRESS-CHANGE implies that the old address is also being stored. Is this necessary? Do we ever want to know an employee's previous address? Not on the outflows listed here, at least.

2. EMPLOYMENT-HISTORY, on the other hand, requires a list of all the different salaries earned by an employee throughout his career with us, so we will need to store a variable number of pairs of salaries and dates, not just SALARY-OLD and SALARY-NEW as is implied in the structure of SALARY-CHANGE.

3. JOB-HISTORY requires a data storage similar to SALARY-HISTORY. But we have no data flow defined for entering changes in JOB-TITLE! As far as we can see from Figure 6.2, a new employee gets a JOB-TITLE when he is hired, and can never change it. Comparing the inflows and outflows has turned up a fairly serious omission from the data flow diagram, which we must immediately put right, together with some minor redundancies in some data structures. A similar comment applies to REVIEW-SUMMARY.

To see the reality behind the data structures, let us draw out a few examples of the contents, as shown in Figure 6.4.

6.1 WHAT COMES OUT MUST GO IN

SAMPLE CONTENTS OF D5-EMPLOYEE-DETAILS

NAME	PERSONNEL NUMBER	ADDRESS	CURRENT SALARY	DATE HIRED	JOB-HISTORY JOB-TITLE	JOB-HISTORY DATE-EFF	SALARY-HISTORY SALARY	SALARY-HISTORY DATE-EFF	REV. SUMM.
John P Jones	26622	15 Corona Dr New York NY 10077	18200	12/01/74	Supervisor Sen. Prog. Programmer	08/01/76 04/15/75 12/01/74	18200 15250 14500 12750	12/01/76 12/01/75 04/15/75 12/01/74	4.0 3.5
Mary A Worth	30604	2221 W 54 Faterson NJ 07070	21250	06/15/73	Manager Shift-Leader Operator Typist	11/01/75 12/15/74 07/01/74 06/15/73	21250 19000 16500 14000 9500 8500	06/15/76 11/01/75 06/15/75 12/15/74 06/15/74 06/15/73	4.5 4.8 4.5
Edward P Dullson	20927	492 East Highway New York NY 10066	8500	01/01/75	Mail-clerk	01/01/75	8500 8250	01/01/76 01/01/75	2.8
. . . . . .	. . . .								

Figure 6.4

6. DEFINING THE CONTENTS OF DATA STORES

Examining the employment histories of these three people, we see that a change in job-title usually (but not always) carries a change in salary. As an example of such an exception, Jones was promoted to Supervisor on 08/01/76 without a raise. We see that each employee gets a review on the anniversary of being hired, and that the results of the review are expressed as a single figure (a composite rating on a 0-5 scale).

6.2 SIMPLIFYING DATA STORE CONTENTS BY INSPECTION

Can this first draft of the structure be simplified? First, let us see what we can do by common sense alone. We can see some duplication of data right away. CURRENT-SALARY is always the most recent value of SALARY-HISTORY. So why need it be stored as a separate data element? Similarly, DATE-HIRED is always the earliest date in both JOB-HISTORY and SALARY-HISTORY, so why need that be held separately? This observation suggests a more important simplification; since several of the dates in JOB-HISTORY and SALARY-HISTORY are the same, we could make a composite data structure out of JOB-HISTORY and SALARY-HISTORY, calling it SALARY-TITLE-HISTORY, and making an entry whenever either the job-title changes or the salary changes. The structure would be defined as:

>   SALARY-TITLE-HISTORY*
>   DATE-OF-CHANGE
>   JOB-TITLE
>   SALARY
>   [REVIEW-SUMMARY]

and for Mr. Jones, its contents would be:

DATE-OF-CHANGE	JOB-TITLE	SALARY	REVIEW-SUMMARY
12/01/76	Supervisor	18200	4.0
08/01/76	Supervisor	15250	
12/01/75	Sen. Prog.	15250	3.5
04/15/75	Sen. Prog.	14500	
12/01/74	Programmer	12750	

6.2 SIMPLIFYING DATA STORE CONTENTS BY INSPECTION

It is fair to say that this simplification in structure is bought at the price of slightly more complex logic in some subsequent processing. For example, Process 20, "Produce salary listing" will have to do more than merely retrieve name, personnel-number, and current salary (as it would with our first draft). It will have to search for the most recent entry in the SALARY-TITLE-HISTORY repeating group, and extract the current salary from there. However, what we lose on the swings, we more than gain on the roundabouts. In return for a little extra process logic, we have gained three advantages:

1. We have a simpler, smaller data store.

2. Each time CURRENT-SALARY changes, we only have to record the fact in one place, instead of two, as in the original draft. This reduces the chance that the two entries will get out of step, either because someone forgets to make both changes (in a manual file), or a hardware/software error causes one to be updated and not the other. As a general principle we note that *redundancy* (duplication, triplication, etc.) *increases the risk of error*.

3. We have a better insurance against change. Consider Process 21, "Produce Individual Profile." The data flow structures in Figure 6.2 imply that the user wants job-history and salary-history reported separately; that's why the first draft of the data store looks that way. But what if the user changes his mind, and wants a combined listing of both job and salary changes in date order? With the original structure, the process logic will have to do the combination and sorting; with the new simplified structure, the processing is almost trivial.

Both for manual and for computer systems, it usually is *easier and cheaper to change the logic of a process than to change the structure of a data store*. Consequently, the *simpler* and *more general* the *structure* of a data store, the *easier and cheaper* it will tend to be to make changes in the the data.

6. DEFINING THE CONTENTS OF DATA STORES

6.3 SIMPLIFYING DATA STORE CONTENTS BY NORMALIZATION

In the previous section, we simplified the data store contents by finding and removing redundant data elements. A further level of simplification can be achieved by reorganizing the contents to remove repeating groups, a process that has come to be known as *normalization*.

After the removal of redundancy, the contents of D5 have the structure:

>   NAME
>   PERSONNEL-NO
>   ADDRESS
>   SALARY-TITLE-HISTORY *
>   DATE-OF-CHANGE
>   JOB-TITLE
>   SALARY
>   [REVIEW-SUMMARY]

How can we get rid of the repeating group, SALARY-TITLE-HISTORY? The only way is to split the structure into two structures, both of which are simpler. We end up with a structure containing name and address (which occurs only once for each employee) and a structure containing each change of title or salary (of which there may be several for each employee). Each structure must contain PERSONNEL-NO, the only data element which specifies each employee uniquely.

```
D5-EMPLOYEE-DETAILS                    EMPLOYEE-HOME
   NAME                                   PERSONNEL-NO
   PERSONNEL-NO                           NAME
   ADDRESS                                ADDRESS
   SALARY-TITLE-HISTORY*
      DATE-OF-CHANGE
      JOB-TITLE                        SALARY-TITLE-HISTORY
      SALARY                              PERSONNEL-NO
      [REVIEW-SUMMARY]                    DATE-OF-CHANGE
                                          JOB-TITLE
                                          SALARY
                                          [REVIEW-SUMMARY]

    Unnormalized Structure              Two normalized Structures
```

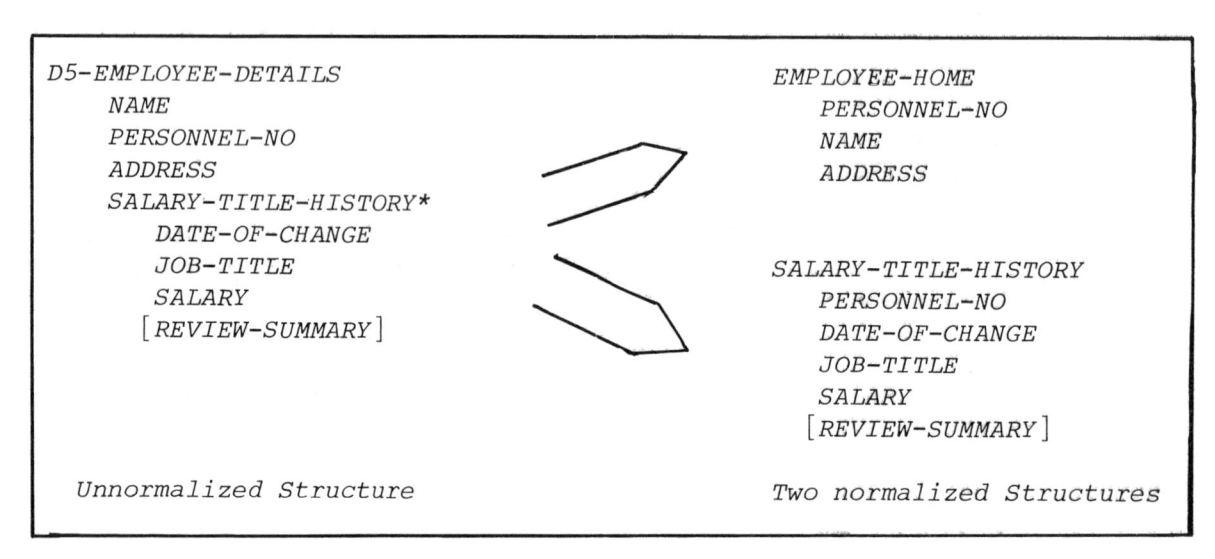

Figure 6.5

6.3 SIMPLIFYING DATA STORE CONTENTS BY NORMALIZATION

```
EMPLOYEE-HOME
        (PERSONNEL-NO,    NAME,              ADDRESS)

        20927          Ed Dullson        492 E. Highway, New York, NY 10066

        26622          John P. Jones     15 Corona Dr, New York, NY 10077

        30604          Mary A. Worth     2221 W 54, Paterson, NJ 07070



SALARY-TITLE-HISTORY
        (PERSONNEL-NO, DATE-OF-CHANGE, JOB-TITLE, SALARY, REVIEW-SUMMARY)

        20927          01/01/76       Mail-clerk     8500      2.8
        20927          01/01/75       Mail-clerk     8250       -

        26622          12/01/76       Supervisor    18200      4.0
        26622          08/01/76       Supervisor    15250       -
        26622          12/01/75       Sen. Prog.    15250      3.5
        26622          04/15/75       Sen. Prog.    14500       -
        26622          12/01/74       Programmer    12750       -

        30604          06/15/76       Manager       21250      4.5
        30604          11/01/75       Manager       19000       -
        30604          06/15/75       Shift-Leader  16500      4.8
        30604          12/15/74       Shift-Leader  14000       -
        30604          07/01/74       Operator       9500       -
        30604          06/15/74       Typist         9500      4.5
        30604          06/15/73       Typist         8500       -
```

Figure 6.6

Contents of normalized structures for D5-EMPLOYEE-DETAILS

6. DEFINING THE CONTENTS OF DATA STORES

Figure 6.6 shows how the data store contents now look for the three employees we showed in Figure 6.4. The reader is invited to verify that all the data flows listed in Figure 6.2 as coming out of the data store, could be make up from structures in the form shown in Figure 6.6, by appropriate selection of data elements.

6.3.1 THE VOCABULARY OF NORMALIZATION

The concepts and techniques of normalization have been developed by Dr. E.F. Codd of IBM [6.1, 6.2] and others, working in terms of the mathematics of sets. For this reason, the reader may meet somewhat different terms from those we have become used to. While this different vocabulary is, strictly speaking, unnecessary, the normalization approach is likely to become increasingly important over the next few years as data bases get more complex and simplicity becomes more vital.

We shall explain the terms, and use them in the rest of this and the next chapter, with the more usual equivalent term in parentheses, where relevant. Each term is explained in the glossary at the end of the book. A more technical discussion is given in Martin,[6.3] and Date, [6.4].

In place of our term "data structure," the word "relation" is used, in the sense of a data structure expressing a relation between data elements. In place of our term "data element," the term "domain" is used, meaning a range of values a data element could take up. Each individual record is called a "tuple," (pronounced to rhyme with "couple.") A tuple (record) with two domains (data elements) is called a 2-tuple; a tuple such as "26622, 12/01/75, Sen.Prog., 15250, 3.5" with 5 domains, is a 5-tuple. The relations (data structures) of which these tuples are part are called relations of degree 2 and degree 5 respectively. Figure 6.7 illustrates these terms. The relation is of degree 5, even though in some tuples, the value of the domain REVIEW-SUMMARY is null.

Every tuple in a relation (like every record in a file) must have a unique key, by which the tuple can be identified. What is the key in the relation SALARY-TITLE-HISTORY in Figure 6.7? PERSONNEL-NO is not good enough, since by the very nature of the way we set the relation up, there are likely to be several tuples for each employee. A unique key can be formed by *concatenating* (chaining together) PERSONNEL-NO and

178

6.3 SIMPLIFYING DATA STORE CONTENTS BY NORMALIZATION

```
SALARY-TITLE-HISTORY
        (PERSONNEL-NO, DATE-OF-CHANGE, JOB-TITLE, SALARY, REVIEW-SUMMARY)

            20927       01/01/76    Mail-clerk      8500      2.8
            20927       01/01/75    Mail-clerk      8250       -

            26622       12/01/76    Supervisor     18200      4.0
            26622       08/01/76    Supervisor     15250       -
            26622       12/01/75    Sen. Prog.     15250      3.5
            26622       04/15/75    Sen. Prog.     14500       -
            26622       12/01/74    Programmer     12750       -

            30604       06/15/76    Manager        21250      4.5
            30604       11/01/75    Manager        19000       -
            30604       06/15/75    Shift-Leader   16500      4.8
            30604       12/15/74    Shift-Leader   14000       -
            30604       07/01/74    Operator        9500       -
            30604       06/15/74    Typist          9500      4.5
            30604       06/15/73    Typist          8500       -
```

Relation (data structure) ↗

Tuple (record) → (row: 30604 11/01/75 Manager 19000)

Domain (data element) ↑ (SALARY column)

null value ↙

Figure 6.7

DATE-OF-CHANGE. You could say, "Get me the record for the status of employee 30604 effective 11/01/75," and be sure of retrieving only one tuple. It is conventional to show the relation's structure and key by writing:

SALARY-TITLE-HISTORY (PERSONNEL-NO, DATE-OF-CHANGE,
            JOB-TITLE, SALARY, REVIEW-SUMMARY)
or
EMPLOYEE-HOME (PERSONNEL-NO, NAME, ADDRESS)

Sometimes, it is not obvious what the key of the relation should be. For instance, at first sight you might think that another possible *candidate-key* would be the concatenation of PERSONNEL-NO, JOB-TITLE, and SALARY. In Figure 6.7 there are no duplications of these three taken together. But PERSONNEL-NO/JOB-TITLE/SALARY would be a poor choice as a key, because of the possibility that an employee could be given a review without a raise or a promotion, creating two tuples with the same key value. We choose the *primary key* of the relation to have as few data elements as possible (obviously one data

6. DEFINING THE CONTENTS OF DATA STORES

element is ideal), and with data elements that have no undefined values. (The "no undefined values" criterion would rule out REVIEW-SUMMARY as part of a candidate key, since it has a null value in several tuples.)

Figure 6.8 summarizes the questions that have to be asked of a candidate key in a relation.

Tests to see whether a candidate key should be chosen as the primary key

1. *Can any data element be removed, and the remaining key still be unique for each tuple?*

2. *Is there any foreseeable situation in which this candidate key would not be unique?*

3. *Does any part of the candidate key have undefined values?*

4. *Of the remaining candidates, which has the fewest domains (data elements)?*

Figure 6.8

6.4 SOME NORMALIZED FORMS ARE SIMPLER THAN OTHERS

Codd has established that there are three types of normalized relations, called, in increasing order of simplicity, first normal form, second normal form, and third normal form. We will define each of these, and discuss how to reduce relations to the simplest form, third normal.

6.4.1 FIRST NORMAL FORM (1NF)

Any normalized relation (a data structure without repeating groups) is automatically in 1NF, no matter how complex its key, or what interrelationships there may be between the component data elements. Relations in first normal form may suf-

6.4 SOME NORMALIZED FORMS ARE SIMPLER THAN OTHERS

fer from two kinds of complexity:

1. If the primary key is concatenated, some of the non-key domains may depend on only part of the key, not the whole key.

2. Some of the non-key domains may be interrelated.

This is easy to say, but hard to visualize. Suppose we were trying to produce a normalized version of the structure, ORDERS, describing book purchases from the CBM Company. We define an order to consist of the customer-name, an order date, the ISBN (standard book number) for the book ordered, the title, the author, the quantity of this title that has been ordered, and the total cost of the order for a given book. We can create a normalized relation:

BOOK-ORDER (<u>CUSTOMER-NAME</u>, <u>ORDER-DATE</u>, <u>ISBN</u>, TITLE, AUTHOR QUANTITY, PRICE, ORDER-TOTAL)

the underlining showing that we have chosen the concatenated key CUSTOMER-NAME/ORDER-DATE/ISBN to uniquely identify each order, a fair assumption provided no customer ever orders the same book twice in one day. As we have noted, this relation is in *first normal form* by virtue of the fact that it contains no repeating groups.

The first complexity we would like to remove is the fact that several of the non-key domains (TITLE, AUTHOR, and PRICE) can be identified with only part of the key, the ISBN. Customer name and order date are irrelevant. In other words, if you are given the ISBN, you know the TITLE, AUTHOR and PRICE, without caring about the CUSTOMER-NAME or the ORDER-DATE. This is a different situation from that of QUANTITY: to know the quantity in any particular order, you have to know all three of the domains which are concatenated to make the key.

In the vocabulary of normalization QUANTITY is *fully functionally dependent* on the whole concatenated key. On the other hand, TITLE is not fully functionally dependent, since you only need to know part of the key, (ISBN) to know the TITLE. A domain is fully functionally dependent if it is dependent on the whole of the key. It is functionally dependent, but not fully so, if the value of the domain can be determined from only part of the key. This concept allows us to define:

6. DEFINING THE CONTENTS OF DATA STORES

6.4.2 SECOND NORMAL FORM (2NF)

A normalized relation is in second normal form if *all* the non-key domains are fully functionally dependent on the primary key. BOOK-ORDER, though in 1NF, is not in 2NF as it stands. To get BOOK-ORDER into 2NF we must get rid of the partial functional dependence. This can be done by taking out the domains which describe the book, and putting them in a separate relation:

ORDER (<u>CUSTOMER-NAME</u>, <u>ORDER-DATE</u>, <u>ISBN</u>, QUANTITY, ORDER-TOTAL)

BOOK (<u>ISBN</u>, TITLE, AUTHOR, PRICE)

ORDER is now in *second normal form*; each of the non-key domains (QUANTITY, ORDER-TOTAL) can only be specified by knowing the full concatenated key, i.e., all the non-key domains are fully functionally dependent on the primary key.

We can simplify the situation even further because ORDER has a complexity still lurking inside it; QUANTITY and ORDER-TOTAL are not mutually independent. For any given PRICE, the QUANTITY determines the ORDER-TOTAL. Therefore, ORDER-TOTAL is functionally dependent on QUANTITY.

6.4.3 THIRD NORMAL FORM (3NF)

A normalized relation is in third normal form if:

a) all the non-key domains are *fully* functionally dependent on the primary key

and also

b) no non-key domain is functionally dependent on any other non-key domain.

So to transform a 2NF relation into a 3NF relation, we examine each of the non-key domains to see that they are independent of each of the other non-key domains, and remove any such mutual dependence. In the case of ORDER, we see that ORDER-TOTAL is in fact a redundant data element, because it can be computed. So we can remove it altogether, and express each order by means of two relations, both in 3NF.

6.4 SOME NORMALIZED FORMS ARE SIMPLER THAN OTHERS

BOOK-ORDER (CUSTOMER-NAME, ORDER-DATE, ISBN, QUANTITY)

BOOK (ISBN, TITLE, AUTHOR, PRICE)

Sometimes we can only get a relation into 3NF by expressing the dependence between non-key domains as a separate relation. Suppose we had a relation like this:

PROJ-ASSIGN (EMPLOYEE-#, PHONE, HOURLY-RATE, PROJECT-#, FINISH-DATE)

The relation is used for storing assignments of temporary employees to projects.

Is this in 1NF?   Yes, it is a normalized relation.

Is it in 2NF?   Yes, since the primary key is only one domain, the non-key domains must be fully functionally dependent on it.

Is it in 3NF?   No, there is a dependence between PROJECT-# and FINISH-DATE: if you know the PROJECT-#, you know the FINISH-DATE for the project.

We can express PROJ-ASSIGN in 3NF only by splitting it into two relations:

TEMP-HIRE (EMPLOYEE-#, PHONE, HOURLY-RATE, PROJECT-#)

PROJECT (PROJECT-#, FINISH-DATE)

Figure 6.9 summarizes the tests and steps to transform unnormalized relations into 3NF relations.

As an example of some 3NF relations, consider EMPLOYEE-HOME and SALARY-TITLE-HISTORY, reproduced in Figure 6.10. The reader is invited to apply the tests and verify that both these relations are in third normal form.

As a practical matter, 1NF and 2NF relations tend to arise when we try to describe two underlying real things in the same relation, as in the case of PROJ-ASSIGN or ORDER. The simplest relations describe one aspect of the real world at a time.

6. DEFINING THE CONTENTS OF DATA STORES

UNNORMALIZED RELATION (data structure with repeating groups)

↓

> Split the relation into one or more relations without repeating groups. Assign one or more domains (data elements) as the primary key: the smallest key that uniquely identifies each tuple (instance of the data structure).

↓

NORMALIZED RELATION IN FIRST NORMAL FORM (without repeating groups)

↓

> For relations whose keys have more than one domain, verify that each non-key domain is functionally dependent on the *whole* key, and not on just part of it. Split the relation, if necessary, to achieve this.

↓

SECOND NORMAL FORM (all non-key domains fully functionally dependent on the primary key)

↓

> Verify that all non-key domains are mutually independent of each other. Remove redundant domains or split the relation as necessary to achieve this.

↓

THIRD NORMAL FORM (all non-key domains fully functionally dependent on the primary key, and independent of each other.)

Figure 6.9

6.5 MAKING RELATIONS OUT OF RELATIONS--PROJECTION AND JOIN

```
SALARY-TITLE-HISTORY
     (PERSONNEL-NO, DATE-OF-CHANGE, JOB-TITLE, SALARY, REVIEW-SUMMARY)

        20927        01/01/76     Mail-clerk    8500       2.8
        20927        01/01/75     Mail-clerk    8250        -

        26622        12/01/76     Supervisor    18200      4.0
        26622        08/01/76     Supervisor    15250       -
        26622        12/01/75     Sen. Prog.    15250      3.5
        26622        04/15/75     Sen. Prog.    14500       -
        26622        12/01/74     Programmer    12750       -

        30604        06/15/76     Manager       21250      4.5
        30604        11/01/75     Manager       19000       -
        30604        06/15/75     Shift-Leader  16500      4.8
        30604        12/15/74     Shift-Leader  14000       -
        30604        07/01/74     Operator      9500        -
        30604        06/15/74     Typist        9500       4.5
        30604        06/15/73     Typist        8500        -
```

Figure 6.10

6.5 MAKING RELATIONS OUT OF RELATIONS--PROJECTION AND JOIN

Once the contents of the data stores have been simplified into a few normalized relations, we need to be able to use these relations to answer queries, as well as for other processing tasks. This means we need a method of building up more extensive relation structures and extracting parts of our stored relations.

   Two operations have been defined which enable us to describe what we want to do. They are called *projection* and *join*.

6. DEFINING THE CONTENTS OF DATA STORES

6.5.1 PROJECTION

A better name for this operation would be "selection"; it consists of "projecting" a whole relation over selected domains to end up with a relation between just those domains. Figure 6.11 shows the projection of:

    SALARY-TITLE-HISTORY   over PERSONNEL-NO
    SALARY-TITLE-HISTORY   over PERSONNEL-NO plus JOB-TITLE

As you can see, projection involves the extraction of the chosen domain(s), followed by the removal of any duplicate tuples. Projection is the logical operation involved in handling queries such as, "What jobs has Employee 26622 had with us," or "Get me a salary history for Employee 30604."

The figure also illustrates the fact that it is perfectly possible to have a relation with only one domain, such as PERSONNEL-NO, just as in theory one can have a data structure with only one data element.

6.5.2 JOIN

This operation is the reverse of projection, putting two relations together to make one larger relation. The two relations to be joined must have a common domain. Suppose we have one relation describing suppliers of produce, say

    SUPPLIER   (SUPP-NO, SUPP-NAME, CITY)

with values like this:

SUPPLIER:	SUPP-NO	SUPP-NAME	CITY
	013	Handy	New York
	061	Readymix	Pittsburgh
	062	Readymix	New York
	063	Readymix	Boston
	095	Friendly	Buffalo

and another relation describing the produce itself, say,

6.5 MAKING RELATIONS OUT OF RELATIONS--PROJECTION AND JOIN

SALARY-TITLE-HISTORY
    (PERSONNEL-NO, DATE-OF-CHANGE, JOB-TITLE, SALARY, REVIEW-SUMMARY)

PERSONNEL-NO	DATE-OF-CHANGE	JOB-TITLE	SALARY	REVIEW-SUMMARY
20927	01/01/76	Mail-clerk	8500	2.8
20927	01/01/75	Mail-clerk	8250	-
26622	12/01/76	Supervisor	18200	4.0
26622	08/01/76	Supervisor	15250	-
26622	12/01/75	Sen. Prog.	15250	3.5
26622	04/15/75	Sen. Prog.	14500	-
26622	12/01/74	Programmer	12750	-
30604	06/15/76	Manager	21250	4.5
30604	11/01/75	Manager	19000	-
30604	06/15/75	Shift-Leader	16500	4.8
30604	12/15/74	Shift-Leader	14000	-
30604	07/01/74	Operator	9500	-
30604	06/15/74	Typist	9500	4.5
30604	06/15/73	Typist	8500	-

PROJECT *over* PERSONNEL-NO

PERSONNEL-NO
20927
20927
26622
26622
26622
26622
26622
30604
30604
30604
30604
30604
30604
30604

PROJECT *over* PERSONNEL-NO, JOB-TITLE

PERSONNEL-NO	JOB-TITLE
20927	Mail-clerk
20927	Mail-clerk
26622	Supervisor
26622	Supervisor
26622	Sen. Prog.
26622	Sen. Prog.
26622	Programmer
30604	Manager
30604	Manager
30604	Shift-Leader
30604	Shift-Leader
30604	Operator
30604	Typist
30604	Typist

REMOVE DUPLICATE TUPLES

PERSONNEL-NO
20927
26622
30604

PERSONNEL-NO	JOB-TITLE
20927	Mail-clerk
26622	Supervisor
26622	Sen. Prog.
26622	Programmer
30604	Manager
30604	Shift-Leader
30604	Operator
30604	Typist

Answer to "What jobs has employee 26622 held?" } 26622 Supervisor, 26622 Sen. Prog., 26622 Programmer

Figure 6.11

6. DEFINING THE CONTENTS OF DATA STORES

PRODUCE (<u>PRODUCT-CODE</u>, <u>SUPP-NO</u>, DESCRIPTION, QUANTITY-SHIPPED)

with values like this:

PRODUCE:

PRODUCT-CODE	SUPP-NO	DESCRIPTION	QUANTITY-SHIPPED
A123	013	Cornmeal	100
B671	013	Maize	32
A123	061	Cornmeal	100
C777	095	Cattlefeed	50
B671	062	Maize	72
A123	095	Cornmeal	100

We can join these two relations over SUPP-NO to get a composite relation PRODUCE-SUPPLIERS as shown in Figure 6.12. The tuples of PRODUCE-SUPPLIERS have been sequenced by supplier number within product code, though strictly speaking a relation does not have to be in sequence.

PRODUCE:

PRODUCT CODE	SUPP-NO	DESCRIPTION	QUANTITY SHIPPED
A123	013	Cornmeal	100
B671	013	Maize	32
A123	061	Cornmeal	100
C777	095	Cattlefeed	50
B671	062	Maize	72
A123	095	Cornmeal	100

SUPPLIER:

SUPP-NO	SUPP-NAME	CITY
013	Handy	New York
061	Readymix	Pittsburgh
062	Readymix	New York
063	Readymix	Boston
095	Friendly	Buffalo

JOIN over SUPP-NO

PRODUCE-SUPPLIERS:

PRODUCT CODE	SUPP-NO	SUPP-NAME	CITY	DESCRIPTION	QUANTITY-SHIPPED
A123	013	Handy	New York	Cornmeal	100
A123	061	Readymix	Pittsburgh	Cornmeal	100
A123	095	Friendly	Buffalo	Cornmeal	100
B671	013	Handy	New York	Maize	32
B671	062	Readymix	New York	Maize	72
C777	095	Friendly	Buffalo	Cattlefeed	50

Figure 6.12

Note that Supplier 063, Readymix of Boston, does not appear in the join because he shipped us no produce. Note also that PRODUCE is not in 3NF. Why not? What form is it in? How could the relations be made all 3NF?

The order in which the domains are listed has no importance in a relation, though it is convenient to have the key at the left, and the non-key domains grouped in a way which makes the most sense to the person reading the relation. This contrasts with our data-structure notation, in which we have to put the members of a repeating group together. If we were to describe PRODUCE-SUPPLIERS as a data structure, it would have to be like this:

```
PRODUCE-SUPPLIERS
    PRODUCT-CODE
    DESCRIPTION
    SHIPMENTS* (1-)
        SUPP-NO
        SUPP-NAME
        CITY
        QUANTITY-SHIPPED
```

6.6 THE IMPORTANCE OF THIRD NORMAL FORM

3NF is the simplest possible representation of data that we can get. It represents a kind of "inspired common sense," in that often, when we boil the contents of data stores down into data structures in 3NF, and look at the result, we find ourselves saying "Of course! Those are the basic data elements which describe a customer (or part, or an employee)." Non-technical users find normalized relations, and 3NF in particular, easy to comprehend--after all, stripped of the special vocabulary, we are representing all the data in the form of a perfectly ordinary table, and what could be more familiar than that?

In addition to these benefits to the analyst and the user, the trend in the physical design of data bases is towards the use of normalized relations, through "relational" data bases, rather than indexes or hierarchies (as described in the next chapter). The basic simplicity of data in 3NF makes it much more flexible and easy to change compared with other methods of organizing a physical data base. Relational data bases are not yet in wide use largely because the hardware required

6. DEFINING THE CONTENTS OF DATA STORES

to process relations at acceptable speeds is still too expensive. As hardware is getting cheaper very rapidly, and as the complexity of data and the need for change are growing, relational data bases will become more and more important. In the meantime, if we can set up files and data bases using 3NF structures, it will make the changeover to relational data bases much easier.

So, as analysts, we can use 3NF to kill three birds with one stone:

1. We can use 3NF relations as the basic building blocks of the data stores we specify (indeed, just knowing about 3NF enables you to specify simpler structures more fluently).

2. We can use 3NF as the standard medium for communicating the contents of data stores to the physical designer, whether the eventual system turns out to be data base-oriented or file-oriented.

3. We can show the logical contents of data stores to interested users in the familiar form of tables.

6.7   A PRACTICAL EXAMPLE OF 3NF

To see how the techniques of normalization work out in practice, we will take part of the CBM system defined in Chapter 3, and analyze the contents of the data stores for that subsystem. Figure 6.13 shows that portion of the system which is concerned with order entry, taking in orders for books, checking those orders to see whether they can be filled, creating a stream of "Shippable Items" for the warehouse, together with "Non-Shippable Items" that need to be ordered. (In Chapter 9, we will suppose that the analyst and designer have decided to make this a physically separate subsystem, and we will discuss the use of the Structured Analysis and Structured Design tools in deriving a good physical design. We will be using the normalized relations from this section as an input to that design.)

6.7 A PRACTICAL EXAMPLE OF 3NF

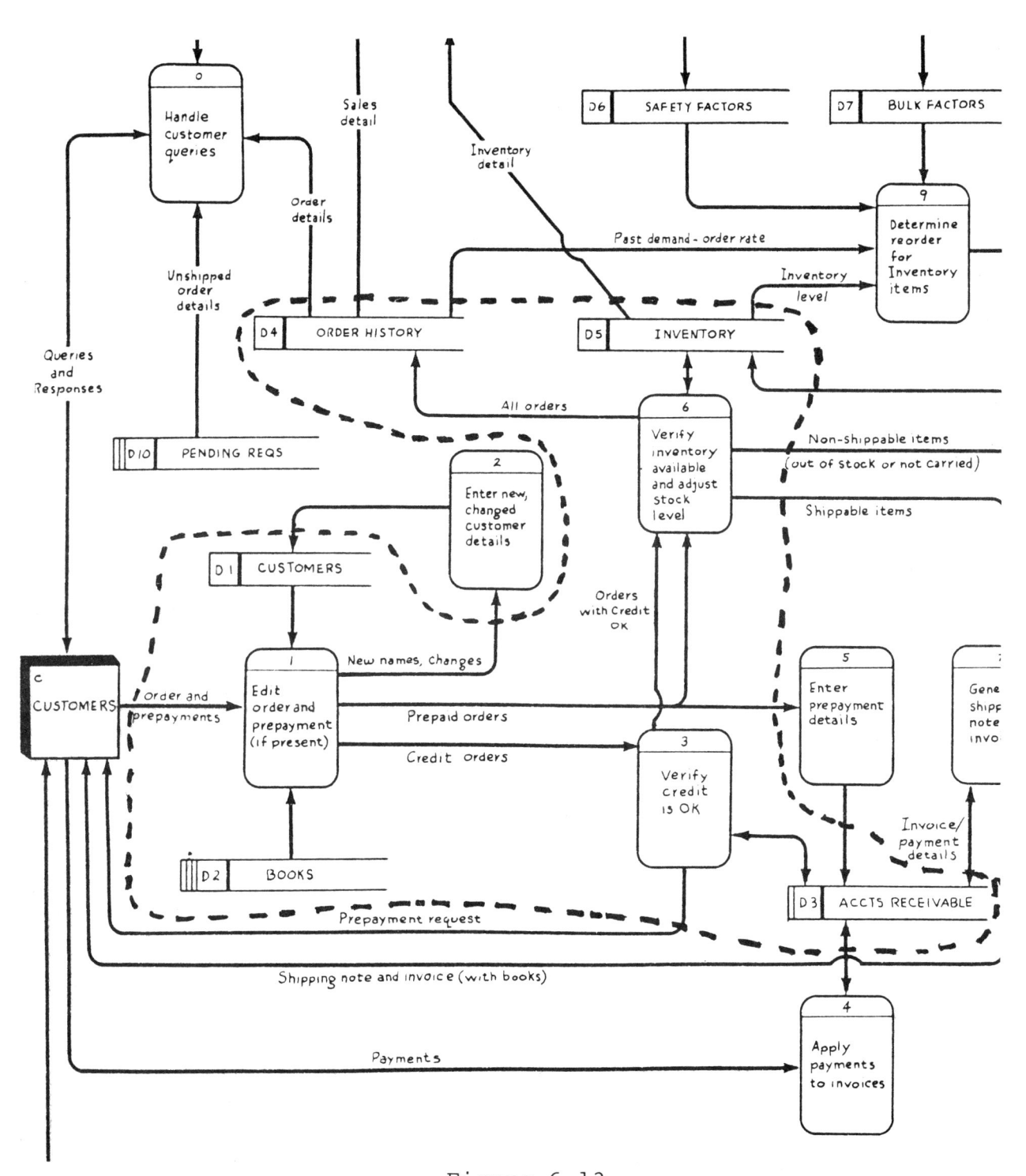

Figure 6.13

The "Order-entry" Subsystem of the CBM Company

6. DEFINING THE CONTENTS OF DATA STORES

As you can see from Figure 6.13, we are concerned with four logical data stores:

> D1: CUSTOMERS
> D2: BOOKS
> D3: ACCOUNTS RECEIVABLE
> D5: INVENTORY

We will assume for the purposes of this example that the raw contents of each of these data stores have been defined in the data dictionary, based on an examination of the data flows into and out of each data store, as we discussed in Section 6.1, "What comes out must go in." Let us take each data store in turn, and normalize it.

6.7.1 NORMALIZATION OF THE "CUSTOMERS" DATA STORE

Suppose the data dictionary entry for CUSTOMERS specifies its structure thus:

> D1: CUSTOMERS
>
> ORGANIZATION-NAME
>
>> ORGANIZATION-ADDRESS*(1-)
>>   STREET-BOX
>>   CITY-COUNTY
>>   STATE-ZIP
>>
>> PHONE
>>   AREA-CODE
>>   EXCHANGE
>>   NUMBER
>>
>> CONTACT*(1-)
>>   CONTACT-NAME
>>   [JOB-TITLE]
>>   [PHONE-EXTENSION]

From this data structure we see that we can expect any organization, say IBM, to have one or more locations, (e.g., While Plains, Palo Alto, Boca Raton), each with its address. At any one location we may have a number of contacts, e.g.,

6.7 A PRACTICAL EXAMPLE OF 3NF

White Plains
    John Jones
    Mary Roe

Palo Alto
    Jim Smith
    Lucy Velez

Boca Raton
    Fred Smith

Is this data structure a relation in first normal form? Definitely not; it contains a repeating group (CONTACT) within another repeating group (ORGANIZATION-ADDRESS). So our first step must be to remove these repeating groups and identify a suitable unique key for each tuple in the new relation. A first attempt might look like this:

CUSTOMERS (<u>ORGANIZATION-ADDRESS</u>, <u>CONTACT-NAME</u>, ORGANIZATION-NAME, PHONE, EXTENSION, JOB-TITLE)

In this relation, the key is the concatenation of the defining attributes of the two original repeating groups. Since it has no repeating groups in it, it is in 1NF. It is slightly unsatisfactory, since if two people, both named John Jones, happened to work for different organizations in a large building, we would get duplicate keys. To be on the safe side, we should include ORGANIZATION-NAME in the key: the relation then becomes:

CUSTOMERS (<u>ORGANIZATION-NAME</u>, <u>ORGANIZATION-ADDRESS</u>, <u>CONTACT-NAME</u>, PHONE, EXTENSION, JOB-TITLE)

We can now identify each contact uniquely, except for the very rare case, where the same *company* has two John Jones at the same address! The price we pay is an unwieldy key; indeed the relation could be said to be almost all key.

Leaving that consideration aside for the moment, is the new relation is second normal form, in addition to being in 1NF? Are all the non-key domains fully functionally dependent on the concatenated key? At first sight, we might say that EXTENSION and JOB-TITLE are specified when we know CONTACT-NAME. If so, that would imply that EXTENSION and JOB-TITLE are not *fully* functionally dependent on the whole key. However, we have set up such an elaborate key, precisely because CONTACT-NAME may be duplicated. Consequently we can

6. DEFINING THE CONTENTS OF DATA STORES

say that the whole key is necessary to specify each of the non-key domains, and the relation is therefore in 2NF.

Are all the non-key domains mutually independent? Yes, there is no sure way of telling any one of PHONE, EXTENSION, or JOB-TITLE given any other. Therefore the relation is not only in 2NF, but also in 3NF as it stands.

Though we have a 3NF relation, we still have this very unwieldy key, which in rare cases is not unique; as a refinement, we might create a new domain to uniquely identify each location of each organization, called ORG-ID perhaps. This might correspond to an account number, in which the first 5 digits represent the organization, and the last 2 digits represent the location within that organization. For example, 10926 might be the basic ORG-ID for the General Distributing Corp; 1092601 would represent GDC in New York, 1092602 would represent the Chicago branch, and so on. Given this domain, we could split CUSTOMERS into two 3NF relations:

ORGANIZATIONS (<u>ORG-ID</u>, ORGANIZATION-NAME, ORGANIZATION-ADDRESS
               MAIN-PHONE)
and

CONTACTS   (<u>CONTACT-NAME</u>, <u>ORG-ID</u>, CONTACT-PHONE,
               JOB-TITLE)

As with all simple keys or account numbers, ORG-ID will suffer from the fact that, unless special provisions are made, the user will need to know the ORG-ID code to get at any of the records describing the organization or the contacts. (We will also assume that if two name-sakes work in the same organization at the same address, *they* would have had enough trouble with mixed mail as to identify themselves uniquely in their name!)

6.7.2 NORMALIZATION OF THE "BOOKS" DATA STORE

In the data dictionary entry for this data store we find the following structure:

6.7 A PRACTICAL EXAMPLE OF 3NF

BOOKS

    ISBN   (International Standard Book Number)
    BOOK-TITLE
    AUTHOR
    ORGANIZATION-AFFILIATION
    PUBLISHER-NAME
    PRICE

There are no repeating groups in this structure; it is therefore automatically in 1NF. One might think that AUTHOR constitutes a repeating group, but a little thought will show that AUTHOR is merely a variable length field. For example. if SMITH and Jones are the authors of "Structured Structures," it is *not* true that there is one book of that name by Smith and another by Jones, as would be the case with a repeating group. It can be argued that there should be a separate ORGANIZATION-AFFILIATION for each AUTHOR: if Smith is from the University of Podunk and Jones is from the General Distributing Corp, we should be able to show that fact. In this case, the analyst has noticed that most collaborating authors have the same affiliation, and specified that, where different affiliations exist, the affiliation of the senior author only should be held.

The ISBN uniquely defines the book; thus we can write the 1NF relation as:

BOOKS (<u>ISBN</u>, BOOK-TITLE, AUTHOR, ORGANIZATION-AFFILIATION,
     PUBLISHER-NAME, PRICE)

Are all the non-key domains fully functionally dependent on the key? Provided we regard ISBN as a single domain, they must be. (In fact, as we noted in Section 2.2, ISBN has subcodes, one of which specifies the PUBLISHER-NAME, but we will ignore this for now.) The relation is therefore in 2NF.

Are all the non-key domains mutually independent? No, ORGANIZATION-AFFILIATION clearly depends on AUTHOR; we must therefore separate these out into a different relation. We now end up with two 3NF relations:

BOOKS (<u>ISBN</u>, BOOK-TITLE, AUTHOR, PUBLISHER-NAME, PRICE)

and

AUTHOR-AFFILIATION (<u>AUTHOR</u>, ORGANIZATION-AFFILIATION)

6. DEFINING THE CONTENTS OF DATA STORES

6.7.3 NORMALIZATION OF THE "ACCOUNTS RECEIVABLE" DATA STORE

The data structure definition is:

```
                    ORG-ID
                    BILLING-ADDRESS
                    DATE-ACCOUNT-OPENED
                    INVOICE*(1-)
                       INVOICE-NO
                       INVOICE-DATE
                       INVOICE-AMOUNT
                    PAYMENT*(0-)
                       CHECK-NO
                       PAYMENT-DATE
                       PAYMENT-AMOUNT
                    BALANCE-OUTSTANDING
                    NUMBER-OF-ORDERS-TO-DATE
```

In this data store we note that there are two repeating groups, representing the multiple invoices that have been sent out, and the various payments received. In practice, this information would only be kept in the main data store for a certain period (usually the last 6 months or the fiscal year); after that period the invoice and payment details will be transferred to an archive where they can be referred to if required, but do not clog up the active data store. This practical consideration, however, does not affect the *structure* of the data store.

As a first cut at removing the repeating groups, let us split the relatively static data elements from the more volatile:

ACCOUNT-MASTER (<u>ORG-ID</u>, BILLING-ADDRESS, DATE-ACCOUNT-OPENED)

INVOICES   (<u>INVOICE-NO</u>, ORG-ID, INVOICE-DATE, INVOICE-AMOUNT)

PAYMENTS   (<u>CHECK-NO</u>, <u>ORG-ID</u>, PAYMENT-DATE, PAYMENT-AMOUNT)

We note that INVOICE-NO is allocated by CBM in sequence as invoices are sent out, and therefore defines each invoice uniquely. CHECK-NO is assigned by each customer, and therefore may not be unique unless combined with ORG-ID. But what about BALANCE-OUTSTANDING and NUMBER-OF-ORDERS-TO-DATE? They

196

change all the time as orders and payments are received, yet do not intuitively seem to belong to INVOICES or PAYMENTS. If we could establish that they were fully functionally dependent on the key of one of the relations, that would give us a good reason to include them in that relation. And, in fact, that is the case; both BALANCE-OUTSTANDING and NUMBER-OF-ORDERS-TO-DATE are functionally dependent on ORG-ID. Therefore we can safely incorporate both domains into the relation:

ACCOUNT MASTER (ORG-ID, BILLING-ADDRESS, DATE-ACCOUNT-OPENED,
                BALANCE-OUTSTANDING, NUMBER-OF-ORDERS-TO-DATE)

even though they are volatile.

6.7.4 NORMALIZATION OF THE "INVENTORY" DATA STORE

In the data dictionary we find this specification for the contents of INVENTORY:

>   ISBN
>   PUBLISHER-NAME
>   QUANTITY-ON-HAND
>   QUANTITY-ON-ORDER
>   REORDER-LEVEL

Since there are no repeating groups, we can define a 1NF relation right away, with ISBN as the key:

INVENTORY (ISBN, PUBLISHER-NAME, QUANTITY-ON-HAND,
           QUANTITY-ON-ORDER, REORDER-LEVEL)

There is only one domain in the key, therefore the relation must be in 2NF; since the non-key domains are independent, it is also in 3NF.

6.7.5 PUTTING THE RELATIONS TOGETHER

Figure 6.14 shows the analysis of the four data stores into their component relations.

6. DEFINING THE CONTENTS OF DATA STORES

```
DATA STORE                  NORMALIZED RELATION

D1: CUSTOMERS

     ORGANIZATION    (ORG-ID, ORGANIZATION-NAME, ORGANIZATION-ADDRESS,
                      MAIN-PHONE)

     CONTACTS        (CONTACT-NAME, ORG-ID, CONTACT-PHONE, JOB-TITLE)

D2: BOOKS

     BOOKS           (ISBN, BOOK-TITLE, AUTHOR, PUBLISHER-NAME, PRICE)

     AUTHOR-
     AFFILIATION    (AUTHOR, ORGANIZATION-AFFILIATION)

D3: ACCOUNTS RECEIVABLE

     ACCOUNT-MASTER (ORG-ID, BILLING-ADDRESS, DATE-ACCOUNT-OPENED,
                     BALANCE-OUTSTANDING, NUMBER-OF-ORDERS-TO-DATE)

     INVOICES        (INVOICE-NO, ORG-ID, INVOICE-DATE, INVOICE-AMOUNT)

     PAYMENTS        (CHECK-NO, ORG-ID, PAYMENT-DATE, PAYMENT-AMOUNT)

D5; INVENTORY

     INVENTORY       (ISBN, PUBLISHER-NAME, QUANTITY-ON-HAND,
                      QUANTITY-ON-ORDER, REORDER-LEVEL)
```

Figure 6.14

Scanning the eight relations in Figure 6.14, we note that two of them have ORG-ID as their key (ORGANIZATIONS, ACCOUNT MASTER) and two have ISBN as their key (BOOKS, INVENTORY). Can we merge these two pairs, and end up with relations still in 3NF? We inspect the domains of each pair to be sure that they are mutually independent. In fact they are independent in both cases, so we could reduce the 8 relations

6.7 A PRACTICAL EXAMPLE OF 3NF

to 6, as shown in Figure 6.15, by using the JOIN operation over the common key.

```
ORGANIZATION-    (ORG-ID, ORGANIZATION-NAME, ORGANIZATION-ADDRESS,
MASTER              BILLING-ADDRESS, MAIN-PHONE, DATE-ACCOUNT-OPENED,
                    BALANCE-OUTSTANDING, NUMBER-OF-ORDERS-TO-DATE)

CONTACTS         (CONTACT-NAME, ORG-ID, CONTACT-PHONE, JOB-TITLE)

BOOK-INVENTORY   (ISBN, BOOK-TITLE, AUTHOR, PUBLISHER-NAME, PRICE,
                    QUANTITY-ON-HAND, QUANTITY-ON-ORDER, REORDER-LEVEL)

AUTHOR-          (AUTHOR, ORGANIZATION-AFFILIATION)
AFFILIATION

INVOICES         (INVOICE-NO, ORG-ID, INVOICE-DATE, INVOICE-AMOUNT)

PAYMENTS         (CHECK-NO, ORG-ID, PAYMENT-DATE, PAYMENT-AMOUNT)
```

Figure 6.15

We might well decide to arrange the physical files differently; it is common practice to have the accounting information separate, for reasons of security and control. But by applying the technique of normalization, we have ended up with the groupings of data elements which describe the six basic entities with which the order entry subsystem is concerned (other than orders, of course):

- organizations
- people in organizations
- books
- people who write books
- invoices
- payments

However the actual files are designed (and we discuss the issues in Chapters 7 and 9), we are starting from a firm bedrock understanding of *what* those physical files represent at the logical level. The 3NF relations are a logical model of the data stores, as free as possible from considerations of physical implementation. Much of the effort and difficulty

6. DEFINING THE CONTENTS OF DATA STORES

of physical file or physical data base design comes from the difficulty of seeing what data elements in the existing files truly *belong together*, what data elements describe *different entities*, current *overlapping* and *duplicate* data elements, and what *omissions* exist. Building a logical relational model goes a long way to get round these problems.

REFERENCES

6.1 E. F. Codd, "A Relational Model of Data for Large Shared Data Banks," *Communications of the ACM*, June 1970.

6.2 E. F. Codd, "Further Normalization of the Data Base Relational Model," in *Courant Computer Science Symposia*, Vol. 6, "Data Base Systems," Prentice-Hall, 1972.

6.3 J. Martin, *"Computer Data-Base Organization,"* Prentice-Hall, 1975.

6.4 C. J. Date, *"An Introduction to Database Systems,"* Addison-Wesley, 1975.

6. DEFINING THE CONTENTS OF DATA STORES

EXERCISES AND DISCUSSION POINTS

1. Explain in your own words, the terms relation, domain, tuple, degree, candidate-key, fully functionally dependent.

2. Assuming that the data store D5: EMPLOYEE-DETAILS in Figure 6.1 were to be physically implemented as two files, each with 3NF structure as shown in Figure 6.6, write the pseudocode for the processes in FIgure 6.1.

3. In what sense is it true that normalization increases redundancy?

4. What projection/join operation would be required to answer the request "Give me Mary Worth's salary history," starting with the relations in Figure 6.6.

5. The Friendly Feed Company has its suppliers file set up with this data structure:

    ```
    SUPPLIER
        SUPPLIER-NO
        SUPPLIER-NAME
        SUPPLIER-ADDRESS
    PRODUCE*
        PRODUCT-CODE
        DESCRIPTION
        PRICE
    ```

    There is no separate file of PRODUCE.

    a) Suppose Bighorn Inc., the only supplier of Produce A679, goes out of business and is deleted from the file. How can we find out what the description of A679 is?

    b) Suppose the management of Friendly decides to begin handling a new feed BETTACOW, which is given the code D302. How can we hold its description on file while we are looking for a supplier?

    Convert the data structure above into 3NF relations. How are situations a) and b) handled with the relational model?

6. From the relations of Figures 6.14 and 6.15, produce a 3NF relation (or relations) to represent the orders for books received from customers.

Chapter 7

Analyzing response requirements

7.1 DESCRIBING THE WAYS DATA IS USED

The outputs we derive from data stores and define in data flow diagrams are of widely varying natures. Consider the following requests for information.

1. "I want to receive each morning a list of all orders received during the previous business day. The report must be correct up to 4:30 P.M., and available to me at 9:00 A.M."

2. "Every time the inventory level of an item goes below the safety level, give me a report of the on-order status of the item. This report must be produced by the following morning."

3. "What is the balance owing on Customer 349's account this morning? I have him on the phone."

7. ANALYZING RESPONSE REQUIREMENTS

4. "How much business has Customer 349 placed with us in the three months up to last Friday? I have a meeting with him the day after tomorrow."

5. "How many times in the past six months has Supplier X missed his promised delivery dates? I have him on the phone..."

6. "I need a report showing what percentage of our widget business over the past five financial years was with firms doing less than $1 million annually. I have to make a presentation to the board next month."

7. "Give me the names and home phone numbers of all our electronic engineers with fluent Arabic who have serviced the Model 999 in the last six months and are not on urgent projects. Put asterisks against the unmarried ones, list them in order of seniority, and make it snappy. I want someone on the plane to Dhahran this evening."

These seven requests span the whole range of those we meet in practice, in very rough order of difficulty and expense to answer. In spite of their apparent diversity, there are four main underlying factors which distinguish them.

1. How predictable is the *timing* of the request? Can it be scheduled, like request 1, or is it triggered by some event, like Request 2, or is it on-demand like Requests 3 through 7? If it is triggered or on-demand, what volume of transactions can we predict?

2. How predictable is the *nature* of the request? While scheduled and triggered requests are predictable by definition, the on-demand requests can vary widely both in the data elements they call for, and the processing needed by those data elements. Request 3, for the balance of an account, could almost certainly be predicted by the user. Requests 4 through 7 are increasingly more difficult to predict, not only in detail, but even in their very nature. Request 6, for an analysis of business based on the customer's historical size, may never have been envisaged until the Chief Executive got enthusiastic about supporting small business.

3. How *fresh* must the data be? Is it enough to have it correct to the end of last month, or should it be correct to the last transaction known to the business? In request 3, for the customer's balance, should the information reflect checks received in the mail this morning?

4. How *rapidly* must the requests be handled? As we observed in Chapter 2, the critical point is whether the answer is required faster than the physical file or data base can be searched from end to end, or sorted. If the user of the information cannot wait for the answer while the entire file is searched or sorted, the physical designer will have to make provision for immediate access to the data. Immediate access, and particularly an unpredictable request requiring immediate access, creates complexity and cost of a different magnitude from predictable, non-immediate access. Request 3 is for immediate access to a customer balance, which we said was reasonably predictable. Many quite modest systems can provide facilities like this. Request 7, requiring access to (probably) several files, and being very hard to anticipate, will be very expensive in machine resources, if (as implied) it has to be answered quickly.

There are a number of other aspects the analyst should investigate, as we shall see, such as security (who has a right to make the request and/or get the answer), and importance to the business of this request (whether it is a "nice-to-have" or a "must-have"). The issues of predictability, immediate access, and freshness of data are the key ones which affect the cost and usefulness of the system.

To see why this is so, we will review some of the physical techniques involved in providing immediate access, before examining tools that the analyst can use to resolve these issues.

7. ANALYZING RESPONSE REQUIREMENTS

7.2 PHYSICAL TECHNIQUES FOR IMMEDIATE ACCESS

7.2.1 INDEXES

Suppose our company buys electrical parts from a large number of suppliers all over the country. We have a file of suppliers, holding their reference number, their name, their location, and the initials of the purchasing agent who is responsible for negotiating with them as well as other information such as address for payments, phone, etc. which we will omit for clarity. The file may look like this:

SUPPLIER-NO	SUPPLIER-NAME	CITY	AGENT
6293	Reliable	St. Louis	ALC
4421	Lightning	Chicago	TDM
6604	Quicktronic	Omaha	ALC
6498	Speedy	Chicago	PAS
2727	Electrico	St. Louis	TDM

and so on.

Let us further suppose that we have the file organized so that we can have random access to it based on SUPPLIER-NO. That is, given a number, say 2727, we can immediately retrieve Electrico's record. SUPPLIER-NO is thus the primary key. We can answer immediately questions such as "Where is Supplier 6604 located?" or "Who is the agent for Supplier 4421?", simply by retrieving the specified record and displaying it. This is true no matter whether the file is held manually, say on a rotary card file in supplier number order, or on direct access disk.

But if we now have to answer questions like "Who are our Chicago suppliers?" or "List the suppliers for which Agent TDM is responsible" life becomes more complex. We essentially have two alternatives, either to sort the file into sequence on the required data element, or to search it by reading every record, and deciding for each one, whether it matches our search criteria or not. Figure 7.1 illustrates these alternatives.

7.2 PHYSICAL TECHNIQUES FOR IMMEDIATE ACCESS

```
"List the suppliers for which Agent TDM is responsible."

        SUPPLIER-NO    SUPPLIER-NAME      CITY          AGENT

            6293         Reliable       St. Louis        ALC
            4421         Lightning      Chicago          TDM
            6604         Quicktronic    Omaha            ALC
            6498         Speedy         Chicago          PAS
            2727         Electrico      St. Louis        TDM
              .
              .

  Either SORT into                                     or SEARCH
  sequence on AGENT

  6293   Reliable     St. Louis   ALC     Read first record
  6604   Quicktronic  Omaha       ALC     Repeat until no more records
  6498   Speedy       Chicago     PAS       IF     AGENT is TDM
                                              THEN write on list
 [4421   Lightning]   Chicago     TDM        ELSE   (not for TDM)
 [2727   Electrico]   St. Louis   TDM          SO no action
                                         Read next record
```

Figure 7.1

No matter whether we sort or search, we have to read every record at least once, and that takes time. Whereas retrieving a record given a SUPPLIER-NO might take seconds, sorting or searching will take a matter of minutes, depending on the size of the file. A typical time to sort 10,000 records of 100 characters each on an IBM 370/158 might be 2-3 minutes, and to retrieve any one of those records given the key, 2-3 seconds. We can estimate how long it might take to search through a rotary card file of 10,000 cards manually; at 1 per second, it's about 3 hours. Worse, if other people want to make immediate accesses using the key, while the sorting process is in progress, it will slow them up, if not shut them out altogether. For this reason, it is very tempting to do all sorting and searching overnight, when no one is making on-line accesses to files. This creates a tendency to divide accesses into "immediate" or "tomorrow morning," though in fact some requests could be answered within minutes if strictly necessary.

7. ANALYZING RESPONSE REQUIREMENTS

If immediate access on the basis of data elements other than the primary key can be predicted and is important enough to the users (and this is up to the analyst to determine), we can provide it in several ways. The simplest way is by constructing an index to the file. An index by AGENT would be:

AGENT	SUPPLIER-NO
ALC	6293, 6604
PAS	6498
TDM	2727, 4421

The index would be set up for random access by agent's initials. Thus to answer the request "List those suppliers for which Agent TDM is responsible," we would:

1. Look up TDM in the index (1 random access) and retrieve the supplier numbers.

2. Look up the two supplier numbers (2 random accesses) in the main file.

Thus the existence of this *secondary index* enables us to replace a search of the whole file with a small number of fast random accesses.

The penalty for having an index is that it takes up more space in the file, and that every time the agent for a supplier is changed, the index has to be changed accordingly.

We can have an index for any of the non-key data elements in the file. Indeed, we could represent the whole file as a set of indexes!

AGENT-INDEX	CITY-INDEX	SUPPLIER-NAME-INDEX
(Agent:Supplier-No)	(City:Supplier-No)	(Supplier-Name:Supplier-No)
ALC 6293, 6604	Chicago 4421, 6498	Electrico 2727
PAS 6498	Omaha 6604	Lightning 4421
TDM 2727, 4421	St. Louis 2727, 6293	Quicktronic 6604
		Reliable 6293
		Speedy 6498

Figure 7.2

An Inverted File

This so-called "inverted" file structure in fact contains all the information that was in the original file, though now there is no "main" file as such; it has disappeared into indexes.

The inverted file structure is good for immediate access to supplier number based on anything you might have in mind; where it is not so convenient is for answering the original request type "Where is supplier 6604 located?" or "Give me all the details of Supplier 2727." In this last case we would have to search all the indexes to find all references to 2727 and then piece the individual record together from there! Truly there is no free lunch.

The most usual solution in practice is to retain the main file, accessible by primary key, and provide a small number of secondary indexes to handle those requests which cannot wait for searching of the whole file.

But which requests are they? They must be predictable in nature, (otherwise we will not know which indexes to set up), be satisfied by data which is only as up-to-date as the index, and be important enough to the business to be worth the trouble and expense of creating a special index. The systems analyst has the responsibility of assessing these factors, so that the best physical choice can be made.

7.2.2 HIERARCHICAL RECORDS

In the previous chapter, we discussed ways of simplifying the structure of records by normalization, removing repeating groups. In the past, many designers have taken the opposite approach; rather than simplify records, they have put as much information and structure as possible into each record. In part this makes more immediate access possible; if we have not only the supplier's name, number, city, and agent, but also the parts he supplies with their names, order and delivery dates, prices, and weights, all in the same record, we can retrieve all that information with, in principle, only one immediate access.

7. ANALYZING RESPONSE REQUIREMENTS

For instance, the record for Electrico might be structured to hold much more information, as shown in Figure 7.3. (In fact all the data elements listed would be simply strung out together in one long record; we have indented them to show the structure.)

Electrico is shown as supplying 2 parts, R23B and C117. For each part, the part number, name, description, and price are given in the part segment of the record. Also each part has a number of order segments associated with it, and also a number of delivery segments. The only way we know Delivery #D614 was for Part R23B is that its segment physically follows the part segment and comes before the next part segment (for part C117).

We can draw a schematic picture of the structure of each record in Figure 7.3 like this:

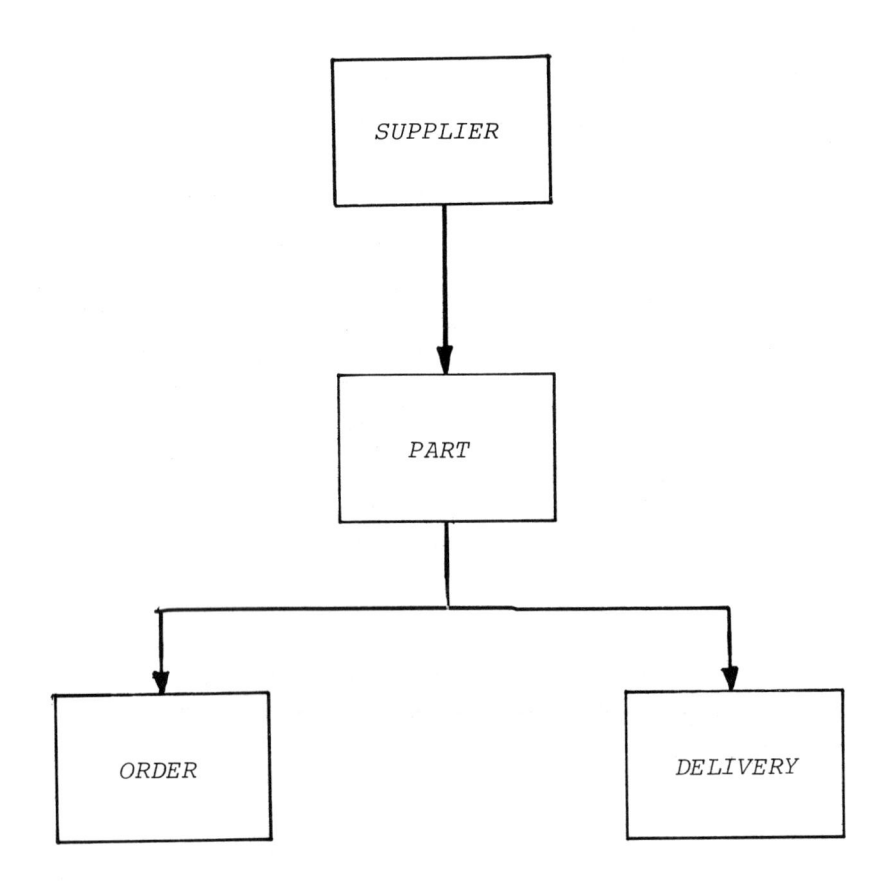

```
        SUPPLIER
           │
           ▼
          PART
         ╱    ╲
        ▼      ▼
      ORDER  DELIVERY
```

7.2 PHYSICAL TECHNIQUES FOR IMMEDIATE ACCESS

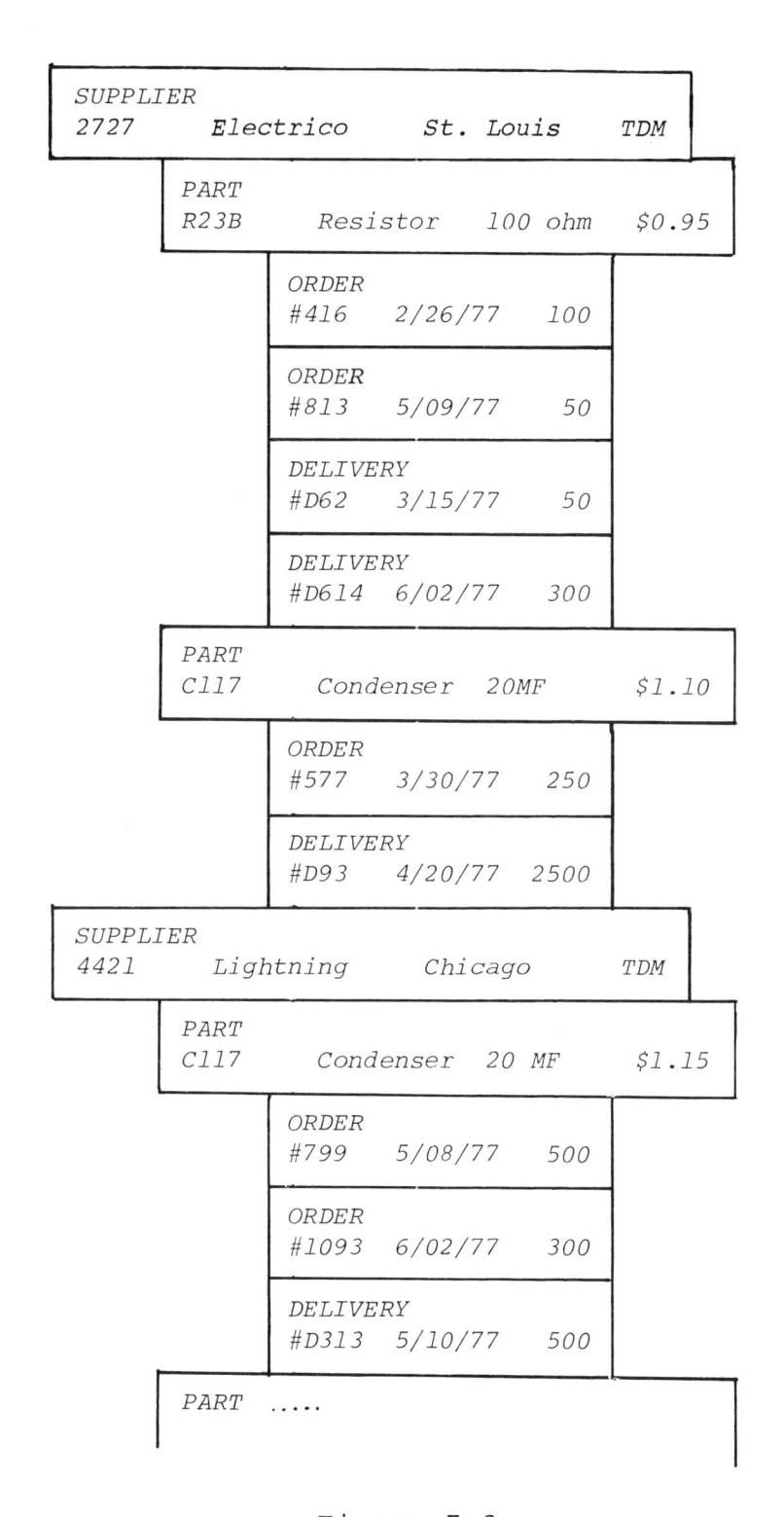

```
SUPPLIER
2727      Electrico      St. Louis      TDM
    PART
    R23B        Resistor    100 ohm     $0.95
        ORDER
        #416     2/26/77      100
        ORDER
        #813     5/09/77       50
        DELIVERY
        #D62     3/15/77       50
        DELIVERY
        #D614    6/02/77      300
    PART
    C117        Condenser   20MF        $1.10
        ORDER
        #577     3/30/77      250
        DELIVERY
        #D93     4/20/77     2500
SUPPLIER
4421      Lightning      Chicago        TDM
    PART
    C117        Condenser   20 MF       $1.15
        ORDER
        #799     5/08/77      500
        ORDER
        #1093    6/02/77      300
        DELIVERY
        #D313    5/10/77      500
    PART   ....
```

Figure 7.3

A Hierarchical Record Structure

7. ANALYZING RESPONSE REQUIREMENTS

The structure diagram implies that each SUPPLIER has one or more dependent PARTs, each PART has one or more dependent ORDERs, and so on. We note that any one part can be supplied by more than one supplier; C117, the condenser, is supplied by both Electrico and Lightning. The term for this supplier-part relationship is "many-many," to distinguish it from, say, a company-employee relationship which is "one-many." Any one company can have many employees, but any employee can have only one company. Hierarchical records can handle both one-many and many-many relationships.

Hierarchical records can thus handle a lot of complexity and detail, and are currently in use in a number of data base management systems. IBM's Information Management System (IMS) is perhaps the best known such system. With hierarchical systems, some types of immediate access are easy. For instance, with the hierarchical records of Figure 7.3, we could say, "Get me the price of Part C117 charged by Supplier 4421." The combination of supplier number and part number specifies *uniquely* the segment which contains the needed information. One of the commands provided by IMS is "Get Unique," which, given the necessary keys, will retrieve the specified segment. We can think of such a command as "coming down the hierarchical path," starting with a specific supplier (the highest level or "root" segment), moving from that segment to a lower segment, then to a lower segment, and so on. The request "How many of Part C117 has Supplier 2727 delivered, and when?" would be answered in such a way.

However, if we want to get at the data other than by following the hierarchy, we may have a difficulty. Suppose we want to answer the question, "Who supplies Part C117?" we have to search the entire data base of all suppliers, checking each part within each supplier to see if it is Part C117. IMS has a "Get Next" command for searches of this type; the command sequence could be:

```
                Get to start of data
Supplier-loop:  Get next SUPPLIER
                Get next PART for which PART-# is C117
                IF      C117 is found
                  THEN  write supplier details on reports
                  ELSE  (this supplier does not supply C117)
                    SO  go to supplier-loop
```

"Get next" allows us to move forward from our current position in the data base (whichever segment we retrieved last) to retrieve the next segment of the kind we specify. For example, in Figure 7.3, if the last segment we retrieved was ORDER #813, "Get next PART," would retrieve PART C117; "Get next SUPPLIER," would retrieve SUPPLIER 4421. It follows that we can use the "get next" command to search a hierarchical file or data base, to answer any information request, (provided the necessary data elements are present, of course). However, we are back with the problem that we met in the previous section, that unless the request can be handled by following the hierarchical path, we effectively have to search through the entire data base to be sure we have found all the answers. These searches can take a long time, especially with large IMS data bases. In practice, queries such as "How many of PART C117 have been delivered since June 1st?" cannot realistically be responded to immediately unless some type of index is set up.

In IMS, there are a number of methods for creating indexes, or their equivalents, all of which involve more or less cost and complexity. The physical design of an IMS data base is a complex job, involving the weighing of many factors of space required for the data, relative frequency of access, growth of the data base, reliability and security requirements, and so on. The data base designer needs the systems analyst to determine what immediate accesses are needed, how frequently each type of access is likely to be needed, and the relative importance of each access. With this information, it is much easier for the data base designer to select the best compromise between retrieval capability and cost.

7.3 GENERAL INQUIRY LANGUAGE CAPABILITY

Even if the analyst has specified the appropriate indexes to a file or data base, it may be hard to predict in detail the precise form in which the user will want the answers to his information requests.

Suppose we have a supplier file, with a secondary index to each part. The user can say, "List all the suppliers who

7. ANALYZING RESPONSE REQUIREMENTS

supply part C117," and can get the answer immediately, in principle. But suppose one day the request is "Give me all the sources of supply for part C117 in California, Arizona, and New Mexico, in descending order of the numbers of the part they have delivered in the past 3 months." There is nothing to stop us retrieving the necessary information from the supplier file, but we will need a program that tests for the state in which each supplier is located, accepts only those in the three states specified, computes the deliveries make by the surviving suppliers over the last 3 months, and sorts the records into descending order on total deliveries before listing them. This program is not complex, but if written in a standard programming language, say COBOL, might take a day or two to code and test. Further, since we didn't predict that this precise type of request would come up, we may not be sure that it will ever come up again. So we are put in the situation of having to write a "one-time-only" program to handle this request; the factor limiting our speed of response to the user is not how long it takes to search the file, but how long it takes to write the program!

For this reason, a number of simple query languages have been developed which enable a wide variety of requests to be programmed rapidly, in many cases by user people without programming knowledge. As an example we will consider IQF (Interactive Query Facility), a software package produced by IBM for use with IMS-type data bases, [7.1,7.2].

IQF allows a user to sit at a terminal and enter requests for information in English, provided the important words in each sentence have been previously defined to IQF. For instance, the user could key in:

> FROM THE PURCHASING DATA BASE, TOTAL THE DELIVERIES OF 20MF CONDENSERS IN THE LAST MONTH, AND LIST THE NAME AND CITY FOR SUPPLIERS OF 20MF CONDENSERS WITH DELIVERY LEAD TIMES LESS THAN ONE WEEK AND BULK DISCOUNTS OF GREATER THAN 10%.

Some of the words used in this sentence such as FROM, THE, OF, would have previously been defined as null words, which the user may put in to make the request more readable and recognizable to him. All the other terms, such as SUPPLIERS, 20MF CONDENSERS, and so on, would need to be defined in IQF's private data dictionary. LIST and TOTAL are IQF commands with obvious meanings. The IQF software translates the terms used into their predefined meanings, works out a method of searching the data base, assembles the resulting records and lists the specified data elements on the user's terminal.

7.3 GENERAL INQUIRY LANGUAGE CAPABILITY

Once the definition of terms has been done (and each user can add his own definitions), IQF essentially enables each user to write his own query programs. This very useful facility has the limitation that the user must use only the defined terms; if he types in VENDORS instead of SUPPLIERS, IQF will not know what is meant, until VENDORS is defined to it. The user must also be able to phrase his request in a logical manner using the IQF commands; some users of general inquiry languages like IQF have a specially trained "Information Assistant" to help them translate their thoughts into the form required by the language.

IQF can be used on-line so that the response to the query can be available in a matter of seconds provided the necessary access paths have been built into the structure of the data base, as we discussed in the previous section. The user has no way of knowing, however, whether the English-language-like query he enters is going to mean one random access, or a search of the entire data base. Indeed, he should not have to know; the access problems should be "transparent" to him! Nonetheless, if 20 users simultaneously originate 20 separate queries, each of which involves searching a reasonable size data base from end to end, they are going to slow each other down considerably, and hamper anyone else who may be using the data base at the time. Consequently, IQF is equipped with a warning facility; if the user originates an inquiry which IQF decides will involve a lengthy search, it will notify him before starting the search and give him an opportunity to submit the query on an overnight basis, when the search can be conducted off-line and involve less interference with on-line users. Some of the time, of course, the user will say, "I don't care how long the search is, or what it costs, get me the answer": this is why general inquiry language facilities can expose the data center to unpredictable heavy loads. On the one hand, we should be glad the users are getting valuable results out of the computer, (and that without taxing the resources of the programming staff). On the other hand, we want to avoid a situation in which the users are burning up computer time, and degrading response time on other systems, because they are having to search the data base to find information, when they could have immediate access to that information, if only a suitable index had been provided. We come back again to the moral of this chapter; *the systems analyst should do everything possible to foresee what types of access the users are going to want to make*, and how valuable the various types are. Only then can a good balance be struck between the cost of providing secondary indexes and the cost of end-to-end searching.

7. ANALYZING RESPONSE REQUIREMENTS

    This moral holds good for all inquiry languages which allow for on-line report generation. As well **as** IQF, IBM also markets GIS (Generalized Information System), which has more extensive and more powerful facilities [7.2].

    GIS is more like a high-level programming language than IQF, allowing the users to build files, modify the data base, do extensive calculations and format reports in many different ways. IQF, with its English-like format can be quickly learned by a clerical person: GIS is more demanding and requires more programming ability, though it can be learned by non-programming people. Other popular general inquiry packages marketed by independent software houses are CULPRIT (Cullinane Corp.), EASYTRIEVE (Pansophic), and MARK IV (Informatics). The addresses of these firms are listed at the end of the chapter.

    It is likely that we will see an expansion in the use of general inquiry software over the next few years, as hardware becomes cheaper and more powerful, and more advanced software makes it easier to provide immediate access to any data element in a file, not just to those for which secondary indexes have been established.

7.4 TYPES OF QUERY

Before trying to analyse the different priorities that users put on queries, it is useful to survey the various logically different types of query that can be made. We can distinguish six basically different types of query [7.3], and note some variations on them.

7.4.1 ENTITIES AND ATTRIBUTES

Things in the real world have various properties; a machine has size, weight, identifying number; a person has name, address, personnel number, date-of-hire, salary, and so on. When describing data, a thing or class of things is referred to as an *entity*; the data elements which describe properties of the things are *attributes* of the entity. We can see that the attributes of the entity "customer" are customer-number, phone, address, etc. One attribute is usually chosen as the key attribute, because it uniquely identifies the entity (though as we said when considering third normal form, it is sometimes necessary to combine more than one attribute to form a unique key).

We can illustrate these concepts thus:

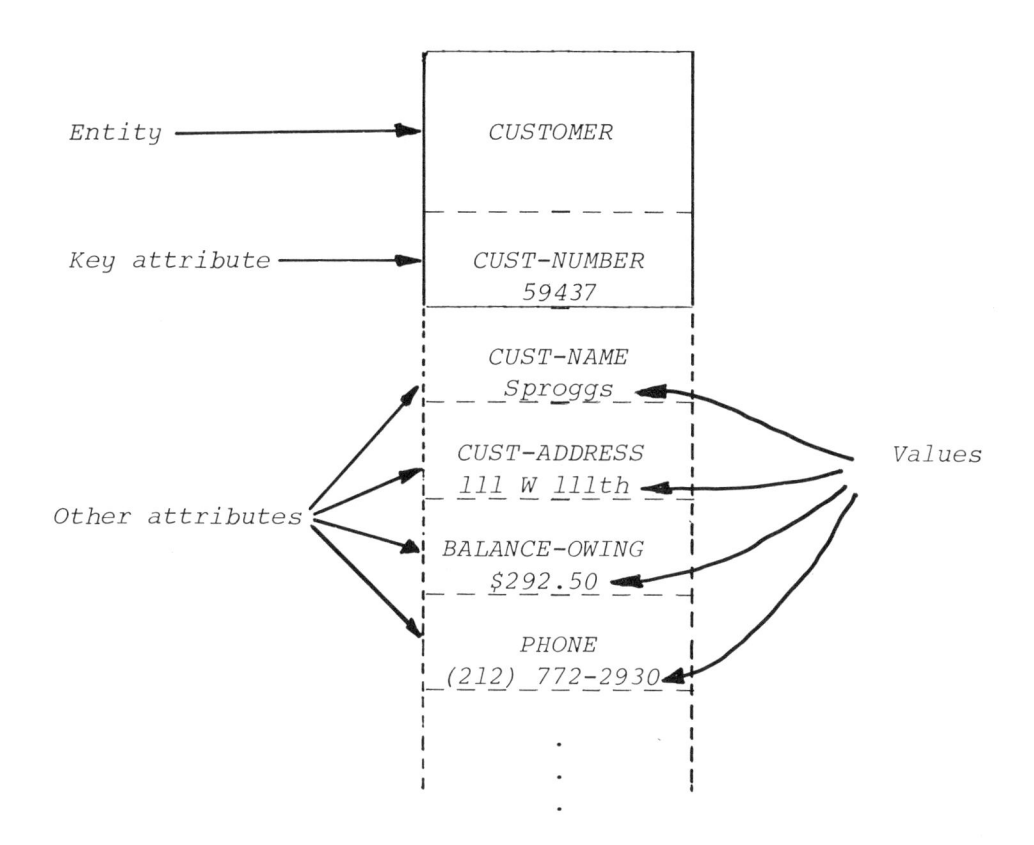

```
Entity ─────────────►  CUSTOMER

Key attribute ──────►  CUST-NUMBER
                         59437

                       CUST-NAME
                        Sproggs    ◄──┐
                                      │
                       CUST-ADDRESS   │
Other attributes ───►   111 W 111th ◄─┤── Values
                                      │
                       BALANCE-OWING  │
                        $292.50    ◄──┤
                                      │
                       PHONE          │
                        (212) 772-2930 ◄┘
```

Sometimes it is convenient to make the distinction between *basic entities*, (which stand for permanent things like customers, employees, suppliers, and products,) and *transaction entities* which stand for some less permanent thing or event, often connecting two basic entities. Thus an invoice is an entity connecting a customer with one or more products; a purchase-order is an entity connecting one or more purchase-items with a supplier.

7. ANALYZING RESPONSE REQUIREMENTS

7.4.2 SIX BASIC QUERY TYPES

We can describe the six basic types of query in terms of entities, their attributes, and values the attributes may take.

> QUERY TYPE 1: *"Given a particular Entity (E), what is the value of a particular attribute (A)?"*

For example,

"Given part number B232, what is the value of its weight?"

The conventional notation for this type of query is:

$$A(E) = ?$$

where A stands for a stated attribute, of entity E (whose key attribute is known), and the question mark stands for the unknown value. The query can be shown thus:

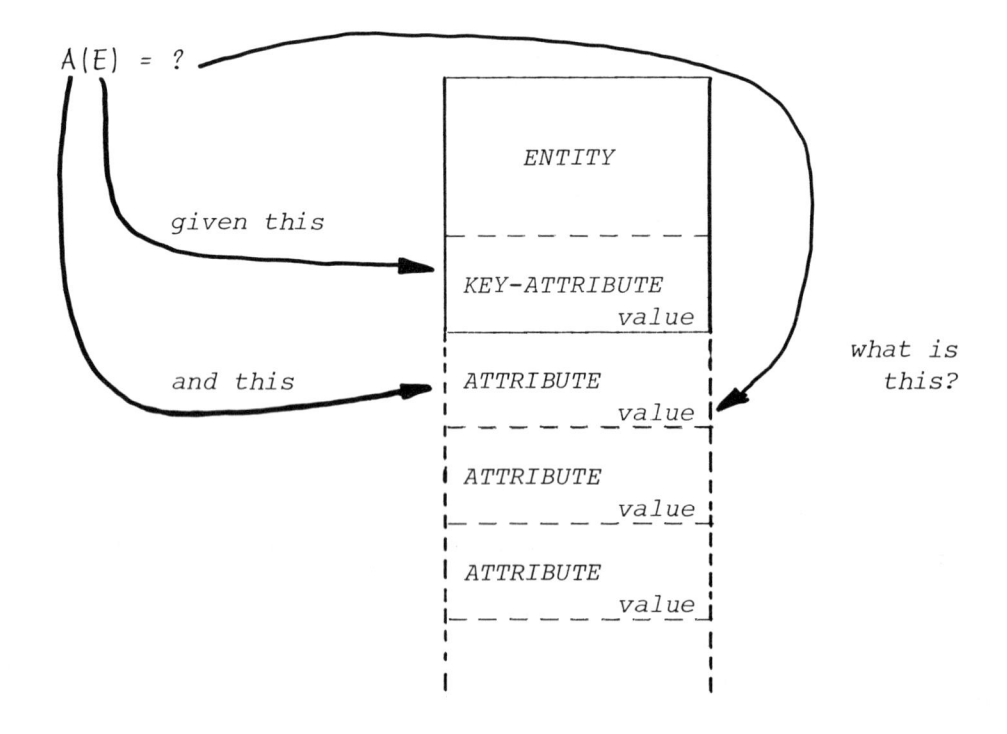

```
A(E) = ?
            ┌─────────────────┐
            │                 │
            │     ENTITY      │
            │                 │
given this  ├ ─ ─ ─ ─ ─ ─ ─ ─ ┤
       →    │  KEY-ATTRIBUTE  │
            │           value │
            │                 │       what is
and this    ├ ─ ─ ─ ─ ─ ─ ─ ─ ┤       this?
       →    │   ATTRIBUTE     │
            │           value │
            ├ ─ ─ ─ ─ ─ ─ ─ ─ ┤
            │   ATTRIBUTE     │
            │           value │
            ├ ─ ─ ─ ─ ─ ─ ─ ─ ┤
            │   ATTRIBUTE     │
            │           value │
            │                 │
```

7.4 TYPES OF QUERY

QUERY TYPE 2: *"Given a value of a particular attribute, which entities are specified?*

For example,

"Given a weight of two pounds, (attribute), which parts (entities) have this value?

or

"Given a grade of ENGINEER, which employees have this grade?"

This query is the inverse of type 1; in fact query type 2 is the sort of "inverted" file situation we discussed in Section 7.2. Type 2 requires a secondary index, if it is concerned with non-key attributes. We can draw a diagram thus:

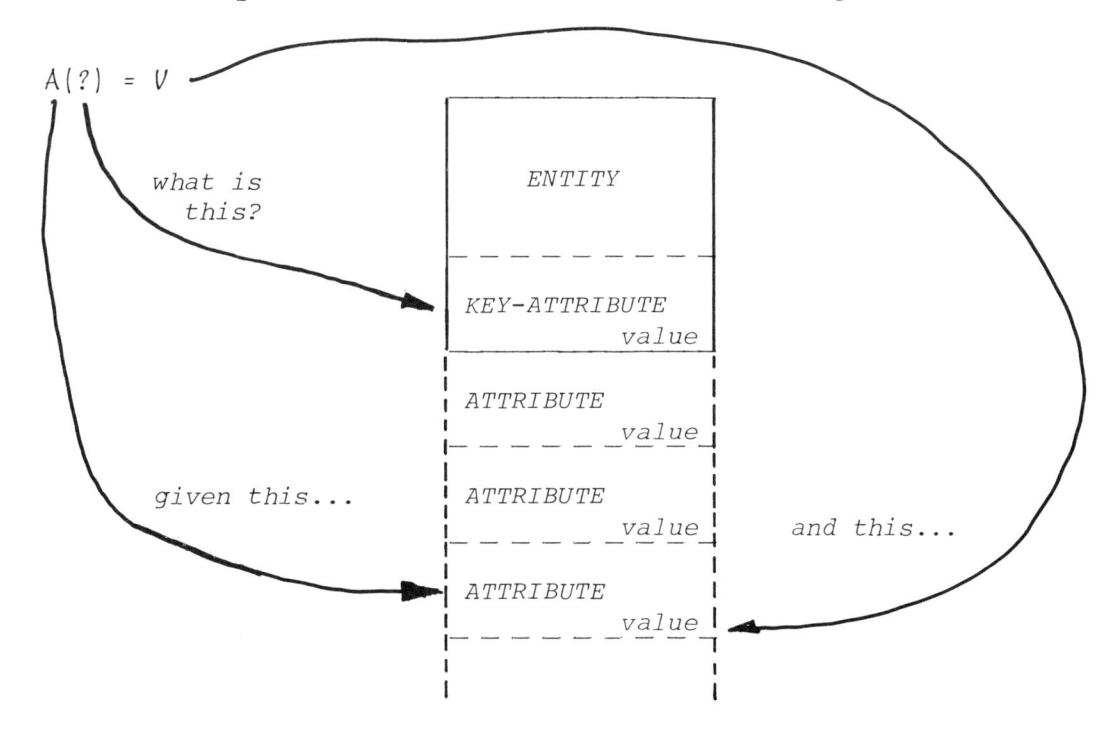

$A(?) = V$

what is this? → ENTITY

KEY-ATTRIBUTE *value*

ATTRIBUTE *value*

given this... → ATTRIBUTE *value*   *and this...*

ATTRIBUTE *value*

We often want to specify more than one value for the search, for example, we might say, "Given a weight of *less than* two pounds, which parts have this range of values?" So the most general form of the type 2 query is:

$$A(?) \begin{Bmatrix} = \\ \neq \\ \geq \\ \leq \end{Bmatrix} V$$

where the symbols have their usual meaning ($\neq$ means "not equal to," $>$ means "greater than," and $<$ means "less than,").

219

7. ANALYZING RESPONSE REQUIREMENTS

QUERY TYPE 3: "Given a particular entity, and a value, which attribute(s) of the entity match that value?"

This is a rarer type of query, most used when several attributes describe the same property at different points in time.

For example,

"Given employee number 26622, and earnings of $15,000, in which years has that value been exceeded?"

"Given product number B232, which dimension (if any) exceeds 12 inches?"

The notation is:

$$?(E) \begin{Bmatrix} = \\ \neq \\ > \\ < \end{Bmatrix} V$$

QUERY TYPE 4: Type 4 is similar to type 1, except that it asks for all values of all attributes, rather than just one attribute.

"List the details of employee number 26622"

"Give me a profile of part number B232"

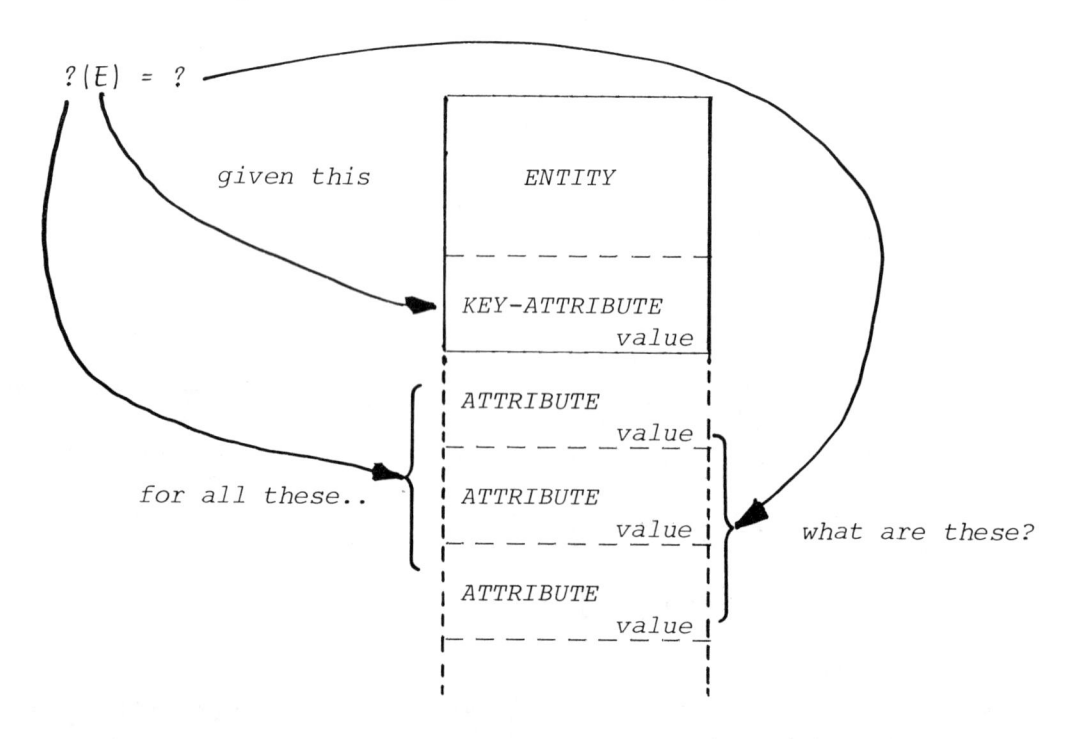

$?(E) = ?$

given this → ENTITY

KEY-ATTRIBUTE value

for all these.. → ATTRIBUTE value
ATTRIBUTE value
ATTRIBUTE value

← what are these?

7.4 TYPES OF QUERY

QUERY TYPE 5: This is similar to type 2 but, like type 4, is also global in that it asks for all values of an attribute, for all entities.

"List the weights of all parts"

"List the salaries of all employees"

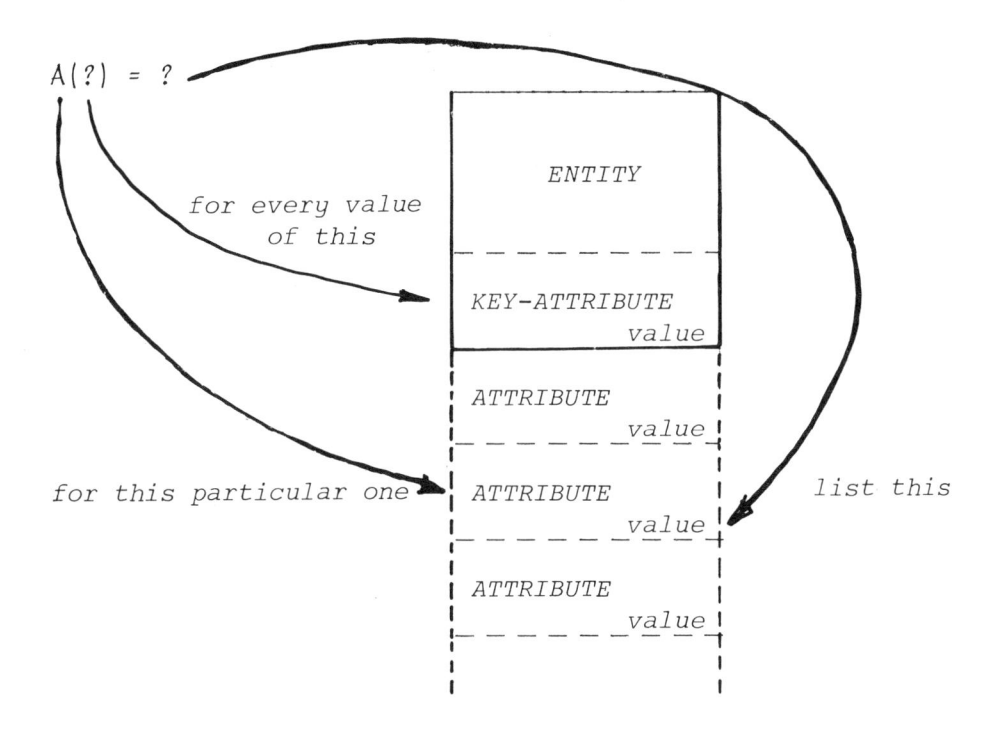

A(?) = ?

for every value of this → ENTITY / KEY-ATTRIBUTE value

for this particular one → ATTRIBUTE value

list this → ATTRIBUTE value

ATTRIBUTE value

QUERY TYPE 6: This is the global equivalent of type 3, asking for all attributes of all entities that have a certain value.

"List every employee who earned more than $25,000 in any year."

"List every part having any dimension greater than 6 inches."

The notation for type 6 is:

$$?(?) \begin{Bmatrix} = \\ \neq \\ > \\ < \end{Bmatrix} V$$

Like type 3, it does not occur in practice very often.

7. ANALYZING RESPONSE REQUIREMENTS

7.4.3 VARIATIONS ON THE BASIC TYPES OF QUERIES

Multiple values. Wherever a test of a value is made in a query, there is always the possibility of multiple tests being combined. We might say, in a type 2 query, "Which parts have a weight of more than 2 pounds, *and* are colored red or blue *and* have a length of more than 30 inches," or "Which employees have a grade of electronics engineer and a language-skill of Arabic-fluent?"

Ordering and quotas. As a further refinement, the response to the query may be needed arranged in some order, either ascending or descending on some attribute. The query may call for only one or a specified quota of entities that fit the criteria. "Give me all the overdue invoices, arranged in descending order of amount." "Give me any one supplier of part B232, located in Chicago."

7.5 FINDING OUT WHAT THE USERS NEEDS AND PREFERENCES ARE

The user's needs for immediate access to data emerge over the whole period of study and system analysis, as part of the iterative process of gathering information, proposing trial solutions, gathering more information, and so on. In some cases, the user managers will not express any need for immediate access to data. This may be for one of two reasons. There may genuinely be little or no business requirement for immediate access, as in the case of an application which is concerned with predetermined historical analysis of past events. More usually, the reason no need is expressed is because the users have got used to functioning with the reports generated by a batch system, and have not grasped the fact that technology now enables them to have a "window into the computer." After all, the last time many managers had immediate access was back in the days of ledger card accounting, when they could walk over to a tub of cards, access a customer's record by name, and inspect the history of transactions. In these circumstances, the analyst has an updating job to do, explaining the current technical possibilities and helping managers to rethink how immediate access could be of value to them.

7.5 FINDING OUT WHAT THE USERS NEEDS AND PREFERENCES ARE

On the other hand, an increasing number of people in business and government are very well aware of the computer's information retrieval capability, and can present the analyst with an elaborate "Wish-list." The analyst is faced with the problem of encouraging the first group of managers to use the computer more effectively, and avoiding promising anything to the second group that can't be delivered cost-effectively.

7.5.1 OPERATIONAL ACCESS VERSUS INFORMATIONAL ACCESS

Many of the immediate accesses to data stores in a system will be to do with the processing of routine transactions, rather than the answering of information requests. Some examples of *operational* accesses are:

1. Retrieving a customer address for use in editing an order

2. Retrieving a customer's payment history for credit checking.

3. Retrieving the inventory level of a product for order processing.

Typically these accesses are needed to implement some strategic design decisions; once it has been decided that the system will have data entry on-line via CRT's in the sales department, it follows that we will need immediate access to the customer file at least once for every transaction. Similarly, if one of the objectives of the system is to keep the inventory level up-to-date during the day, immediate access to the inventory file is implied. The analyst is concerned with the nature of operational accesses, their likely volumes (because the designer will need to know them), and any security considerations involved. He is not specifically concerned with the value of each operational access to the business; either the access will be provided or it won't.

On the other hand, many of the accesses we have discussed in this chapter are valuable, but not mandatory. Their value lies in helping the various levels of management make better decisions faster, or providing faster response to customer queries. If these *informational* accesses are not immediate, the business won't come to a halt, it just won't be conducted so well or so smoothly. It is with informational as opposed to operational accesses, that the analyst can make a great

7. ANALYZING RESPONSE REQUIREMENTS

difference to the cost-effectiveness of the final system; if the informational accesses can be identified and the most valuable ones built into the design, by secondary indexing or otherwise, then the user community as a whole will get the greatest value out of the system. If the analyst is not able to predict the immediate accesses the user will want to make, the information needs may still be provided, but not so quickly, and at greater cost.

7.5.2 GETTING A COMPOSITE "WISH-LIST"

As we mentioned, in initial contacts with the various management and supervisory people who will be users of the system, the analyst will encounter a greater or less demand for immediate access to data. Suppose, in place of the CBM Corporation, the analyst is involved in the early stages of a marketing information system, this time for the Sunray Engineering Company, which manufactures solar energy installations for the construction industry. Sunray salesmen contact contractors and architects to discuss their construction projects, and as a result prepare costed proposals for solar equipment based on standard components which the company manufactures. Some proposals result in orders and are delivered to the customers for installation in buildings. The information system to be developed will keep track of these proposals, the subsequent orders, and the deliveries, feeding the accounts receivable system with the information necessary to bill customers once a product has been delivered. As a result of initial discussions with the various members of management, the analyst has the following notes on probable uses of the data stores:

Comptroller

> *Needs ability to analyze the value of orders due to be delivered in any specified future period, to predict amount that will be billed, and so the cash flow.*

Sales Manager

> *Wants to know orders and outstanding proposals for any customer, or any salesman, and commissions for any salesman.*

7.5 FINDING OUT WHAT THE USERS NEEDS AND PREFERENCES ARE

Product Planning Manager (A recent graduate of Harvard Business School)

    Wants on-line access to order history based on order-size, month (placement and delivery) of orders, component, type of end-use, and geographical region.

Sales Administration Supervisor

    Wants to be able to find customer number given name, and answer customer queries on the progress of orders towards delivery.

These requirements are over and above the normal data flows of entering proposal data, pricing proposals, entering orders, printing shipping notes, and so on. We shall assume that, based on the data flow and data store content analysis of the type we described in Chapters 3 and 6, data stores describing the CUSTOMER, ORDER, PROPOSAL and SALESPERSON have been identified, with attributes as shown in Figure 7.4. Note that PROPOSAL and ORDER have composite keys (each with two data elements) and that there are several repeating groups, e.g., COMPONENTS

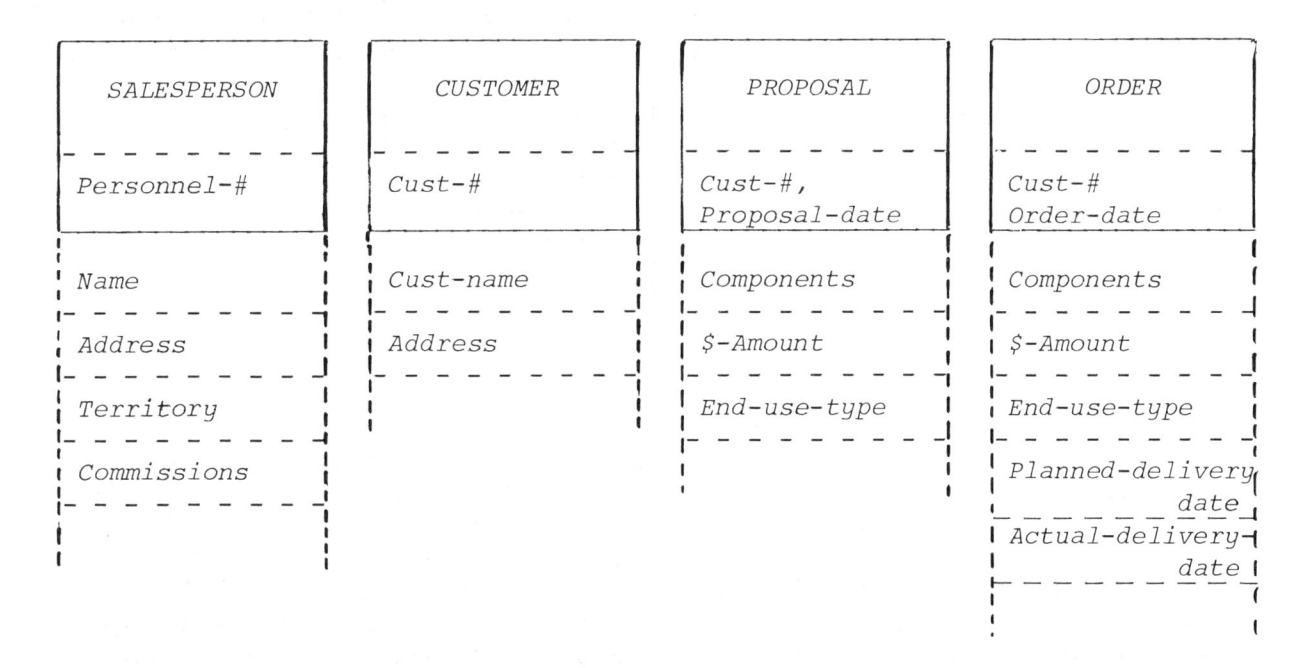

SALESPERSON	CUSTOMER	PROPOSAL	ORDER
Personnel-#	Cust-#	Cust-#, Proposal-date	Cust-# Order-date
Name	Cust-name	Components	Components
Address	Address	$-Amount	$-Amount
Territory		End-use-type	End-use-type
Commissions			Planned-delivery date
			Actual-delivery date

Figure 7.4

Entities and Attributes in SUNRAY ENGINEERING

7. ANALYZING RESPONSE REQUIREMENTS

We can indicate the accesses that various managers want to make by drawing arrows from entity to entity. For example, the Sales Manager wants to access ORDERS given a CUSTOMER, and ORDERS given a SALESPERSON. We can show these accesses as in Figure 7.5, where the arrows represent immediate access.

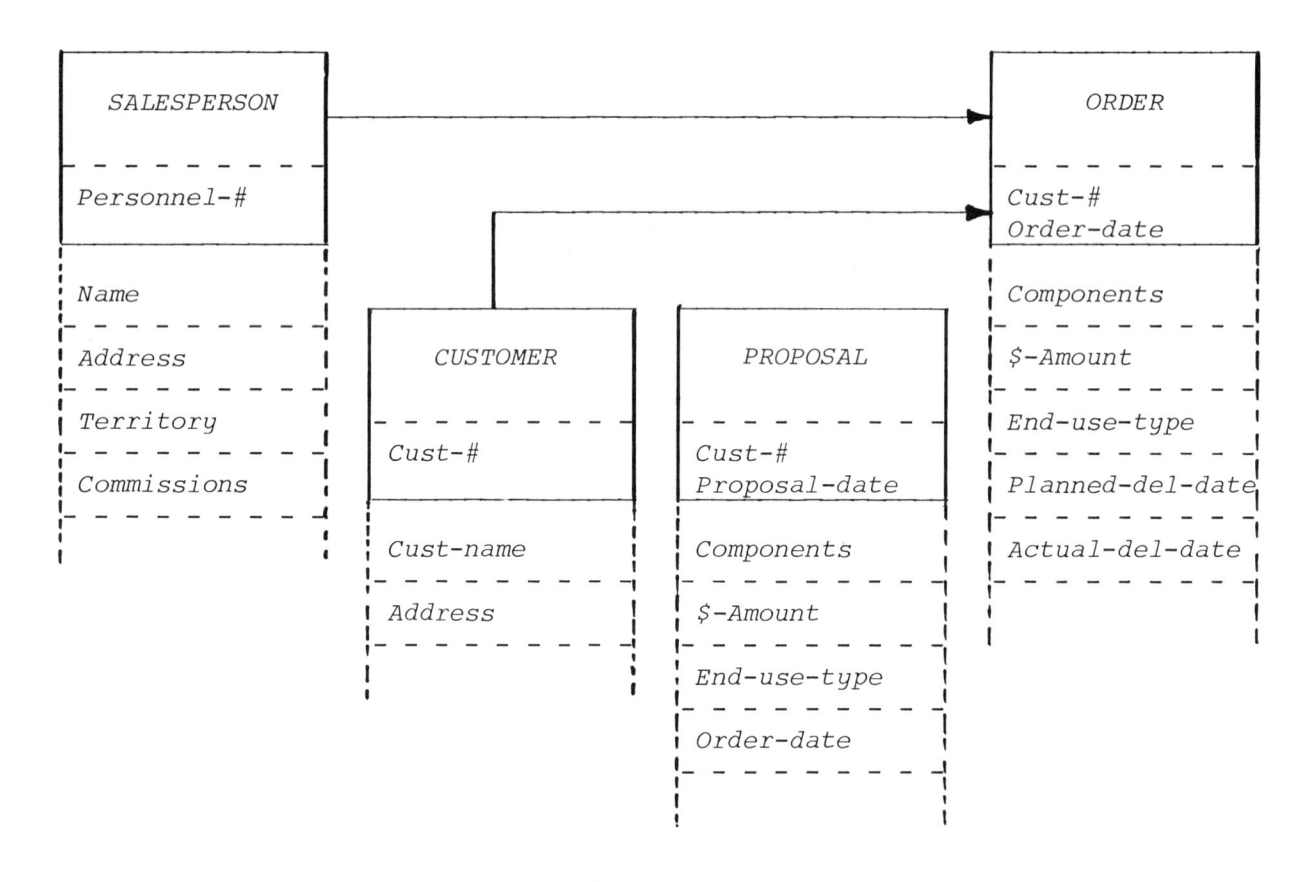

Figure 7.5

These arrows do not mean any kind of flow or any kind of control or hierarchy, just a requirement to be able to access an entity given an instance of another entity. Diagrams like Figure 7.5 are *data immediate access diagrams (DIAD)*, as we described briefly in Chapter 2. They are very useful for summarizing all the immediate access requirements of a group of managers, before coming to any decision as to what the physical file or data base structure will be. We can note

7.5 FINDING OUT WHAT THE USERS NEEDS AND PREFERENCES ARE

that in Figure 7.5, the immediate access from CUSTOMER to ORDERS looks pretty simple logically since CUST-# is part of the key of ORDER. It may not be physically very convenient, since some file handling software systems cannot retrieve all the records that match a data element which is only part of the key. On the other hand, the retrieval would be simple in IMS provided the segments were arranged suitably. The point is that *we don't care* about physical implementation yet; we just want to pin down all the requirements for immediate access at this stage, and worry about the best form of implementation when we know values and volumes.

We can also note that there appears to be no obvious way to access ORDERS from SALESPERSON. We could include the key of each ORDER as as attribute of SALESPERSON, or we could include the SALESPERSON's number as an attribute of each ORDER, and create a secondary index. We don't care yet which.

Some accesses the user asks for involve only retrieval of an entity based on its key attribute. The Sales Manager's requirement to access commissions for each salesperson is like this; given a salesperson's number, the specified record can be retrieved, and it includes the commissions. We do not need to indicate a requirement like this specifically; it follows from the presence of the attribute.

On the other hand, where an entity must be retrieved based on a non-key attribute, (the secondary index situation we discussed in Section 7.2) we want to indicate that fact, by creating a new entity out of the attribute. For example, suppose the Sales Manager wished to access each salesperson by name instead of number, we would indicate that fact as shown in Figure 7.6.

The Comptrollers request can be represented in this way also since it amounts to accessing ORDERS based on a non-key attribute of ORDERS. Figure 7.7 shows the addition of this immediate access path.

Figure 7.8 shows the full set of requests representing the sum-total of the "wish-list" for the Comptroller, Sales Manager, Product Planning Manager, and Sales Administration Supervisor.

7. ANALYZING RESPONSE REQUIREMENTS

SALESPERSON NAME

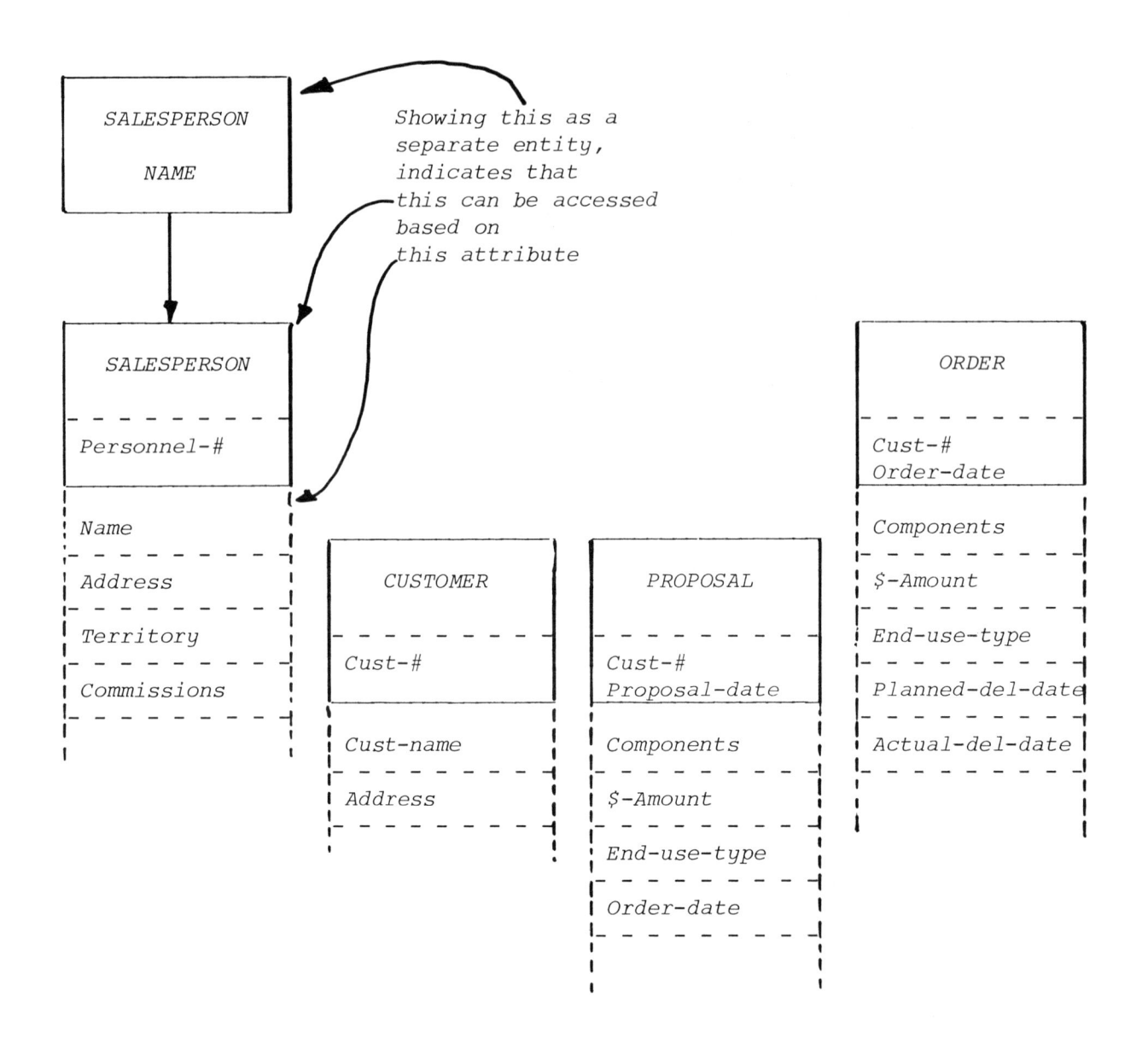

Showing this as a separate entity, indicates that this can be accessed based on this attribute

SALESPERSON
- *Personnel-#*
- *Name*
- *Address*
- *Territory*
- *Commissions*

CUSTOMER
- *Cust-#*
- *Cust-name*
- *Address*

PROPOSAL
- *Cust-#*
- *Proposal-date*
- *Components*
- *$-Amount*
- *End-use-type*
- *Order-date*

ORDER
- *Cust-#*
- *Order-date*
- *Components*
- *$-Amount*
- *End-use-type*
- *Planned-del-date*
- *Actual-del-date*

Figure 7.6

A DIAD showing part of the Sales Manager's requirements

228

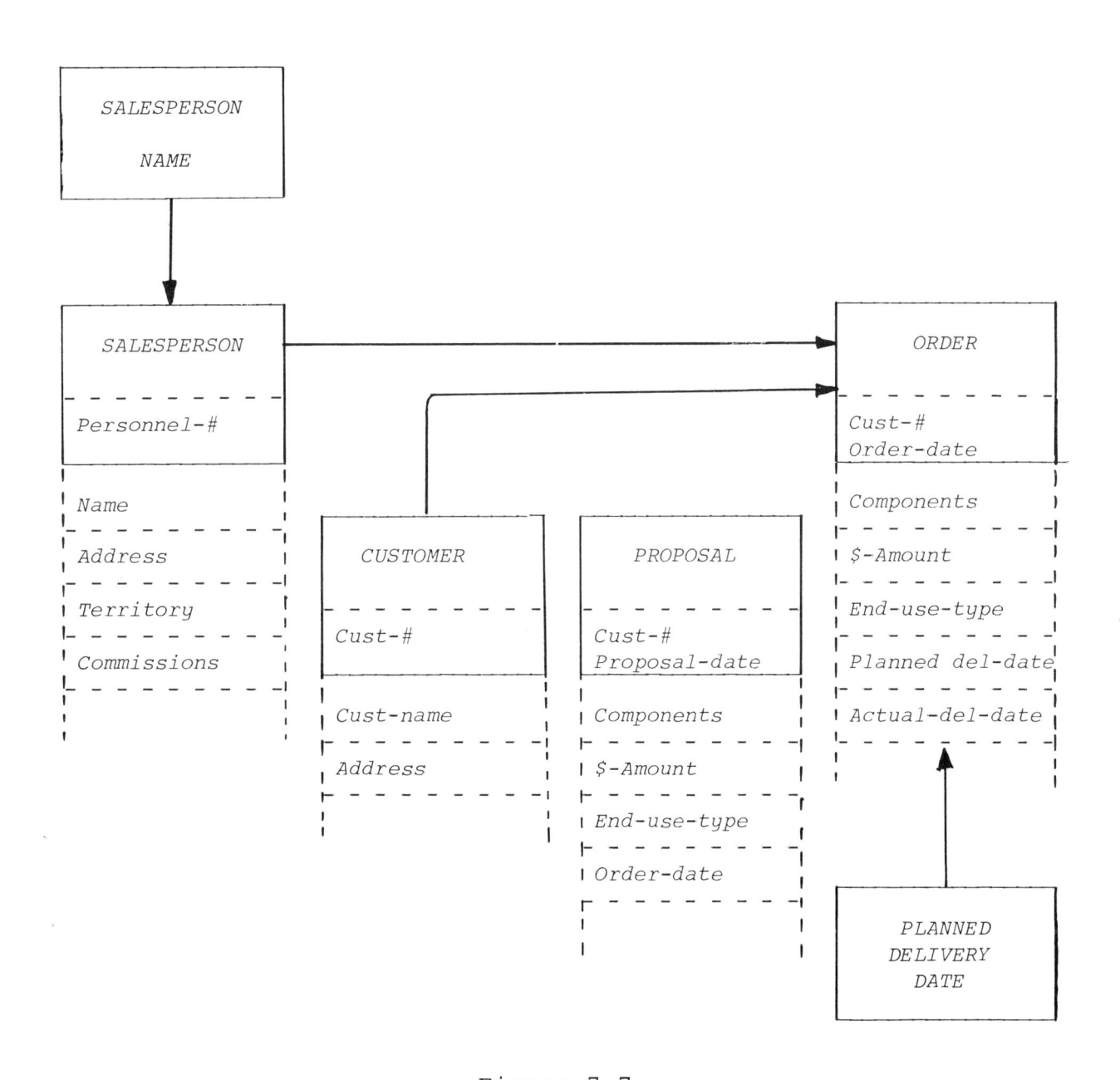

7.5 FINDING OUT WHAT THE USERS NEEDS AND PREFERENCES ARE

Figure 7.7

The DIAD with the Comptrollers requirements added in

7. ANALYZING RESPONSE REQUIREMENTS

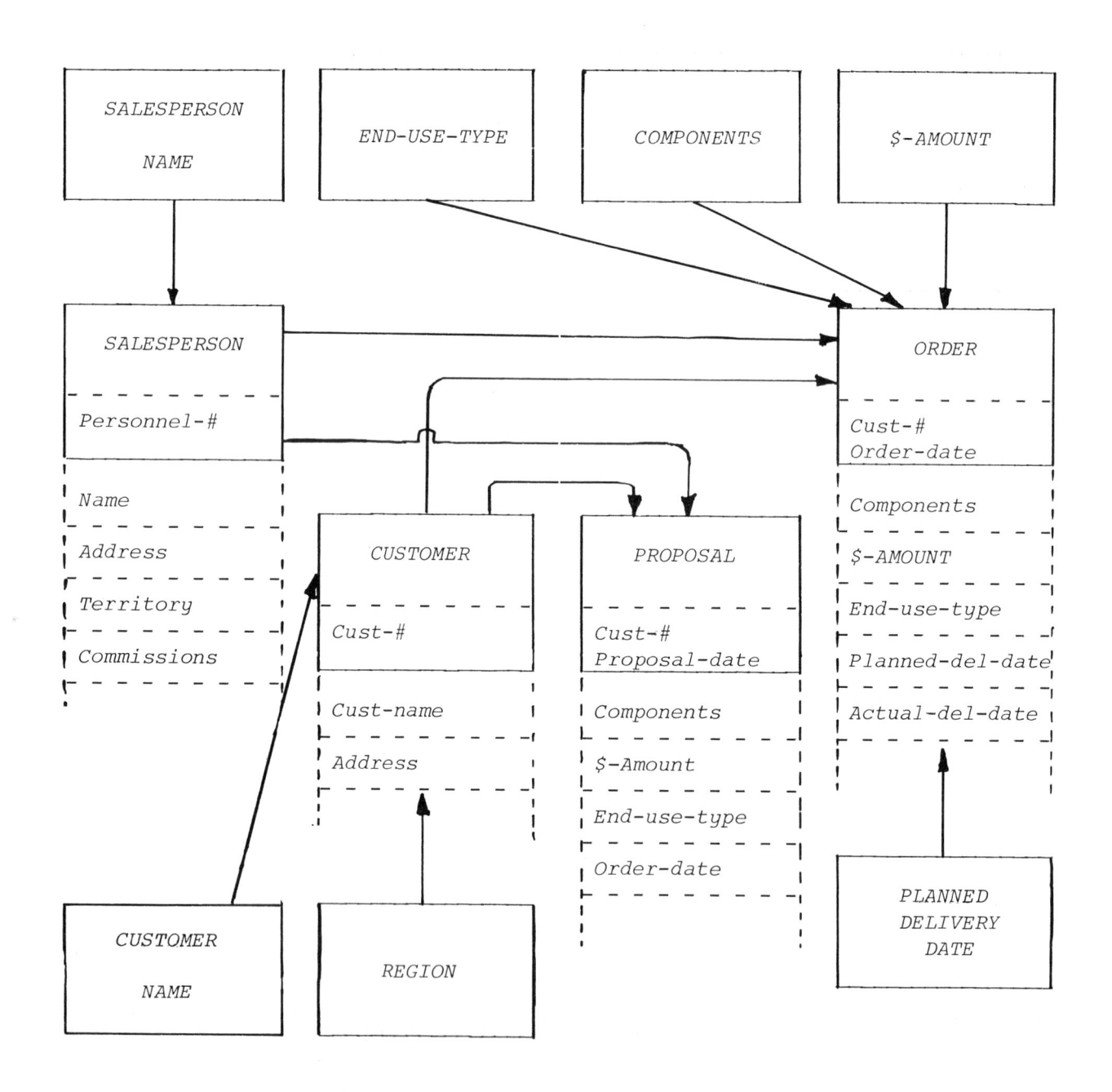

Figure 7.8

The Composite DIAD for Sunray Engineering

7.5 FINDING OUT WHAT THE USERS NEEDS AND PREFERENCES ARE

7.5.3 REFINING THE "WISH-LIST"

If the project budget is no object, and if each user manager expresses a strong need for immediate access the only further information the analyst would need is the likely frequency of each access. Often, however, not all the wishes can be granted cheaply. Then the analyst is faced with the problem of finding out which accesses are least valuable and can be downgraded to an overnight response. One way to do this is to present the immediate access diagram to each user in turn, discussing the meaning of all the various accesses. Even if some of them are not relevant to any one manager, it is of value to show the big picture to all of the decision-makers, explaining that each additional access puts the cost of the system up. This presentation, if tactfully done, stimulates the manager's imagination as to possible uses of the system, while minimizing the chance of friction and disappointment if the final system happens not to give the executive his entire "wish-list."

The presentation of the immediate access diagram may be followed up with a questionnaire, normally filled in by the analyst in conversation with each manager. Figure 7.9 shows representative briefing notes for an analyst to explain such a questionnaire, and Figure 7.10 shows a sample page of the questionnaire for Sunray Engineering.

The questions are framed to make it as quick and easy as possible for each manager to put a value on the immediacy of each response and give an idea of likely frequency. We are not trying to compare the importance of one response with another, just the relative value of immediacy. All the same, managers will tend to give higher values to the more important accesses; they may add comments such as " I don't need it very often, but when I need it, I need it fast, and it's beyond price" (e.g. an analysis of recent proposal success rate, as a guide to competitive bidding).

With accesses which are concerned with data which only changes slowly, such as retrieving a customer number given a customer name, the alternative is really between having to look up the answer in a listing, compared with having the answer displayed or printed on a terminal. With 50 customers the terminal access may seem a luxury; with 5,000 customers the searching of the listing could be quite time-consuming. It is for each manager to decide the relative value of the facility to him and his people.

7. ANALYZING RESPONSE REQUIREMENTS

<u>Briefing notes for the Immediate Access Questionnaire</u>

"We are in the process of deciding on the cost-effectiveness of various features of the integrated marketing system. From the information that you and other managers gave us, we have drawn up a list of all the possible information requests that might be made of the system. I'd like to go through them with you, see which ones you would find useful, and what sort of use you may make of this facility when the system is available.

Each information request has been specified in terms of what would have to be provided to the system (on a request form or keyed into a terminal), and what the system would come back to you with. For example, you might give the system "Customer number" and want it to come back to you with "Details of last invoice." For each request, I would like you, if you can, to give me an idea of how *often* you might want to make the request, and an indication of how valuable it would be to you to have the request answered on the spot, as opposed to getting a report tomorrow morning, or as opposed to having a listing to consult (say of Customer names and numbers).

As I'm sure you know, it costs much more to come right back with an answer to a request in seconds compared with working the answer out overnight and getting you a report the next day, or providing a static listing. We want to be sure we're not building features into the system which are "nice-to-have," so for each request I'll ask you how much you think it is worth to you to have it answered on the spot, supposing that you could have the answer tomorrow morning at a cost of $1, (I'm not saying that's what it will cost, but just looking for a measure of the value to you). Of course, if you *have* to have the information on-the-spot to do your job, the question is irrelevant; please tell me if this is the case."

Figure 7.9

7.5 FINDING OUT WHAT THE USERS NEEDS AND PREFERENCES ARE

MANAGER'S NAME _____ POSITION _____ DATE OF INTERVIEW _____ ANALYST _____

Based on this item that you know.....	You want the system to tell you.....	What is the likely average frequency of this type of request as far as you are concerned?*	If you could have this information tomorrow for $1, roughly what is it worth to you to have it right away?
Salesperson's name	Undelivered orders for that salesperson		
Salesperson's name	Outstanding proposals for that salesperson		
Customer name	Undelivered orders for that customer		
Customer name	History of delivered orders for that customer		
Customer name	Outstanding proposals for that customer		
Any two dates	All orders with planned delivery dates in that interval		
Component number	Orders using that component analyzed by month		

* >1 per min = v. hi, >10/hr = hi, >10/day = med, >10/week = lo, ≤ 10/week = v. lo

Figure 7.10

7. ANALYZING RESPONSE REQUIREMENTS

When the results of the questionnaire interviews with each manager are collated, a good indication of the overall value of each immediate response can usually be seen together with an indication of the overall likely frequency of use. If the questionnaire cannot be answered with any clarity, because the variety of uses managers may make of the data is just too great, a general inquiry language facility may be the answer.

7.6 SECURITY CONSIDERATIONS

While the question of security of data against theft, destruction or alteration concerns the analyst and designer at each stage of design,[7.6], it is particularly important when analyzing immediate access requirements. Confidential documents, generated from a computer, can be kept under lock and key, and shredded when no longer required. Confidential data in a data store to which we have provided immediate access via a terminal, can, unless we are careful, be read by unauthorized people, with no trace left of the fact.

Consequently, the analyst should establish *who* is entitled to make each access. For example, the Sales Manager will have access to the commission earnings of individual salespeople. Should the Sales Administration supervisor also be able to make these accesses? If access is available to outstanding proposals and their dollar amounts, what is to stop some person in the pay of competition from reading this information off a CRT?

The immediate access diagram and questionnaire, once drawn up, are helpful as a tool for considering each access separately and deciding on the security aspects of each access. This may well be a matter for the analyst to clear with senior management, explaining that it is possible, given the right software, to allow certain people to "see" part of a data store, but not the whole of it. Thus it might be that all terminals could access a salesperson's name, address, phone number, and territory, but only someone using the terminal in the Sales Manager's office, equipped with a password known only to the Sales Manager, the President and the Comptroller could see the individual commission amounts. Senior management might decide that a wide group of people could have access to the statistics of proposal dollar amounts, but only the sales person for the account should have access to the specific amount for any specific proposal.

APPENDIX

GENERAL INQUIRY PACKAGES

CULPRIT

    Cullinane Corporation,
    20, William Street,
    Wellesley, MA 02181

EASYTRIEVE

    Pansophic
    709 Enterprise Drive,
    Oakbrook, IL 60521

MARK IV

    Informatics
    65, Route 4,
    Riveredge, N.J. 07661

This is not an exhaustive list: Consult the trade press for other packages.

7. ANALYZING RESPONSE REQUIREMENTS

REFERENCES

7.1 *IQF General Information Manual*, GH20 - 1074, IBM Corp., White Plains, N.Y. 10604.

7.2 J. Martin, "*Principles of Data-Base Management,*" Prentice-Hall, 1976, Chap. 17.

7.3 J. Martin, "*Computer Data-Base Organization,*" Prentice-Hall, 1975, Chap. 5.

EXERCISES AND DISCUSSION POINTS

1. In the terms of this chapter, what relationship does your local Yellow Pages bear to the White Pages of the telephone directory? What other inversions of the data in the telephone directory are possible?

2. Review the systems in your organization which sort or search files to produce reports. Are there any systems where it would be desirable to have a secondary index and provide immediate access to the data?

3. If any of the systems in your organization use secondary indexing, find out how big the index is compared with the main file. How often is the index reorganized? Is the index used for operational or informational purposes? How many accesses are made in the course of an average day?

4. If you have access to a hierarchical file or data base, draw a diagram of its structure. What provision is made for access other than down the hierarchical path?

5. Familiarize yourself with one of the General Inquiry facilities mentioned in the Chapter. In organizations which have used it, how much has it eased the load on the programming department? What factors limit its wider use?
   How many man-months did your organization spend last year on writing one-off reporting programs?

6. Take a system that you know of which has multiple accesses to its stored data, and describe each access in terms of the types in Section 7.4.

7. Draw a data immediate access diagram for the system (or any other system which has multiple immediate accesses). Gather what information you can about the value of the various accesses to the business.

8. Talk to one of the data base design experts in your organization, and ask him/her what information they would ideally like from the systems analyst, in order to produce the most cost-effective design.

Chapter 8

Using the tools: a structured methodology

In Chapters 3 through 7, we have examined the various tools and techniques of Structured Systems Analysis in detail. In this chapter we shall review the use to which they can be put in the early part of systems development, beginning with the point where the question of a system development effort is first raised, through the study and analysis phases, and into physical design. (The techniques of Structured Design, which use the product of our analysis, are described in the next chapter.) In doing so, we sketch out a *structured systems development methodology*, that is, a general approach to the building of commercial data processing systems, built around Structured Analysis and Design.

8.1 THE INITIAL STUDY

As computers get cheaper and cheaper, many more organizations are finding that they can gain advantages from automation, even those with as few as 50 employees. If an organization does not have a computer at present, the tools of Structured Analysis are useful for analyzing the existing manual system, understanding it clearly prior to deciding whether to instal a computer, and deciding what parts of the system to automate.

8. USING THE TOOLS

The sequence of activities set out in this chapter applies to the development of any information system, whether automated or not. We should point out, however, the methodology has been developed primarily for situations in which a computer system already exists, interfacing with manual systems, clerical procedures, and maybe other computer systems, and some improvement has to be made in this mixture of systems.

The questions that should be answered by an initial study are:

- What's wrong with the current situation?
- What improvement is possible?
- Who will be affected by the new system?

In an organization of any size, there is usually a continual stream of requests from managers for improvements in data processing service. While some of these requests can be met by better *operations*, such as improved response time on an existing inquiry facility, and some can be met by *enhancement* of existing systems, say by providing a new report from existing data, many involve the *development of a new system*. We are primarily concerned with these new developments.

Many organizations find that the demand for new systems is several times greater than their ability to build them. Indeed, as more and more systems get implemented, the task of making desirable enhancements in them takes up more and more of the time of the available people, in some cases as much as 50-60% of the entire development staff. Consequently, the systems that are chosen to be built should be the ones that will show the greatest payoff for the most managers in the organization, and hopefully, fit into an overall plan for the development of data processing in the organization. The initial study (sometimes called "Request Evaluation") is in part a screening process to weed out the development requests that are not going to be worth while and to do so relatively quickly and cheaply (since doing a study involves scarce people-time in itself).

This initial study may take from 2 days to 4 weeks (exceptionally, longer). The analyst should study the request (if written), and meet with the manager(s) making the request to get background on the situation and to begin to assess the likely value of the new system. He should ask questions like "Can you put a figure on the revenue we are losing because of the deficiencies of the present system?" "What costs are we incurring that could be avoided by a better system?"

8.1 THE INITIAL STUDY

"Is it possible to put a dollar estimate on the improved service we could give with a better system?" Where the system is needed to comply with a statutory requirement, he should determine the date by which compliance is required, and any penalties for late compliance. Managers should be invited to define the intangible benefits that they see flowing from an improved system, in very general terms, and to identify the other areas of the organization that would be affected. If necessary, the managers of these other areas should be interviewed to get similar information.

It is useful to bear in mind the underlying reasons why new systems are developed. The organization is either taking an opportunity (to increase revenue, avoid cost, or improve service) or is reacting to a pressure, either statutory or from the competition. For example, where a bank is the first in an area to instal a facility which allows customers to pay bills by touch-tone telephone, entering a code for the payee followed by the amount of the bill, they are clearly taking an opportunity to increase revenue by an improved service. Competition is not forcing them to do so, neither is the government. System justification must be based on management estimates of the value of the new service in attracting deposits. The second bank in the area to instal such a system will do so in response to competitive pressure, perhaps as a result of seeing depositors move accounts to the first bank. In governmental agencies, the revenue motive is less common, and cost avoidance, plus improvement of service to the electorate are the more usual justifications. The acronym IRACIS (Increase Revenue, Avoid Cost, Improve Service) has been suggested as a summary of these overall business objectives for systems.

Within these overall categories, there are several *ways* the new system can contribute:

1. Providing existing information, but more rapidly, e.g., providing immediate access to data which now appears in a weekly report.

2. Providing more up-to-date (fresher) information, e.g., providing data on sales as of last night instead of to the previous month-end, or providing up-to-the-minute account balances, instead of close-of-business-yesterday balances.

3. Providing more accurate information, either in terms of arithmetical accuracy, as in the case of a report which is presented subject to adjustment when unprocessable transactions have been dealt with, or in terms of being

8. USING THE TOOLS

more closely representative of the real world, as when inventory figures are reconciled to the actual quantities on-hand and on-order. Some users speak about more accurate information and mean more up-to-date information.

4. Providing information based on more data elements, as when a system that has never captured employee training history is to be replaced by one which does hold details of courses attended and experience gained. This implies the creation of a new data store, or an addition to the contents of an existing one.

5. Providing information based on new logical functions. When a sales trend projection is computed from existing monthly sales figures using time series analysis, the data stores remain the same but a new logical process is introduced.

Of course, frequently a requirement is expressed as a combination of these types: the user wants to capture more information than at present, keep it more up-to-date, analyze it in more complex ways, and have the answer faster! That's when it is useful for the analyst to be aware of these underlying dimensions, so that he can break down a complex requirement into its components.

As well as interviewing appropriate managers, the analyst should get hold of and review any documentation on the limitations of the current system. He should review the DP development plan, (if one exists!), to see how this area fits into it, and he should find out as much as possible about what similar organizations are doing in this area.

By the end of the initial study, the analyst should be reasonably confident about the magnitude of the benefits that could result from a new system. He will know whether the request originates with one manager's flash of inspiration but could save only $10,000 a year, or at the other extreme, whether it is something that could boost a $100 million billing by between 1% and 3%. He will probably be in a position to draw an overview data flow diagram of the existing system and its interfaces; as he gets to know more about the present system during the initial study, the logical data flow diagram is a very useful tool to fit all the fragments together. He should be able to estimate the time and cost involved in doing a detailed study (the next phase of development), and he may have a very tentative idea of what some possible new

8.1 THE INITIAL STUDY

systems might cost. He should be very cautious how he phrases any total project estimates; there is great pressure from users and senior management to establish project budgets as early as possible. Yet all our experience indicates that a firm price cannot be given until the number and complexity of modules in the system is known, that is after a firm physical design has been produced. The best course is to indicate a comfortably wide bracket of cost and time for the whole project, and only commit to a firm figure for the next phase. Thus, boiled down to a paragraph, the results of an initial study could be expressed as:

> "An initial 2 week investigation of the present purchasing system indicates that, based on estimates by manager X and manager Y, we are paying between 1% and 2½% more for our raw materials than we would if we could centralize requirements and take maximum advantage of bulk and prompt payment discounts offered by vendors. Our projected expenditure on raw materials this year is $20 million, indicating a potential annual saving possibility of between $200,000 and $500,000. In addition, an improved purchasing system could have a number of other benefits which we cannot yet quantify, including a reduction in rush purchases and a reduction in inventory. Based on other companies' experience, we estimate that to develop such a system would cost between $300,000 and $750,000, and take between 15 and 30 months. No estimate of operating costs can be made on the information we have at present. Though these figures are necessarily very tentative, we conclude that this project has enough potential to warrant a detailed study which will:
>
> 1. Refine our estimates of potential benefits
>
> 2. Establish objectives for a new system
>
> 3. Define the functions of a new system and how they will fit together
>
> 4. Refine our estimates of development cost.
>
> This detailed study will require 2 analysts for eight weeks, at a fully loaded charge-out cost of $16,000."

8. USING THE TOOLS

8.2 THE DETAILED STUDY

The outcome of the initial study will be reviewed by the appropriate level of management. Many organizations have a Steering Committee of senior managers to provide direction on the development of data processing. Depending on a blend of priorities, politics, and the facts established by the initial study, a detailed study may be authorized. It should be clear to all concerned that the detailed study does not represent a commitment to the implementation of the project. Management is saying, in effect, "Looks promising; tell me more." The detailed study builds on the facts produced by the initial study to document the limitations of the current system in more detail and with greater confidence, and to gain an understanding of the functions of the present system to the level needed to specify a replacement. The activities of the detailed study divide into the following areas:

8.2.1 DEFINING IN MORE DETAIL WHO THE USERS OF A NEW SYSTEM WOULD BE

The user community can be thought of at 3 levels:

1. The senior managers with profit responsibility whose areas will be affected, and who see themselves as paying for the system. This group has been called the "commissioners" since they commission the development of the system from the DP function, as well as possibly using its information outputs.

2. The middle managers and supervisors whose departments will be affected.

3. The clerical and other people who will work directly with the system by using terminals or filling out input forms, and interpreting output to do their jobs.

   Systems commissioners such as the President, Vice-President of Sales, or Vice-President of Production, are able to consider the medium-term, 3-5 year picture and make what is essentially an investment decision, for or against a significant project, on that basis. They can evaluate intangible benefits, even

8.2 THE DETAILED STUDY

if they can't put dollar figures on them. They tend to be very interested in aspects of a system which increase their real or apparent control over the business, since one of the stressful things about being a senior executive is having to take responsibility for some activity with whose detail you can no longer stay in touch.

Where systems are considered which will serve more than one major functional area, the top managers may want different things from the system, and may pull the system in different directions. It is traditional for the Sales Manager to want short delivery times, because that enables him to give better customer service and maximise sales, which is what he's paid for. It is equally traditional for the Production Manager to want long delivery times because that makes for better production scheduling and lower production cost, which is what *he's* paid for. If the analyst has bad luck, he and the system can get drawn into a power struggle, which shows itself in neither manager ever being satisfied with the draft statements of objectives and function. Clearly, the analyst cannot resolve a situation like this; if he suspects that he is trying to serve two masters who cannot agree, he must seek the support of his own management, who must, if necessary, take the situation to the Chief Executive Officer to get a resolution.

Middle managers and supervisors tend to be less concerned with the strategic picture and more concerned with the short-term performance of their department against budget. The analyst comes under pressure to explain the cost and personnel implications of a new system, and must be prepared to deal with resistance from middle managers who fear that the development may take power and autonomy away from them. Middle managers are often plagued by the problems of getting and keeping capable staff, and if the analyst can show how the new system might ease this problem, he may win allies.

As far as the clerical people whose work will be impacted, the analyst should do everything possible to make their jobs more pleasant and interesting under the new system. During the analysis and design stages, people should be reassured about the impact of the system on their jobs by talks and demonstrations. As part of the detailed study, the analyst should become familiar with the clerical jobs involved in the study. If time allows, and it is feasible, the analyst should perform one of the jobs (say telephone inquiries clerk) himself, for a short period. This will give him a wealth of first-hand information which would be hard, if not impossible,

8. USING THE TOOLS

to get via interviews.

Where the analyst has not met with the top managers during the initial study, he should try to get their views on objectives and preferences during the detailed study. Ideally, each of the affected middle managers whould be interviewed, but usually time rules this out, especially where they are geographically spread out. At least the analyst should acquire or prepare an up-to-date organization chart of the relevant departments, their managers/supervisors and their functions. He should determine the numbers of clerical staff, their natural attrition rates, and an understanding of their jobs.

8.2.2 BUILDING A LOGICAL MODEL OF THE CURRENT SYSTEM

From the information in the initial study, and from the information gained in defining the user community, the analyst can now produce a draft logical data flow diagram of the current system. But what exactly is "the current system?" The user's concerns often are to do with business results in a specific area, say, purchasing. As the analyst investigates he may find several interlocking systems are involved, some manual, some automated. In these cases, it can be a problem to *define the system boundary*, deciding which functions are to be considered as part of the system study, and which functions are not. The logical data flow diagram is of great use here; it enables the analyst to reduce each of the interlocking systems to a common terminology, and see how they fit together. Some organizations require their analysts to draw data flow diagrams of each system that interfaces with the area under study, especially the clerical systems. This requires additional time and effort, but has the benefit of showing up duplicated and redundant functions, giving the analyst greater confidence that he is drawing the system boundary in the right place, and giving greater understanding of the clerical jobs that may need to be changed.

Where a data flow diagram of an existing automated system has to be developed, the analyst usually has little difficulty identifying the nature of the functions and the inputs and outputs. There can be a problem specifying the detailed logic of a process if it is implemented in a poorly documented mnemonic language such as 1401 Autocode. (Yes, Virginia, there are 1401 programs still running!) Here the analyst has to choose between the "grease-monkey" approach of extracting

8.2 THE DETAILED STUDY

the logic from the program, (which is often not an economic use of time), and the "cold-turkey" method, of redefining the logic of the function by comparing the input data flows with the output data flows and consulting with a knowledgeable user as to the nature of the external (business) logic which is used to transform those inputs into outputs. The "cold-turkey" method is not quite as bad as its name implies, because we are not concerned with the internal logic of the undocumented program, (the way in which it sets and tests switches, accumulates counters, and executes loops), just with the external business policies, rules, and procedures that are incorporated in the program.

Since a commitment to build a system may not have been made at this point, the analyst may decide not to investigate the detailed logic of the current system functions at all, and just to identify functions at the "Produce Invoices" level. This approach should also be taken with any functions which the analyst knows will not be included in any new system. A similar common-sense decision should be made with respect to the data dictionary. In the course of drawing up the data flow diagram, the analyst will identify many data elements and data structures. If the file structures of the current system are well documented, they are an additional source of data definitions. If these definitions are easy to establish, the analyst should use whatever manual or automated facilities are available to capture them. If defining the details will require an inappropriate amount of work, the analyst should identify and name the data flows and processes with brief descriptions of each, leaving the details until later. As the logical model is built, the analyst should also be on the lookout for any inputs which no longer appear to be used, and any reports which are no longer of value, or could be replaced by exception reports or inquiries. Such redundant outputs and their associated functions can be described briefly.

8.2.3 REFINING THE ESTIMATES OF IRACIS

The initial study showed the order of magnitude of Increased Revenue / Avoidable Cost / Improved Service that could result from a new or improved system. Where the potential benefits of a new system are obvious and great, or the penalty for not having a new system is clearly unacceptable, it may not be necessary to spend any more time in this area. More usually, though, it is worthwhile taking a second look at the limita-

8. USING THE TOOLS

tions of the present system, and the potential benefits, especially in the light of projections of growth in volume of transactions.

For example, in the initial study, we indicated potential savings of between 1% and 2½% on raw materials purchases which are $20 million this year. What is the trend of increase in raw materials purchasing? Is there a way of cross-checking the potential savings? Now that we have met with more of the relevant managers, can we quantify any of the intangible benefits referred to in the initial study? If similar systems have been installed elsewhere, can we find out what benefits they realised? How do the benefits break down between the effect of centralized purchasing in greater bulk, and the effect of taking prompt-payment discounts? It may be important to know the relative sizes of the benefits that will come from various aspects of a new system, because this can be an important pointer to the aspects which should be implemented first. Obviously, if 1½% savings can be realized from taking prompt-payment discounts which appears a relatively simple system to implement, and another 1% would be attributable to centralized bulk purchasing, which in addition to being a more complex system involves major organization changes, we want to take these facts into account in our later planning.

By the end of the detailed study, then, the following information should be available:

- *a definition of the user community for a new system*

  - name and responsibilities of senior executives
  - functions of affected departments
  - relationships between affected departments
  - descriptions of clerical jobs that will be affected
  - number of people in each clerical job, hiring rates and natural attrition rates.

- *a logical model of the current system*

  - overall data flow diagram (including the interfacing systems, if relevant)
  - detailed data flow diagram for each important process
  - logic specification for each basic process at an appropriate level of detail
  - data definitions at an appropriate level of detail

8.2 THE DETAILED STUDY

- *a statement of Increased Revenue / Avoidable Cost / Improved Service that could be provided by an improved system* including:

  - assumptions
  - present and projected volumes of transactions, and quantities of stored data
  - dollar estimates of benefits where possible
  - account of competitive/statutory pressures (if any)

- *revised estimates of probable replacement system cost and a firm cost/time budget for next phase (defining a "menu" of possible alternatives)*

This information may be presented as a report to management, in which case a short summary should be prepared and made available separately from the documentation of the details we now have. It is usually better to make a face-to-face presentation to the relevant management group, organizing the presentation around the overall data flow diagram of the present system. It is frequently convenient to have the overall diagram drawn up as a large visual aid, because it gives the user managers such a clear idea of what the situation is and how the parts fit together. A typical comment is, "That's the first time I've understood what this system is all about." The analyst should also be prepared for more criticism when presenting a data flow diagram than with the traditional approaches; since it is easier to understand, it is also easier for the users to see any misunderstandings and errors in the data flow diagram, (which after all, is one of the purposes of the presentation).

As a result of this presentation, a decision will usually be made to continue the project to the next phase, or to shelve it. Sometimes the management group will say, "Build the best new system you can for $n." They should, if possible, be asked to defer setting a budget until the results of the next analysis phase, the definition of alternatives, is completed. Partly for this reason, the detailed study and definition of alternatives are sometimes treated as one phase, and no user management checkpoint is taken at the end of the detailed study.

8. USING THE TOOLS

8.3 DEFINING A "MENU" OF ALTERNATIVES

In the past it has been common practice for analysts and designers to study the current physical system, understand some of its limitations and then go on right away to devise a new physical system dealing with some of those limitations. In part, this approach was taken because of the difficulty of producing a logical model; the only way to put down one's thoughts about a system was to draw physical flowcharts, and it was such an effort to produce *any* new solution, that there was little energy to spare to investigate alternatives. Partly this production of a single solution was due to the feeling of a "doctor-patient" relationship between the data processing professional and the user. After study of the patient's symptoms, the doctor prescribes the one right solution which will cure the problem. We take the view that a better model of the relationship is that between the hungry person in a restaurant which cooks everything to order, and the maitre d'hotel. After consulting the diner as to his preferences, the maitre d' will suggest a few items, with varying costs and time to prepare, and let the diner choose between them. By analogy, the analyst (in consultation with the physical designer) should produce several possible new systems, which would have differing sets of benefits and involve differing investments, and *invite the user to choose* between these investment alternatives. (Too often we have dished up a gourmet meal to a user who was just hungry and in an hurry, or slapped a hot-dog in front of someone in the mood for a steak.)

Until the advent of Structured Analysis, the offering of a menu has not been a very practical proposition, because the visualization of alternatives, and their presentation to users in a meaningful way, has been so difficult. With Structured Analysis and its tools, the presentation of a menu of alternative solutions involves the user managers in making a business decision between investment alternatives, and gives them a feeling that the system is "their" system.

The activities involved in the development of the "menu" include:

8.3 DEFINING A "MENU" OF ALTERNATIVES

8.3.1 DERIVING OBJECTIVES FOR THE NEW SYSTEM FROM THE LIMITATIONS OF THE CURRENT SYSTEM

It is useful to distinguish *organizational objectives*, such as increased revenue, lowered cost, or improved service, from the *system objectives*, what the system will do to help management achieve the organizational objectives. Systems objectives come under the headings that we noted in Section 8.1, including:

- providing information more rapidly
- providing fresher information i.e. more up-to-date
- providing more accurate information
- handling more data elements
- providing new logical functions

We should note that achieving a system objective does not necessarily achieve an organizational objective. If we provide a manager with more, fresher, better organized, faster information, (a system objective), it does not automatically follow that his costs will be reduced. The manager has to act on the information or capability provided by the system. Sometimes managers get irritated by our assumptions that the new system will automatically make more money for them. Occasionally a system may save money directly, as in the case of a system which monitors power consumption and minimizes energy wastes; more usually, the system just *makes it possible* for managers to achieve their IRACIS objectives.

We should note also that objectives can be strongly or weakly stated. A strongly-stated objective minimizes the chance of disagreement when it is supposed to be achieved. Suppose we set up as a system objective "Produce the monthly Statement of Condition more rapidly," and in fact succeed with the new system in producing the report by the 8th of the month instead of the 10th. It avails us little if what the President really had in mind was a Statement of Condition on the 2nd! Each objective should be reviewed for ambiguities of this kind, and revised to make it as well-stated as possible. Figure 8.1 shows some representative weakly-stated objectives with some strongly-stated candidates.

8. USING THE TOOLS

	Weakly-stated	*Strongly-stated*
1.	Improve the timeliness of reporting	Produce the Daily Sales Report by 9 A.M. each morning
2.	Make the billing less error-prone	Reduce the number of invoices returned by customers as incorrect, to less than 0.5% of those sent out in any given week
3.	Improve the quality of the statements	List the paid checks in check number order
4.	Provide more comprehensive analysis of sales	Report a 3-month rolling average of sales by salesperson, product group and industry.

Figure 8.1
Weakly-stated and Strongly-stated objectives

Well-stated objectives do not necessarily have to be quantified. In cases 3 and 4 of Figure 8.1, the objectives are "all-or-none;" either the facility is provided or it isn't. What matters is that the objective be stated in such a way that there be the least possible chance of an argument when the analyst says to the user, "Here is the system that meets your objectives."

We should recognize that one weakly-stated objective may imply a family of strongly-stated ones, and that each strongly-stated objective can involve very different physical systems. Consider the situation shown in Figure 8.2.

8.3.2 DEVELOPING A LOGICAL MODEL OF THE NEW SYSTEM

Working with the statement of objectives, and with the logical data flow diagram of the current system, the analyst should produce a logical data flow diagram of the new system. The new system may involve handling:

8.3 DEFINING A "MENU" OF ALTERNATIVES

Provide more up-to-date information on available cash

Produce a reconciled bank balance as of close of business Friday, by Wednesday noon

Produce a statement of the available (cleared) balance at close of business each day, by 10 A.M. the following morning

Make the balance of available funds, available on an inquiry basis corrected to take account of transactions reaching the bank up to 30 minutes ago.

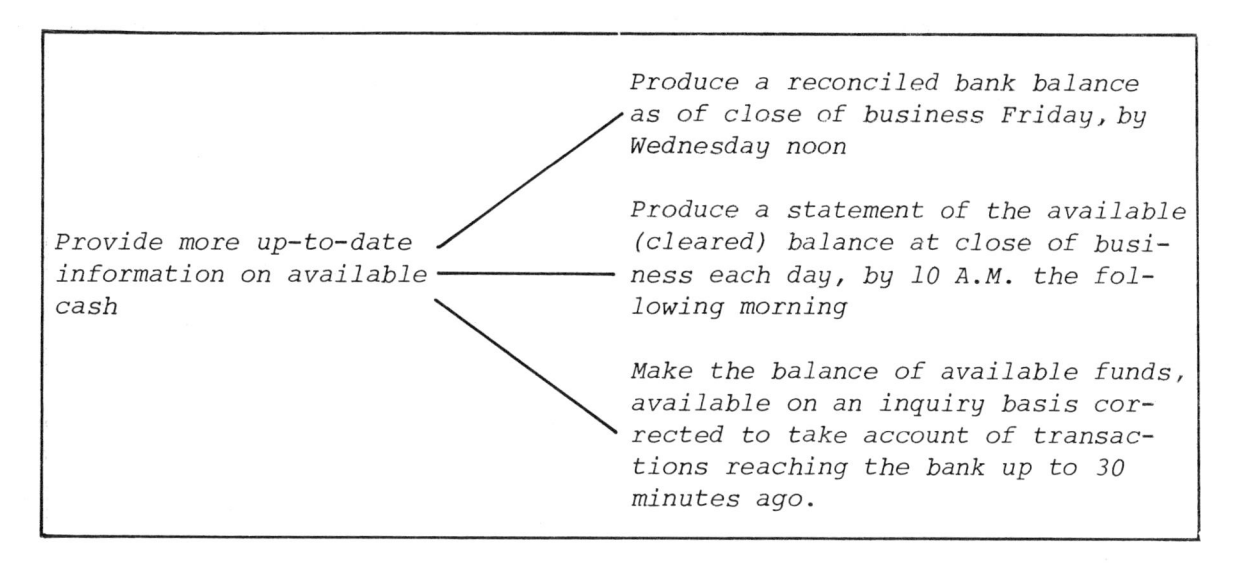

Figure 8.2
A family of strongly-stated objectives

    - new data flows
and/or
    - new transactions coming along existing data flows
and/or
    - new data stores
and/or
    - more data elements being held in existing data stores
and/or
    - new or changed logical processes

Where new data flows and data structures are involved, they should be added to the data dictionary.

Where new data elements are to be included in existing data stores, the elements must be defined, and the data store definitions updated.

Where new processes are involved, their logic should be specified to a level of detail needed to estimate the probable effort involved in programming them. For instance, if the new function is "Prepare report of capital expenditure," with the incoming and outgoing data flows specified, a summary statement of the logic would be adequate. If the new function is "Determine Optimum Routing for Delivery Trucks," a more extensive and detailed account would be needed.

8. USING THE TOOLS

In developing this logical model, the most demanding of the various strongly-stated objectives should be chosen. At this stage, the analyst is trying to define a system that will satisfy every wish of every user.

Reality begins to enter when the analyst carries out an immediate access analysis for each of the data stores to which any immediate access has been requested. The analyst draws up an immediate access questionnaire from the draft immediate access diagram, described in Chapter 7, and uses it to get estimates from the relevant members of the user community of the relative value and likely frequency of the various informational accesses. He consolidates this information, to get an overall picture of the value and frequency attaching to each immediate access.

8.3.3 PRODUCING TENTATIVE ALTERNATIVE PHYSICAL DESIGNS

By this point the analyst will have decided whether or not he is going to take responsibility for the physical design. The organization's philosophy may be to have the same person do both analysis and design, (and possibly coding), or alternatively to have the analyst produce a logical functional specification of the kind we have described, and hand that over to a separate person, with appropriate walkthroughs, for physical design. In our view, the second approach is likely to become more widespread now that techniques are available for presenting precisely "what" the system is required to do, without getting involved in "how" it is done.

Even if the same person does design and analysis, it is good practice to "wear the analyst hat" up to the point we have just reached in the development, (the logical functional specification), and then "put on a designer hat," and review the functional specification as though it had been produced by someone else, to see how it can most cost-effectively be implemented.

The analyst and the designer, then, whether one head or two, work together to devise several alternative systems which enable various versions of the objectives to be achieved, for varying investments of cost and time. They may draw tentative boundaries around different sets of functions on the proposed logical data flow diagram. They may consider implementing all of the immediate accesses requested, or only

8.3 DEFINING A "MENU" OF ALTERNATIVES

the most valuable, or none at all. Design considerations and techniques are discussed in more detail in the next chapter.

Some common types of tentative solution which may be considered are as follows:

1. A batch system, replacing a manual or sequential tape system, with one in which the files, though still possibly sequential, may be placed on disk. Relatively quick and cheap, the benefits are usually improved turnaround time, the addition of some extra data elements to the new files and reports, and more changeable programs, (since this type of solution is often used to replace early second generation systems).

2. A source data entry, overnight update system, in which transactions are entered via terminals in the user area, and edited on-line, building a file of transactions during the day which is used to update the main data base during an overnight run. More expensive than Type 1, the usual benefits include greater accuracy (because errors can be corrected at the time they are entered, by people who understand what they are doing), improved timeliness (because reporting runs don't have to wait for errors to go through repeated "edit, rejection, correction, edit" cycles), and ability to inquire on the state of the business as at the close of business yesterday.

3. An on-line data entry, immediate update, on-line inquiry system, in which each transaction is edited and corrected on-line, and, once accepted, used immediately to update the data base. Such systems are relatively expensive, in terms of hardware required, the development and maintenance of the software, and the security and recovery precautions that need to be taken to prevent corruption of the data base. The benefits, of course, can be considerable; not **only** can errors be corrected at the time of entry, but the data in the files is as fresh as it can be. In an airline reservation system, the seating available for a flight reflects the last booking which may have taken place only seconds ago. In an on-line banking system, if a customer has $100 in his account, and withdraws $80 through an on-line cash dispenser, his balance is immediately shown as $20, and any further attempt to withdraw more than that is refused.

    This type of system, in general, allows operational

8. USING THE TOOLS

people to respond to the changes in the business picture that take place from minute to minute through the day. It gives management the ability to inquire on the situation as of a few minutes ago, rather than as at the close of business yesterday.

4. A distributed system of the kind we described in Section 4.7, in which local minicomputers or intelligent terminals allow source data entry and a limited amount of editing against data held in the minicomputer or terminal files. Operational reports such as invoices may be produced by each local processor; from time to time (often each night) the local processors send information to a central host computer. The central computer uses the information (which may be full details of transactions, or just summary data) to update its data base, and may in turn send back information to the local computers which they use to update their local files.

   Such distributed systems offer many of the benefits of the Type 3 centralized system. Each local manager has access to his own data, and can update it immediately if justifiable. The central management has access to data correct to the close of business last night, (or more recently, if the computers exchange information more frequently). Because minicomputers and terminals are so cheap, a distributed system tends to be less expensive than a centralized system of similar power. Where an organization is widespread, the centralized system needs expensive and vulnerable telephone lines to allow source data entry and access. Also, if a centralized system goes down, all processing stops. If a local minicomputer, or the central host of a distributed system goes down, the effects are much less drastic.

5. A system which uses a dedicated minicomputer rather than sharing the existing central CPU. Such a system would not necessarily be distributed, but would take advantage of the cheapness of minis to implement a system which does not go down when the central system goes down, and does not suffer in response-time or throughput when the central system gets heavily loaded. Such a system is often appropriate where one manager has an application that doesn't need much communication with any other application, but requires a high service level.

8.3 DEFINING A "MENU" OF ALTERNATIVES

6. Lastly, we should bear in mind the possibility that one alternative is to design an improved manual system, with or without an automated system. Frequently, the data flow diagram of the clerical systems shows up redundancies and duplications, which have arisen over the years, and which can be eliminated to advantage. We may even conclude (preferably before extensive gathering of detail), that the users objectives do not require any change in the information system at all, but can be met by organizational or personnel changes. Perhaps the reason that bad debts are so high, is not that the information needed to reduce them is not available, but because even when the Credit Manager refuses an order, the Sales Manager goes to the President and overrules him. In such a situation we can use whatever ability as management consultants we may have, perhaps to suggest that the Credit Manager documents the cost of these decisions and quietly makes the information available to the President.

Naturally, these six possible alternatives are only suggestions, and are not to be followed slavishly. They are useful to consider, as a catalog of model vacation-homes is useful to review, before discussing requirements for your own custom-built home. Frequently, the analyst and designer will select a feature from one alternative, and a feature from another, and so on, to arrive at a tentative physical design.

The tentative physical designs for a new system ideally are selected as:

1. a low-budget, reasonably quickly - implemented system which meets only the most pressing of the user's objectives, though hopefully it can also be built on to make a more elaborate solution later
    (the "hamburger" solution).

2. a medium-budget, medium-timescale, system which achieves a solid majority of the objectives, though not the most ambitious ones
    (the "fried-chicken" solution).

3. a higher-budget, lengthy project which will achieve all of the user's objectives and have a major impact on the business,
    (the "chateaubriand steak" solution).

8. USING THE TOOLS

For each possible tentative alternative, the analyst and designer should work out rough estimates of the likely cost, and list the benefits, quantifying estimates as far as possible. Cost and time estimates may be based either on the known costs and duration of similar projects, or on estimates of the cost of the hardware and software components that each system will require, preferably both.

Thus, at the end of this step, the analyst is in a position to say to the user community something like:

"For about $50,000 - $70,000 and about six to nine months work, we can develop System A. This will give you the reports you get at present, but within 3 days of month-end rather than 10 days as at present. It will speed up the billing cycle by sending out invoices with an average time lag of only two days after shipment, and provide for automatic reminders, which should enable the average collection period to be reduced to 20 days from the current 38 days. Overall this should improve our cash position by $500,000, with annual savings of bank interest of $35,000. This new system will allow analysis of customers by size and by industry, which we cannot do at present. It will be much easier to change and extend than the present system.

For about $150,000 - $250,000 development involving about 12 to 18 months work, and using a $50,000 - $70,000 minicomputer, we can develop System B. This will allow our telephone order clerks to check the customer's creditworthiness and check the inventory picture, correct to the previous night, before accepting an order, and provide management with the ability to retrieve any customer's account on an immediate basis. In addition to the benefits of System A, the additional benefits of System B are a reduction in bad debts by an estimated 50% (equivalent to $20,000 last year), same-day shipment of orders when in stock, intangible improved customer service, and intangible improved sales.

For about $600,000 - $900,000 and about 2½ - 4 years work, we can develop System C, which will integrate order processing and inventory control, so that a salesman, using a portable terminal, can dial our computer from the customer's office, get a delivery date and price for the items the customer wants, enter the order details on the spot, and get a confirmation for the customer then and there. The VP of Sales estimates that this facility will make possible increased sales of between 2% and 5%, and will allow for reassignment of some 30-35 clerical people in Sales Administration and

8.4 USING THE "MENU" TO GET COMMITMENT FROM USER DECISION-MAKERS

> *Purchasing. Subject to management control, System C will automatically reorder components and raw materials when the inventory available falls below a specified safe level."*

Obviously, the features and benefits of each of the alternatives have to be spelled out in greater detail, but, though crude, the three alternatives above present the type of investment decision that businessmen can justifiably be asked to make. We note that this is an investment decision (how much shall we invest, to get what benefits over what timescale) and not a technical decision. We are not asking the users to decide between the merits of IBM's latest product announcement and the Digital Datawhack 17-bit improved architecture. The analyst and designer have done the work of translating features of each tentative design into *benefits* to the users.

8.4 USING THE "MENU" TO GET COMMITMENT FROM USER DECISION-MAKERS

Once the menu has been formulated, it must be presented to the manager(s) who have to make the investment decision. There may be a formal group in existence, perhaps called the Data Processing Steering Committee, or the Automation Policy Group; if no such group exists, the analyst falls back on the user decision makers that were identified as part of the user community in Section 8.2.1. It is worth taking pains to get this group together in the same room at the same time if at all possible, and to make a formal presentation of the alternatives to them using the Structured Systems Analysis tools as visual aids. If the decision makers are spread over a wide geographical area, or have conflicting schedules, the analyst may find it worth while to make a series of presentations to each one. If that is not economically justifiable then a report must be written and circulated for comments, recognizing that this is the least satisfactory way of arriving at a consensus on the alternatives.

Assuming that the decision makers can meet as a group, the analyst, accompanied by his manager, should make a presentation to the group covering the following points:

8. USING THE TOOLS

1. The current system (if one exists), walking through the data flow diagram.

2. The limitations of the current system or situation. If the same group of managers received a presentation at the end of the detailed study, the limitations should be summarized. Otherwise, it is worth spending 5 - 10 minutes to explain the IRACIS figures and the facts behind them, as this is an important input to the user decision making.

3. The logical model of the new system, walking through the logical data flow diagram that represents the most extensive alternative on the menu, pointing out the new functions that would be incorporated.

4. Each of the alternative systems that make up the menu, explaining for each one:

    - the parts of the overall data flow diagram that would be implemented

    - the way the system would look to the users, in terms of terminals, reports, and query facilities (the immediate access diagram can be used as a visual aid). As little jargon as possible should be used here.

    - the estimated benefits of this alternative

    - the current best estimates of cost and time to implement this alternative

    - a statement of the element of risk involved

5. A request for direction as to which of the alternatives represents the best cost-benefit trade-off in the user's eyes.

The management group will have a number of questions arising out of this presentation; it may be possible for them to reach a consensus decision on the alternative to select right away, or they may require the analyst to evaluate some compromise alternatives. They may require a report to consider and review themselves; in any event, a dialogue takes place, ending in the reaching of a consensus as to the nature of the system that should be built, and the assignment of the analyst and designer to come up with a firm budget of cost and time (within the range of estimates already quoted).

As we noted at the start of Section 8.3, this process involves user managers very directly in the selection of objectives for the project, and in the commitment of resources. There is a much greater likelihood that the user community will "own" the project as a result of this process, rather than seeing it as "another data processing boondoggle", something the computer people are putting together for their own purposes which may or may not serve the needs of the organization. This commitment to the project and involvement of senior user management has been identified many times as one of the key factors in success or failure for data processing projects.

8.5 REFINING THE PHYSICAL DESIGN OF THE NEW SYSTEM

Once a commitment has been made by the user decision makers, the analyst and designer work together to translate the logical model, and the tentative physical design, into a firm physical design. This process involves four overlapping activities:

8.5.1 REFINING THE LOGICAL MODEL

Typically the logical model by this stage consists of the overall data flow diagram, logical level data dictionary entries for each major data flow, data structure, data store, and process, and an immediate access diagram for each data store, where relevant. Usually the detailed logic of each process has not been specified, unless it is critical to the cost-estimation. Detailed data flow diagrams must be developed, dealing with error and exception handling, and any other processes not yet specified, as described in Chapter 3. The contents of the data dictionary entries must now be reviewed, and completed where necessary, as described in Chapter 4. Report and screen formats may be "dummied up" drawing on data elements defined in the data dictionary. Process logic must be specified at an "external" logic level, specifying the rules and procedures which have to be built into the system as described in Chapter 5. The contents of logical data stores must be analyzed and simplified as described in Chapter 6.

8. USING THE TOOLS

This logical model may be reviewed in detail with users. Often a user liaison representative, typically a middle manager with knowledge of the application, is appointed, and given the responsibility for approving the detailed specifications. It has been noted by several organizations that the more senior the user representative is, and the more time he can give to the project, the greater the likelihood of eventual success. For example, if the Assistant Sales Manager can be assigned half-time to help the analyst define the detailed requirements of the new system, his involvement and authority will make it likely that everyone in the user community will do a thorough job of thinking through their requirements and supplying the analyst with information that is prompt and complete. Some organizations have gone so far as designate this senior user manager as "Project Manager," assigning the necessary DP staff to him as resources to get the system developed. In any event, anything that can be done to get the active involvement of senior user people, should be done.

8.5.2 DESIGNING THE PHYSICAL DATA BASE

From the logical data store contents, the information held in the data dictionary about the volumes of transactions, and the immediate access analysis, the physical designer must make a near-final commitment as to the contents and organization of the physical files and/or data base. A tentative file/data base design will have been done for the "menu" exercise, and this must now be refined and checked against the logical model. If the system is to use an existing file or an existing data base, the designer must check that the existing contents and organization are known and fit the data flows planned in the logical model.

The subject of physical file/data base design is an extensive and complex one and beyond the scope of this book to treat in detail; some of the trade-offs are discussed in the next chapter. The best recent book on the subject is by James Martin [8.1]. As analysts, rather than physical designers, our concern must be to ensure that we have specified all the information needed for the design process, so that the physical designer can do the most cost-effective job.

8.5 REFINING THE PHYSICAL DESIGN OF THE NEW SYSTEM

8.5.3 DERIVING THE HIERARCHY OF MODULAR FUNCTIONS THAT WILL BE PROGRAMMED

Once the physical files have been specified, the processes and data flows between the files fall into physical subsystems. Thus once a decision has been made to edit incoming transactions on-line, and to build a file of accepted transactions as input to an overnight update of the main data base, it follows that all the processes and data flows involved in editing transactions and building the transaction file can be regarded as the "order-entry subsystem." This subsystem may end up being implemented as one on-line program, or it may be implemented as several programs. Before we make this decision as to what physical "packages" will be involved, we need to structure each subsystem as a hierarchy of modules, in which each module carries out a clearly defined function. This concept, and the techniques for transforming a data flow diagram into a modular structure, are described in the next chapter.

8.5.4 DEFINING THE NEW CLERICAL TASKS THAT WILL INTERFACE WITH THE NEW SYSTEM

The clerical tasks that the new system will require are determined by:

a) where the automated system boundary is drawn on the data flow diagram,

and

b) the physical choice of input and output that is made.

Thus if the editing of transactions is to be done on-line, the clerical tasks of entering input into a terminal (probably a CRT), understanding the terminal responses, and dealing with them, will be implied. If the editing of transactions for completeness and the vetting of customer credit is to be performed manually, followed by keypunching of transaction data from an input form, a completely different set of clerical tasks is involved.

8. USING THE TOOLS

Each clerical task has to be specified, and the necessary reference manuals and training prepared for the clerical people. Some organizations define a "personnel subsystem" consisting of all the data flows and processes which are to be implemented manually, and assign a manager or documentation/training analyst to be responsible for it. Many organizations make the systems analyst responsible for the design and implementation of this personnel subsystem. Especially where the clerical users have not worked with computers before, the personnel subsystem is a critical component, which can threaten the success of the whole system. The clerical procedures must be designed so that they are possible for the people to carry out, which implies that the procedures must be thoroughly tested, just like the software. No matter how good the analysis or the software design, a system which requires factory-floor workers to type long messages into terminals, or asks managers to remember numerical codes for every state and city in the country, is almost certainly doomed to disuse and failure. For details of personnel subsystem design, consult Gilb and Weinberg, [8.2], and Martin [8.3].

In designing the procedures and preparing the user reference manuals, the data dictionary is valuable as a source of codes that have to be used and their meanings. Decision trees and "tight English" are valuable tools for documenting the logic which has to be followed in, say, responding to an error message.

8.5.5 A NOTE ON ESTIMATING

Each of the four activities above should be pursued to the point at which it is possible to give a firm estimate of the cost of developing and operating the new system. Major components of these costs are:

1. The professional time and computer test-time required to develop the modules that have been defined. As people become more expensive and hardware gets cheaper, this element of the system cost is rapidly becoming the major factor, amounting to 70-80% of the cost in some minicomputer projects.

8.5 REFINING THE PHYSICAL DESIGN OF THE NEW SYSTEM

2. The CPU speed, memory size, and auxiliary store (disk), needed to handle the volume of transactions and the planned physical data base(s) and meet the throughput and response-time requirements, as specified in the system objectives.

3. The terminals, data communications costs (leased lines, dial-up charges), and other peripheral equipment needed to implement the system.

    Where the hardware costs are concerned, they may be expressed as either a one-time purchase price, or a monthly equivalent cost, often chosen as 1/40th to 1/60th of the purchase price, since computer equipment is often depreciated over 40, 50, or 60 months.

4. The professional time required to develop the user documentation and deliver the user training. This is usually a minor, but distinct, item.

5. The time of the clerical people who interact with the system. Since many systems enable existing clerical staff to be freed up, the reduced clerical work load is sometimes expressed as a benefit of the system rather than the remaining work load being classed as an operating cost. Clearly, if the present system uses 30 people, and the new system will require 10, at an average cost of $1,000 per month in both cases, the change can either be expressed as a saving of $30,000 per month with a cost of $10,000 per month, or a net saving of $20,000.

6. The professional time of the people required to maintain or enhance the system during its lifetime.

Different organizations have developed different costing standards for putting a dollar value on these various components. Representative examples are:

1K of real memory per hour (including all operation and support costs)	− 50 cents
1 x 3330 disk spindle on-line per hour (100 million bytes)	− $6 ($1,000 p.m.)
1 hour of professional time	− $30

8. USING THE TOOLS

To take a *very* simple example, based on these figures, a system which requires 60 man-months to develop, using 4 hours a day testing time for the last 6 months of the project, and which runs in a 150K region 8 hours a day with four 3330 disk spindles permanently on-line, needing two people full-time to maintain the programs, costs as follows:

Development:

    Professional time: $290,000 approx.

    60 man-months x 20 days/month
    x 8 hours/day x $30

    Computer test time: $ 36,000

    6 months x 20 days x 4 hours/day
    x 150K x $0.50

                                                                 $325,000 approx.

Operation (per month):

    CPU: $ 12,000

    8 hours /day x 20 days
    x 150 K x $0.50

    Disk: $ 4,000

    4 spindles x $1,000
                                                                 $ 16,500 p.m.

Maintenance (per month):

    2 persons x 20 days x 8 hours x $30 $ 9,600 p.m.

This analysis ignores cost headings 3, 4, and 5 above, of course, but is indicative. As we point out in the next chapter, by the time you read these words, the cost of professional time will have gone up, and the cost of hardware will have gone down substantially. Each analyst must establish realistic figures for his own organization and estimate the trend of operations costs over the life of the project.

These estimates, of course, take as their starting point the estimates of man-power, memory size, and run-time. Where

8.5 USING THE TOOLS TO IMPROVE ANALYSIS AND DESIGN

do these estimates come from? There are a number of more or less well-established schemes for estimating the man-power needed for a programming project (see, for instance, Aron's work [8.4]), but they all assume that *the estimator knows how many programs have to be written, and how large they are*. IBM has built a data base of 60 projects, ranging in size from 4,000 lines of source code to 467,000 lines, and studied the effect of many factors on the effort required and speed of development of the system, [8.5]. The median project produced 20,000 lines of source code in 11 months, with an average of six people, for a total effort of 67 man-months, with a testing cost of some $36,000. From the data base it is possible to predict, with some confidence, the average number of lines of code that will be produced per man-month on a project, based on factors such as the use of Structured Programming, the experience of the professionals, the ease of communicating with the customer, and more than 20 other variables. So once the project manager knows the number of lines of code to be produced, he can predict the productivity to be expected from the people he has available, and hence the duration and cost of the project.

This is where the building of the logical model, and the hierarchical design techniques described in the next chapter, are so useful. The techniques enable us to identify the functions and modules that will be needed to be programmed; the designer can then use his experience to estimate the number of source lines that will be needed to implement each module, and so get an estimate of the total programming task.

It is also, of course, very helpful to have a summary of the costs, effort, and duration required by previous similar projects, so that an estimate can be made by comparison, and checked against the estimate built up module by module. Indeed, we suspect that the difference between someone who can estimate well and someone who is not so good, lies mainly in the data base of past project costs and durations that they carry in their heads. A number of organizations, like IBM, are now seeking to formalize their past experience, and make it available in easily usable form to everyone who needs to be involved in estimating. If you have such a "Sears-Roebuck Catalog" of past projects available to you, (describing each project, the nature of the reports and on-line query facilities provided, the people and machine-time costs, the experience of the people with similar projects and techniques, and

8. USING THE TOOLS

other relevant variables), it can be extremely valuable in "bracketing" the project you are trying to estimate. For each case described in the "catalog," you can compare it with the problem or system you are considering, and decide whether it is simpler, more complex, or about the same. This will give you a fair indication as to the relative cost and duration that your present project will probably involve, give or take, say, 20-30%. Indeed, during the initial and detailed study, such comparative "bracketing" may be the only method of estimating the scope of the project, other than "sticking a wet finger in the air."

Since users, quite understandably, put a lot of pressure on analysts and project managers to produce estimates of total project costs quite early in the project, and since, as we noted, it is not really possible to produce a firmly based estimate until the hierarchical design is done, it is remarkable that more organizations have not produced such a "Sears-Roebuck catalog." If you do not have one, we commend it as a potentially valuable project.

8.6 LATER PHASES OF THE PROJECT

Since this is a book on analysis, and not primarily on design or implementation, we shall not carry the structured methodology past this point. Structured Design and implementation techniques are discussed in the next chapter to see how they relate to Structured Analysis. For the sake of completeness, we can distinguish the following phases in subsequent development of the system:

- drawing up an implementation plan, including plans for testing and acceptance of the system

- the concurrent development of the application programs, the personnel subsystem, and the data base/data communications functions (where relevant)

- the conversion and loading of the data base(s)

- the testing and acceptance of each part of the system

8.6 LATER PHASES OF THE PROJECT

- exercising the system under realistic loads to ensure it meets the performance criteria of the system objectives, in terms of response time and throughput

- commitment of the system to live operation

- measurement of system performance to identify bottlenecks and tuning to deal with those bottlenecks

- comparison of overall system facilities and performance to original objectives, and action to resolve any differences, where possible

- analysis of enhancement requests, prioritization of enhancements, and placing the system in "maintenance" state.

The analyst frequently acts as agent for the users in these later phases, much as an architect will monitor the construction of a building to ensure that the plans are being followed, and that materials of acceptable quality are used. The analyst may be involved in the specification of acceptance tests, and possibly in the generation of test data. Where an automated data dictionary has been used in the project, it can be a very convenient tool for generating test data, since acceptable and unacceptable values of each input data element are already defined in the data dictionary, and from these values acceptable and unacceptable transactions can be composed.

The logical model should be kept up to date through design and implementation, especially the data flow diagram. It will serve as a prime tool for planning enhancements especially those which involve new functions.

8. USING THE TOOLS

REFERENCES

8.1 J. Martin, *"Computer Data-Base Organization,"* Prentice-Hall, 1975.

8.2 T. Gilb and G. M. Weinberg, *"Humanized Input: Techniques for Reliable Keyed Input,"* Prentice-Hall, 1976.

8.3 J. Martin, *"Design of Man-Computer Dialogues,"* Prentice-Hall, 1973.

8.4 J. D. Aron, *"The Program Development Process,"* Addison-Wesley, 1974.

8.5 C. E. Walston and C. P. Felix, "A Method of Programming Measurement and Estimation," *IBM Systems Journal*, Vol. 16, No. 1, 1977.

EXERCISES AND DISCUSSION POINTS

1. Review the normal approach to development projects that is used in your organization (whether formal or informal). How does it differ from the methodology set out in Chapter 8? How does it resemble it?

2. Produce a set of strongly-stated objectives for a recently-installed system with which you are familiar.

3. Produce a statement of actual IRACIS for the system in Exercise 2, comparing it with the system which it replaced.

4. Develop an overall logical DFD for the system in Exercise 2, and evolve tentative designs for:

    a) a cheaper system with fewer benefits than the actual system

    b) a more extensive system with greater benefits than the actual system.

5. In what circumstances would the "menu" approach be inappropriate.

6. Produce a statement of the development cost and current operating cost for the last few systems to be implemented in your organization, as described in Section 8.5.5. Use people-hours and computer-hours if you can't get dollar figures.

    Produce data immediate access diagrams for each of the systems which allow any immediate access. Is there any relationship between system costs and amount of immediate access?

7. Can you see any connection between the value of a system and the degree of user participation in its development? What has been your organization's experience in getting user's participation? What, in your view, are the reasons why users do not participate more fully in system development?

Chapter 9

Deriving a structured design from the logical model

What do we mean by design? What is *structured* design? In this chapter we will try to answer these questions, show how a design can be produced from our logical model of a system, and see how improved designs enable systems to be developed more easily through so-called "top-down development."

There is a lot of confusion between analysis and design, partly because until it was possible to produce the sort of logical model we have described, it was very difficult to separate analysis ("what" the system has to do) from design ("how" it is going to be done). We would define "Design" as *the (iterative) process of taking a logical model of a system, together with a strongly-stated set of objectives for that system, and producing the specification of a physical system that will meet those objectives.*

A great deal of progress in system design techniques has been made in the last few years. From a more or less intuitive approach to design, where it was almost a miracle to see *any* way of deriving the required outputs from the given inputs, some fairly clear procedures and guidelines have emerged. In our view, the most useful of these sets of guidelines was originated by Larry Constantine [9.1], as a result of work at IBM, Hughes Aircraft, and elsewhere, in the late 1960's and refined by Myers [9.2], and Yourdon [9.3]. This

9. DERIVING A STRUCTURED DESIGN FROM THE LOGICAL MODEL

is a particular approach known as *Structured Design*; the logical tools we have discussed in this book draw heavily on the concepts of Structured Design. To see why Structured Design (as opposed to the "top-down" design advocated by IBM, or as opposed to bottom-up design, or random design) is the most useful approach, we shall review the objectives of design and relate the Structured Design approach to those objectives.

9.1   THE OBJECTIVES OF DESIGN

The most important objective of design, of course, is to deliver the functions required by the user. If the logical model calls for the production of paychecks, and the design does not produce paychecks, or does not produce them correctly, then the design is a failure. But given that many correct designs are possible, there are three main objectives which the designer has to bear in mind while evolving and evaluating a design.

- *performance*, how fast the design will be able to do the user's work given a particular hardware resource,

- *control*, the extent to which the design is secure against human errors, machine malfunction, or deliberate mischief,

- *changeability*, the ease with which the design allows the system to be changed to, for example, meet the user's needs to have different transaction types processed.

   Though it is not always true, it generally happens that these three factors work against one another; a system with very secure controls will tend to have its performance degraded, a system designed for very high performance may not be easy to change, and so on. We shall review some of the factors behind each of these design objectives, looking at the nature of changeable systems in some detail (Section 9.2). Then in Section 9.3, we will consider the trade-offs between the factors.

9.1.1 PERFORMANCE CONSIDERATIONS

Performance is usually expressed in terms of:

- *throughput*, (transactions/calculations per hour)

or

- *run-time* (for a batch job, where the same amount of work is processed on each run)

or

- *response-time* (the time between pressing the "enter" key on a terminal and the beginning of the computer's response appearing on the terminal)

As the majority of computer systems are primarily involved in data manipulation or data reduction (commercial systems) implicit in each of these measures is the amount of real memory used, since it is almost always possible to improve throughput or response-time by assigning a larger region or more real memory to the program(s).

This is not true for algorithmic-oriented systems, such as real-time process control, where the performance is more dependent on the speed of the central processor, and on the instructions used for the process. Those readers involved in such "number-crunching" systems are invited to skip to Section 9.1.3, since our discussion of performance and control is applicable mainly to data-oriented systems.

On today's computers, the designer is usually concerned primarily with throughput and secondarily with memory size. An exception to this arises with small minicomputers and microcomputers where the overriding performance objective can be to get the maximum amount of function into 4K bytes or so.

We identify five or six factors which affect throughput performance of a data-oriented system at a gross level, and we can arrange them roughly in descending order of importance, as follows:

9. DERIVING A STRUCTURED DESIGN FROM THE LOGICAL MODEL

1. *The number of intermediate files in a system*

   An intermediate file is one which is used to "park" data between programs, rather than a file which holds a data store of attributes about some important entity. An intermediate file is written and then read again by the next program, after which it could be thrown away without hampering the successful processing of the data. Punched card installations and second generation computers with magnetic tapes but no disks were forced to run systems that were designed with intermediate files in a program-file-program-file-program sequence. The physical system flowchart of Figure 9.1 shows such a typical system.

   Of course, with larger computers and disk files this system structure is no longer necessary; the creation and rereading of all the unnecessary files tends to produce an inherently slow-running system, even if the files are taken off tape and put on disk. However, the habit of creating intermediate files dies hard, and such files are often specified, unnecessarily, by people who gained their initial system design experience in second generation days.

2. *The number of times a given file is passed*

   Suppose a file contains a mixture of records of three types, for transaction type A, transaction type B, and transaction type C. The designer has a choice between either reading through the file for type A and processing type A, then reading through for type B, then for type C, or specifying a program which reads the file once only and handles each type of transaction as it is found. Whether the file is on tape or disk, the second strategy will give the faster throughput, other things being equal. Sorting a file, or searching it from end to end to answer an information request (as we described in Section 7.2) amounts to a pass of the file; we have already commented that sorting or searching takes a considerable amount of time, and may degrade the performance of other systems using the same file at the same time.

9.1 *THE OBJECTIVES OF DESIGN*

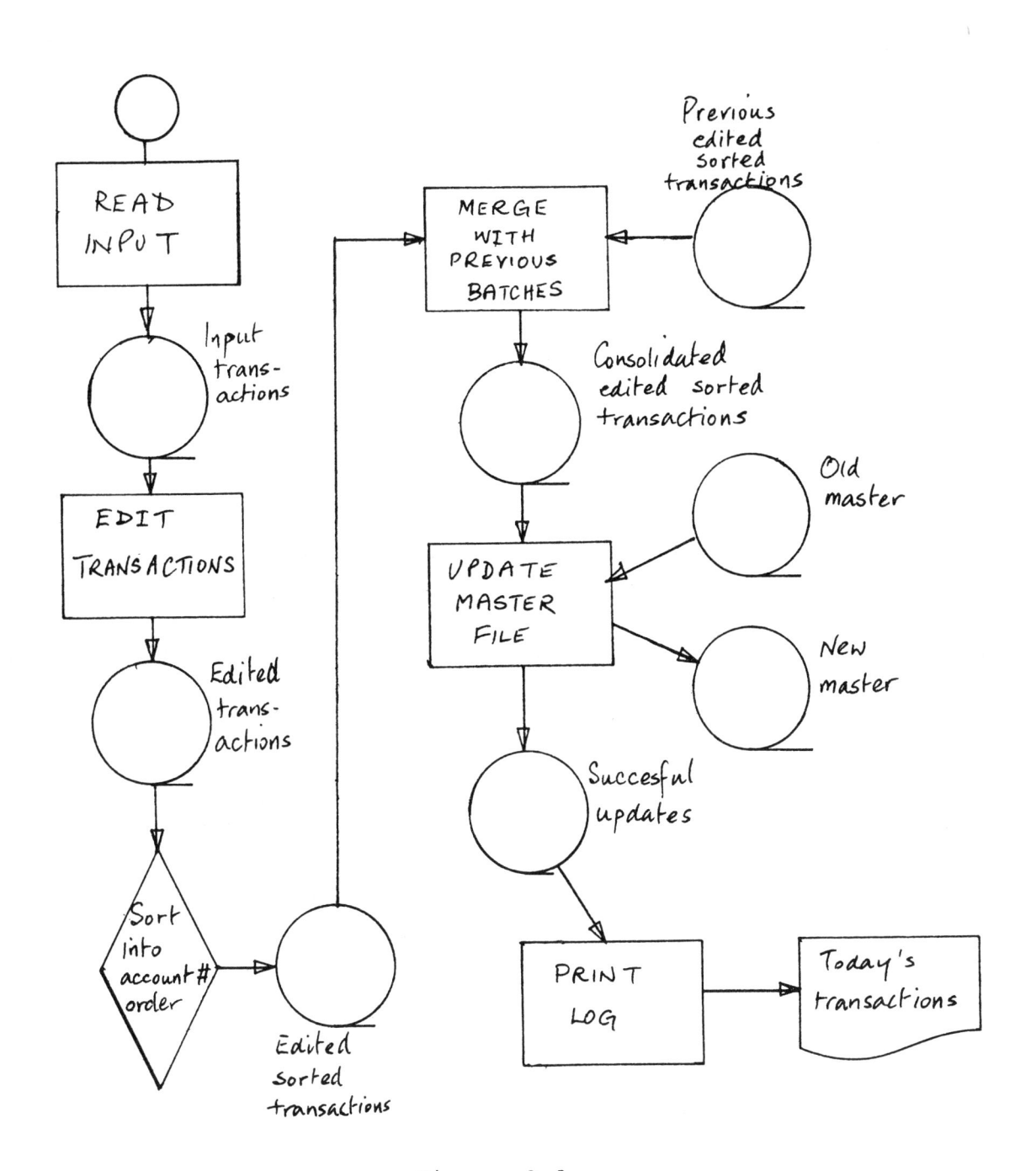

Figure 9.1

9. DERIVING A STRUCTURED DESIGN FROM THE LOGICAL MODEL

3. *The number of seeks against a disk file*

Given that a design has the minimum number of intermediate files, and passes its files the minimum number of times, the next limitation on its throughput is usually the number of seek movements made by the arm of a moving-head disk. Whenever the file is to be read or written, the disk management software must determine where on the disk the appropriate record is held and cause the read-write head to be moved, from wherever it happens to be, to the correct place for reading the record. This seek movement can take no time at all (if the arm just happens to be in the right place), up to 250 or 500 milliseconds (¼ to ½ second) if the arm has to move right across the disk on a slow model of disk drive. Further, a given read or write may involve looking in several different places on the disk. To take a simple case, if the records in the file are found from an index, the read-write head must be moved first to the index, then to the location of the desired record as found in the index. If the specified location contains, not the actual record, but an overflow address saying where the record is to be found, yet a third seek will be necessary. With some disk access methods and file layouts, finding a record can involve 15 to 20 seeks!

Some of the factors giving rise to seeks are outside the designer's control, because they are part of the disk management software that is used. The designer can, by understanding the software, be sure that the file layout does not create excessive seeking; for example, the designer may be able to specify regular reorganization of the physical layout to keep overflow seeking down to a minimum. It may be possible to specify that the file index is kept in the middle of the file extent, rather than at one side, which will cut down the average distance the read-write head has to move. Both of these measures are essentially tactical though; the most important thing the designer can do is structure the system so that the file is read the minimum number of times. This means that, if performance is the main object, when the customer master file is read to check on address, all the data in any one record should be held in the system until the need arises to check on phone, or balance outstanding, or history for that customer, or any other data held in the record. This consideration sometimes conflicts with other design objec-

9.1 THE OBJECTIVES OF DESIGN

tives, such as simplicity or changeability, but the designer should review his design with a view to minimizing reads if he can do so.

4. *The time spent in calling programs and other system overhead*

   A system is built up of a series of programs, each of which may be composed of subprograms. On most computers, each main program is invoked by job control language statements, which themselves form a "mini-program" invoked by the console operator. We can draw the "invocation structure" of most systems as a hierarchy:

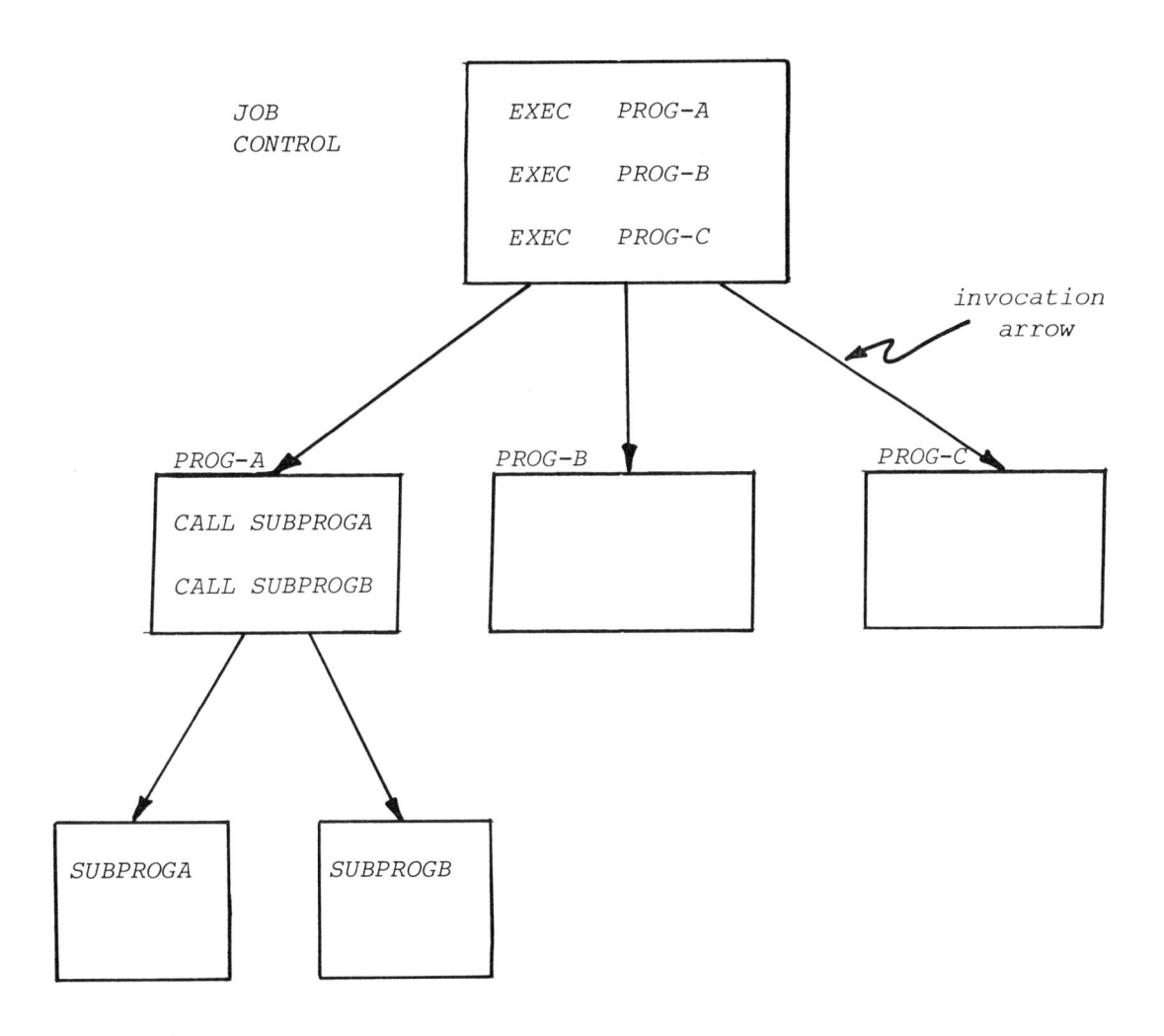

```
                    ┌─────────────────────┐
       JOB          │  EXEC   PROG-A      │
       CONTROL      │                     │
                    │  EXEC   PROG-B      │
                    │                     │
                    │  EXEC   PROG-C      │
                    └─────────────────────┘
                     /        |        \         invocation
                    /         |         \  ←←←    arrow
                   ↓          ↓          ↓
         PROG-A  ┌──────┐  PROG-B ┌──┐  PROG-C ┌──┐
                 │CALL  │         │  │         │  │
                 │SUBPROGA│       │  │         │  │
                 │CALL  │         │  │         │  │
                 │SUBPROGB│       └──┘         └──┘
                 └──────┘
                  /    \
                 ↓      ↓
              ┌─────┐ ┌─────┐
              │SUBPROGA│ │SUBPROGB│
              └─────┘ └─────┘
```

Figure 9.2

9. DERIVING A STRUCTURED DESIGN FROM THE LOGICAL MODEL

The invocation arrow, unlike the arrow in a flowchart or a data flow diagram, implies that control is passed from program to program, but also passed back again, when the invoked or called program has finished execution. This type of control hierarchy, in which numbers of manageably small modules (programs or sections of programs) are combined together to make a system, has many advantages, as we shall see when we come to discuss changeability of designs. From the performance point of view, each invocation takes a certain length of time, either because the computer (and operating system) has to work out where the invoked program is in memory, and transfer control to it, or more seriously, if the invoked program is not in memory, the operating system has to go and fetch it from some library either via a dynamic CALL, or an overlay fetch, or a page-in operation in a virtual memory system. Generally speaking, invocations within a program (e.g., a PERFORM in COBOL) involve less overhead than invocations of one program by another (e.g., a CALL in COBOL). For this reason the designer may decide to "package" two modules together in the same program, if they call one another frequently.

5. *The time taken to execute the actual program*

After the needed data has been read in, and after the appropriate program has been invoked and has gained control, the actual machine instructions, generated by the compiler from the source program, are executed. This generated code execution often represents the smallest proportion of all the time spent by the system, and so should be the last place a performance-conscious designer or programmer should spend time on improvements. Many technicians do not appreciate the relative unimportance of efficient machine-code in these days of large machines, and powerful operating systems. It is hard to remember that if the average seek takes 30 milliseconds, a lengthy machine instruction may take 30 *micro*seconds, one thousand times less. Paying attention to generated code efficiency *used* to be important when machines were small and slow; it is still occasionally important on mini- or micro-computers, and in CPU bound real-time systems where fractions of a second in response-time are important. In general, though, the designer and programmer will achieve more improvement in performance by spending time on the first four factors, than by worrying about the number of assembler instructions generated by some statement. To quote Ed

9.1 *THE OBJECTIVES OF DESIGN*

Yourdon:

"....one must always smile a little at complaints of inefficiency from COBOL programmers working under an operating system such as OS/360. If we were truly interested in efficiency, we would be coding in octal on machines without an operating system..." [9.4]

In fact, unless we can simulate the operating system overhead in our minds, it is pointless to "pre-optimise code" until we see where the major time is spent.

9.1.2 CONTROL CONSIDERATIONS

Depending on the nature of the system, and the amounts of money at stake, the designer will need to build in controls of various kinds. Obviously a system where large sums of money are handled, or where important secret information is stored, needs stronger controls against error or mischief than would our book distribution system.

Some common control aspects are:

1. *The use of check digits on predetermined numbers*

    We noted that the International Standard Book Number has a check digit as the last of its ten digits. Many banks assign the last digit of each person's account number as a check digit. At the time the account number is assigned, the check digit is computed, based on the other digits of the number and some suitable formula. Then each time the account number is processed in the future, particularly during data entry, the digits of the number as entered are used to recompute the check digit, and the answer is compared with the check digit as entered. If the two do not agree, one or more of the digits has been entered incorrectly, and the transaction must be rejected. The use of a check digit requires a little extra processing, but is amply worthwhile in terms of the errors it prevents.

2. *The use of batch control totals*

    Where transactions are entered into the system in batches, a total (perhaps the total amount of all the orders in the batch) is computed by hand before entry. As part of the editing of that batch, the designer will

9. *DERIVING A STRUCTURED DESIGN FROM THE LOGICAL MODEL*

specify that the computer work out the batch total and compare it with the hand-produced total. If they do not agree, it may be that a transaction has been dropped, or an amount entered incorrectly.

3. *The creation of journals and audit trails*

It is desirable that management or internal auditors should always be in a position to tell what transactions the system has processed, and what it did with them. The designer may specify that every transaction that goes into the system be written on a log or journal file, which can be read by the auditors at their leisure. Sometimes a "file-action" journal will also be kept, recording every change to every file. Creating these journals will slow the system down to some extent and require more hardware resources; the designer has to trade this off with the need for control and security. Indeed, he may not have much choice in the matter of controls, since they may be dictated by the auditors.

In on-line systems, a journal file is created for another reason. If the system should malfunction, some of the transactions that have been entered from remote terminals may be lost; it may be necessary to take a previous copy of the master file and run the day's transactions against it from the journal file. The taking of back-up copies of files or of checkpoints, also represents a designed use of hardware resources for security and control.

4. *The limiting of access to files*

Much of the security and control aspect of design is concerned with answering the questions:

"Who can have access to this data?"
"Who is allowed to change this data?"

Access may be limited by passwords, which have to be entered before a person may use a terminal, or by software which prevents unauthorized users from using certain programs, or which prevents certain programs from reading specified data elements in a file. It may even be controlled by physical possession of the file itself, as when the payroll master tape is locked in a safe. For a detailed description of these issues, and the whole question of security and accuracy in design, see Martin [9.5].

9.1 THE OBJECTIVES OF DESIGN

In general we can say that control and security costs money, in terms of either additional hardware and/or software, and may impact performance. The designer, guided by the analyst, has to make the best trade-off between these factors.

9.1.3 CHANGEABILITY CONSIDERATIONS

We talk about "the system" as though it were a fixed static thing, but of course nothing could be further from the truth. Since the system processes data about the real world, whenever the real world "outside" changes, the system may need to change. New lines of business are introduced or dropped, pricing and salary schemes change, user managers change and the new users have different ideas about their information requirements, policies change, new laws are passed.

As well as these externally induced changes, the technology of data processing is in a ferment. More powerful, cheaper hardware, new operating systems and languages, new data communications and data base techniques, all may mean that the system must be changed in some way. Further, no matter how careful the testing that the system receives, there will be bugs that only show up once the system is in production status; these bugs must be found and removed. Finding and removing a bug involves very much the same activities as making a change to benefit a user; the programmer must locate the part of the system that needs to be fixed or changed, make the fix, and be sure that the fix works. (Indeed, the three types of changes are often lumped together and all called "maintenance," though making changes which provide users with additional or better functions might more properly be called "enhancement.")

Is there a difference between removing bugs after the system is put into production, and removing them *before* the production cutover? Apart from the impact of the bugs in the production environment, the activities involved are identical. So we conclude that the *changeability* of a system is a very important thing; if we could find a way to design the most changeable system, it would be easy to debug it, easy to adapt it to changing hardware and software, and easy to incorporate user changes and enhancements.

By changeability we mean a measure of the time it takes to make any change in the system, whether removing a bug, or

9. DERIVING A STRUCTURED DESIGN FROM THE LOGICAL MODEL

installing an upgrade. Suppose a given system were to be designed in two different ways, and the same set of changes made in each system. If design A required 20 person-hours for the changes, and design B required 60 person-hours to instal the same changes, we could reasonably say that design A was three times as changeable as design B.

Recent figures show just how important a factor changeability is in the overall cost of systems. Figure 9.3 summarizes representative figures, which indicate that, for systems currently being developed, the vast majority of their lifetime cost is going to be incurred in changing those systems, and that of the "visible" manpower budget for development, about half will also be spent for the changes involved

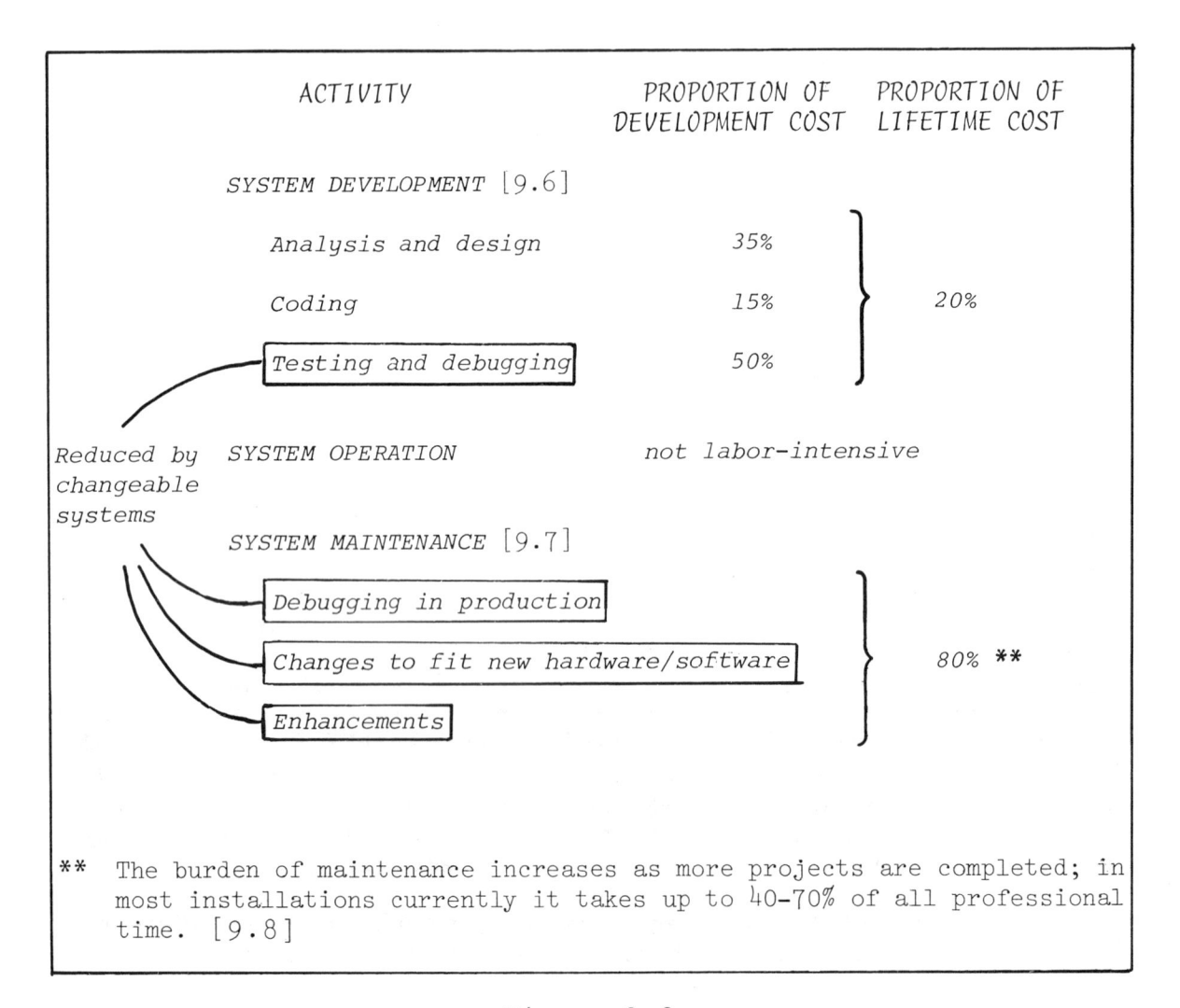

ACTIVITY	PROPORTION OF DEVELOPMENT COST	PROPORTION OF LIFETIME COST
SYSTEM DEVELOPMENT [9.6]		
Analysis and design	35%	
Coding	15%	20%
Testing and debugging	50%	
SYSTEM OPERATION	not labor-intensive	
SYSTEM MAINTENANCE [9.7]		
Debugging in production		
Changes to fit new hardware/software		80% **
Enhancements		

Reduced by changeable systems (arrows pointing to: Testing and debugging, Debugging in production, Changes to fit new hardware/software, Enhancements)

** The burden of maintenance increases as more projects are completed; in most installations currently it takes up to 40-70% of all professional time. [9.8]

Figure 9.3
Summary of distribution of manpower costs across project life cycle

9.1 THE OBJECTIVES OF DESIGN

in testing and debugging. It is no wonder that in many installations, the effort involved in keeping the existing systems "on the air" interferes with the development of new systems, even when the business badly needs them.

Debugging and changing software is heavily labor-intensive; there is not a great deal of help to be gotten from the use of computer power, other than with interactive debugging techniques. While, as Figure 9.4 shows, the cost of hardware is declining at a remarkable rate, the cost of professional labor is going up all the time. There is thus *great pressure, which will increase with each month that goes by, for systems to be designed to be as changeable as possible.* After all, if we could produce systems which are twice as changeable as those at present, we could potentially reduce their lifetime cost by 40%! More importantly to many managers, we could cut their development cost by 25% (half the testing and debugging cost). These are important numbers.

	EQUIVALENT PURCHASE PRICE FOR 1K MEMORY	AVERAGE CHARGE PER HOUR FOR PROGRAMMER/ANALYST
1964	$2,000 (360/30)	
1969	$1,000 (S/3)	$20
Sep 1975	$ 260 (370/158)	
May 1976	$ 170 (")	
Apr 1977	$ 110 (")	$30
TREND:	DOWN 20% p.a.	UP 7% p.a.

Figure 9.4

9. DERIVING A STRUCTURED DESIGN FROM THE LOGICAL MODEL

9.2 STRUCTURED DESIGN FOR CHANGEABILITY

9.2.1 WHAT MAKES FOR A CHANGEABLE SYSTEM?

There is general agreement that the easiest systems to change are those built up from manageably small modules, each of which are as far as possible independent of one another, so that they can be taken out of the system, changed, and put back in without affecting the rest of the system.

>What do we mean by "module"?
>What is "manageably small"?
>What is "independence"?

A logical module is a defined function with a name that expresses the purpose of the function; the processes we have defined on data flow diagrams such as "Verify Shipment Quantity," "Compute Order Total," "Generate Payment," can all be regarded as logical modules. A physical module is a particular implementation of a logical module, usually in terms of some group of statements in a programming language, which can be referenced by a name. Thus a physical module may be a program, a subprogram, a subroutine, a COBOL section, or even a COBOL paragraph. Where job control language allows sets of statements to be given a name and stored as a cataloged procedure, that procedure is a physical module. Strictly speaking, a set of instructions to a console operator, or to a clerk, could be regarded as a physical module. Obviously, modules often invoke other modules, which invoke other modules, and so on, as Figure 9.2 showed in outline. Within a system or within a program, the modules naturally fall into a hierarchy of boss, sub-boss, sub-sub-boss, and so on, down to "worker" modules which perform particular tasks.

When we say we want a module to be manageably small, we mean that a competent person would be able to take the listing of a module, read it, and keep a picture of its internal function in his head while deciding how to change it. Some modules may be as small as 10 lines of code, some (where the function is simple in principle, but lengthy in execution, such as a linear programming problem) may involve several hundred lines. 50 - 100 lines is a common size for well-formed "worker" modules.

9.2 STRUCTURED DESIGN FOR CHANGEABILITY

The modules that make up a system cannot be completely independent of each other, or there would be no system, just a heap of isolated modules. The designer's job is to form the modules and design their interconnection to minimize the chance of the dreaded "ripple-effect." [9.9] The "ripple-effect" occurs when the designer (or implementer) has allowed the various parts of the system to be connected in obscure ways; perhaps two modules will share the same temporary storage, or one piece of the program reads a switch which is set in a quite different part of the program. If many of these subtle inter-module couplings exist, the maintenance programmer will have a very hard life. Suppose a user needs a change made which affects module A. The programmer works out the details of the change, alters module A, and puts the system back in production. Alas, the change to module A in some way causes a bug in module B which is not noticed until the middle of the following night. The programmer traces the bug, and changes module B to avoid it. But that change affects module C and fixing module C blows up module D..... The effect of the original change "ripples" around the system, like the ripples caused by throwing a stone into a pool. Some systems are so interconnected as to be unmaintainable; the poor programmers follow the bugs round and round the system without ever finally removing them.

The design guidelines of Structured Design are of great help in producing systems in which the ripple-effect is kept to a minimum.

Another factor making for changeability is the extent to which the designer achieves *isolation of function* to as few modules as possible. This is fairly obvious, once stated; if the users want to change the discount calculation policy, and the discount is calculated in one module (which is changeable without setting up a ripple-effect), the change will be relatively easy, quick and cheap to make. If different parts of the discount calculation are spread through several modules, or if discounts are calculated independently within several modules in the system, the changes will be correspondingly harder to make.

Finally, we want our functions to be contained within "black boxes," where the function will always produce predictable results from a set of data passed to it. By black box, we specifically mean that:

1. it produces predictable results as viewed from the invoking module

9. DERIVING A STRUCTURED DESIGN FROM THE LOGICAL MODEL

2. we do not need to know the specific *internal* code of the module to know its function

Figure 9.5 summarizes the factors we have discussed as making for changeable designs.

System composed of hierarchy of "black box" modules

Each module manageably small

Each module changeable without creating a ripple-effect

User functions isolated to as few modules as possible

Figure 9.5

9.2.2 DERIVING A CHANGEABLE SYSTEM FROM THE DATA FLOW DIAGRAM

Experience with Structured Design, and examination of systems known to be changeable, suggests that the modules in a changeable system often resemble the units of a military organization. Each module has its own job, which it performs only when given orders from above; it communicates only with its superior officer and with its subordinates, to whom in turn it will issue orders. Figure 9.6 illustrates this analogy.

In this modular hierarchy, the "System Commander" module invokes the "Input Commander" with the implicit order "Get me a good transaction." The "Input Commander" in turn invokes a "Reader" module with the order "Get me a transaction." The "Reader" performs its sole function in life, and then returns control to its superior officer, saying "Here is a transaction," or "There are no more transactions." Ignoring the end-of-file possibility for the moment, the "Input Commander" then "wakes up" an "Editor" module, passes it the raw transaction and says, "Is this any good?" The "Editor" performs its function and returns with the answer. If the transaction is not good, the "Input Commander" will take a decision as to whether it should be fixed up in some way, or whether it should be rejected, and the "Reader" ordered to obtain another one. The sum total of this process is that the

9.2 STRUCTURED DESIGN FOR CHANGEABILITY

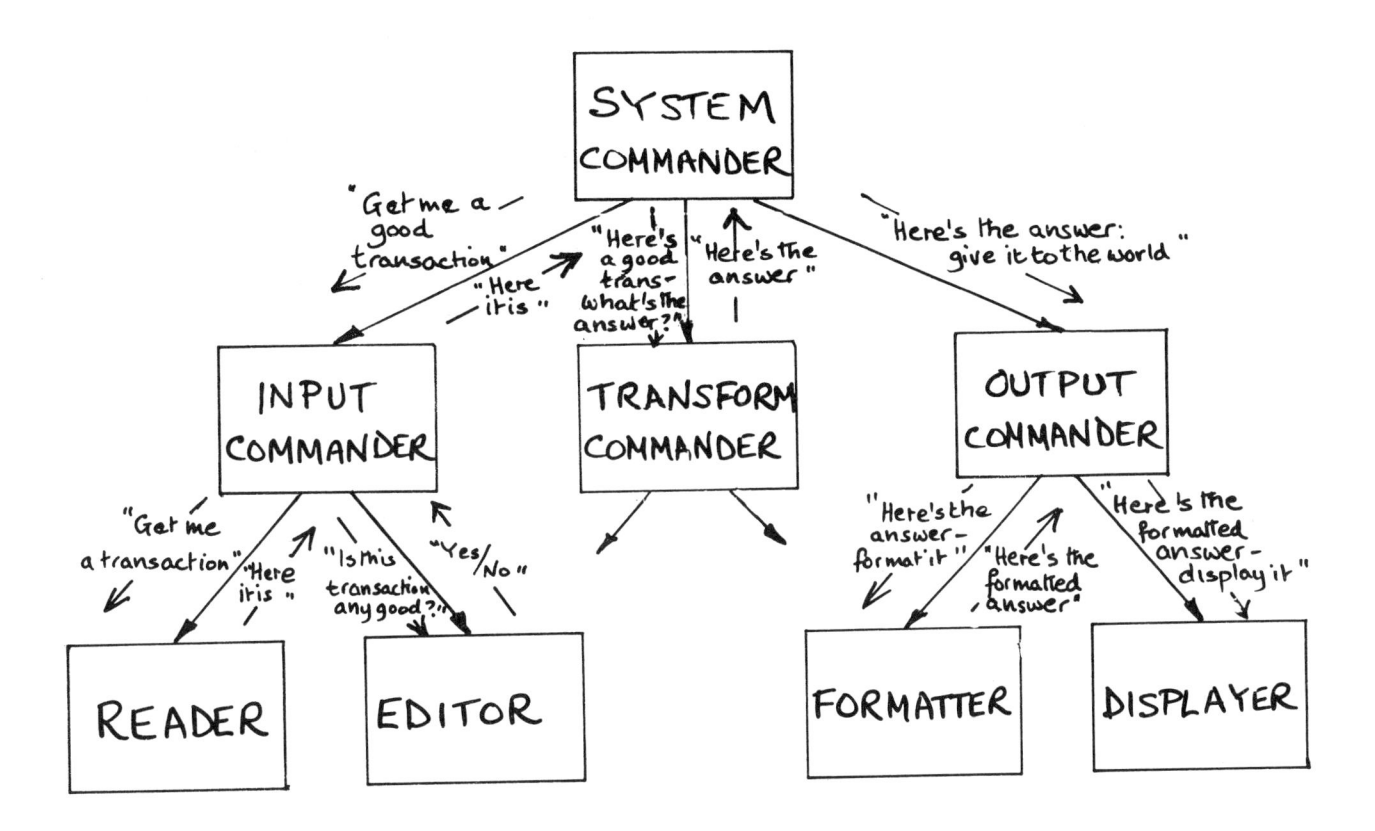

Figure 9.6

"Input Commander" will return control to the "System Commander" and pass back a good transaction. The "System Commander" then passes that good transaction to a "Transform Commander," who, using its subordinates each to perform their special function, computes the result that is to become the output corresponding to the original good transaction. The transformation might be computing net pay and deductions from a pay record, or determining the optimum flight plan for an aircraft from loading and weather parameters, or determining the quantity of goods to order from usage and discount information. Whatever the details, the "Transform Commander" will return to its boss with the result, which will be handed over to the "Output Commander" for disposition. Note that the "worker" modules do not communicate directly with one another, but only with their commanders. This is one way of simplifying inter-module coupling, and making it easier to understand the behavior of the system.

9. DERIVING A STRUCTURED DESIGN FROM THE LOGICAL MODEL

Systems with this type of structure, in which an input leg handles all input functions, a transform leg takes good input and produces a result, and an output leg handles all output of that result, are called *transform centered* systems, for obvious reasons. Their hierarchical structure can be derived fairly directly from the data flow through the system, which could be shown as in Figure 9.7.

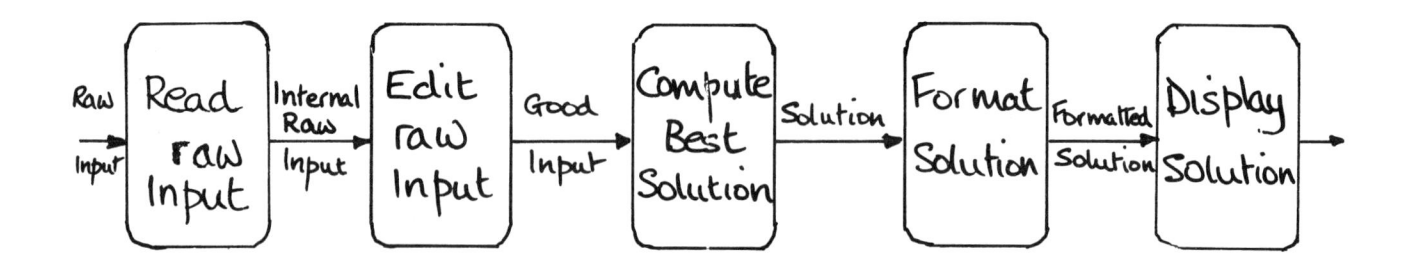

Figure 9.7

Hypothetical data flow

To get from the data flow to a hierarchical structure, we start with the rawest form of input and trace it through the data flow until we reach a point at which it can no longer be said to be input. Likewise, we take the final output, and trace it back into the system until it can no longer be thought of as output. Figure 9.8 shows the data flow with these points marked on it.

Once these "points of highest abstraction" have been identified, we can always create a transform centered design by specifying a system which calls for the most abstract form of input, transforms it into the most abstract form of output, and puts out the output. Figure 9.9 shows the "top-level" of such a hierarchy.

Note that in this case we have named the modules according to their functions, or the "orders which they obey," rather than calling them "Input Commander" and so on. This is precisely similar to the way in which we have described

9.2 STRUCTURED DESIGN FOR CHANGEABILITY

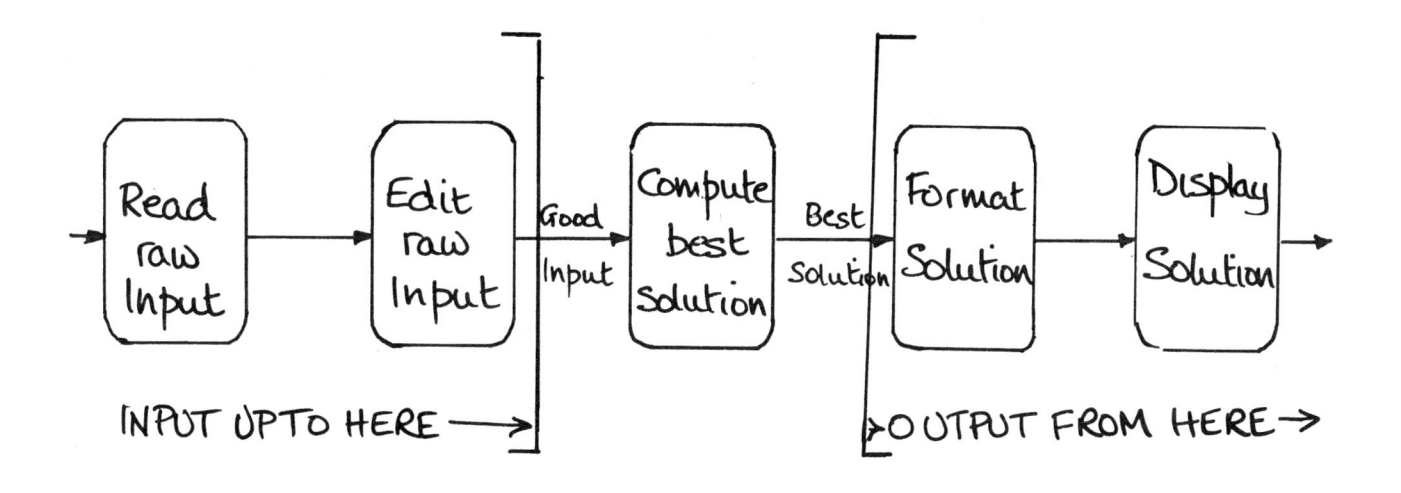

Figure 9.8

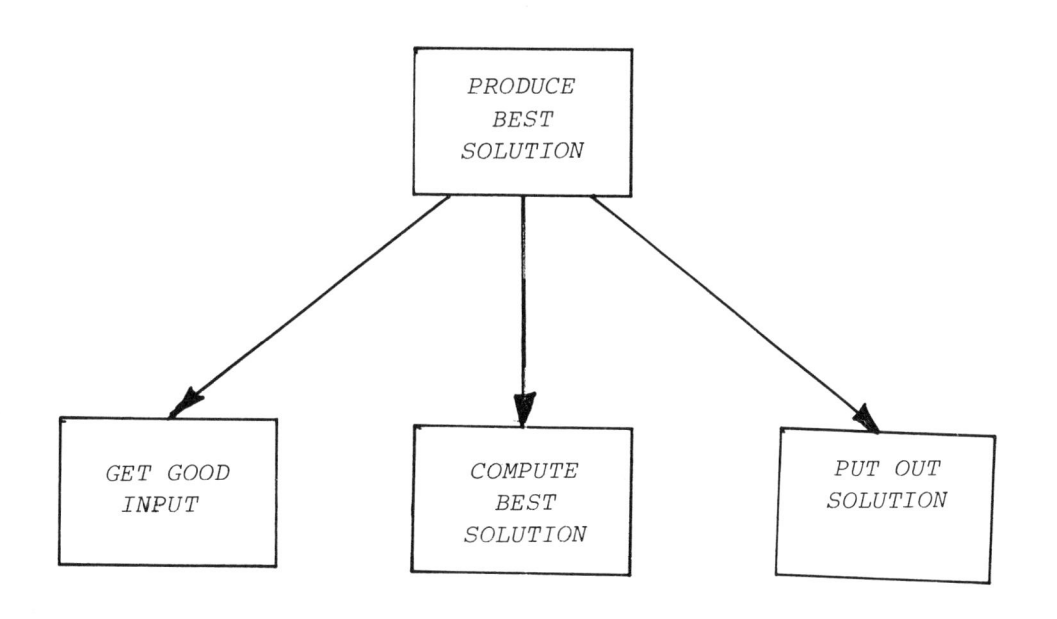

Figure 9.9

9. *DERIVING A STRUCTURED DESIGN FROM THE LOGICAL MODEL*

logical functions on our data flow diagram, using an active verb and a single object.

We can fill in the lower level modules and mark on this so-called "structure chart" the data that is passed between modules. It is conventional to show each data structure or data element with a small arrow having an open circle on its tail: o⟶. Where one module passes a control flag or switch to another module, telling the receiving module what happened, or what to do, the control element is shown with a filled in circle: •⟶.

Figure 9.10 shows our hypothetical system structure using these conventions:

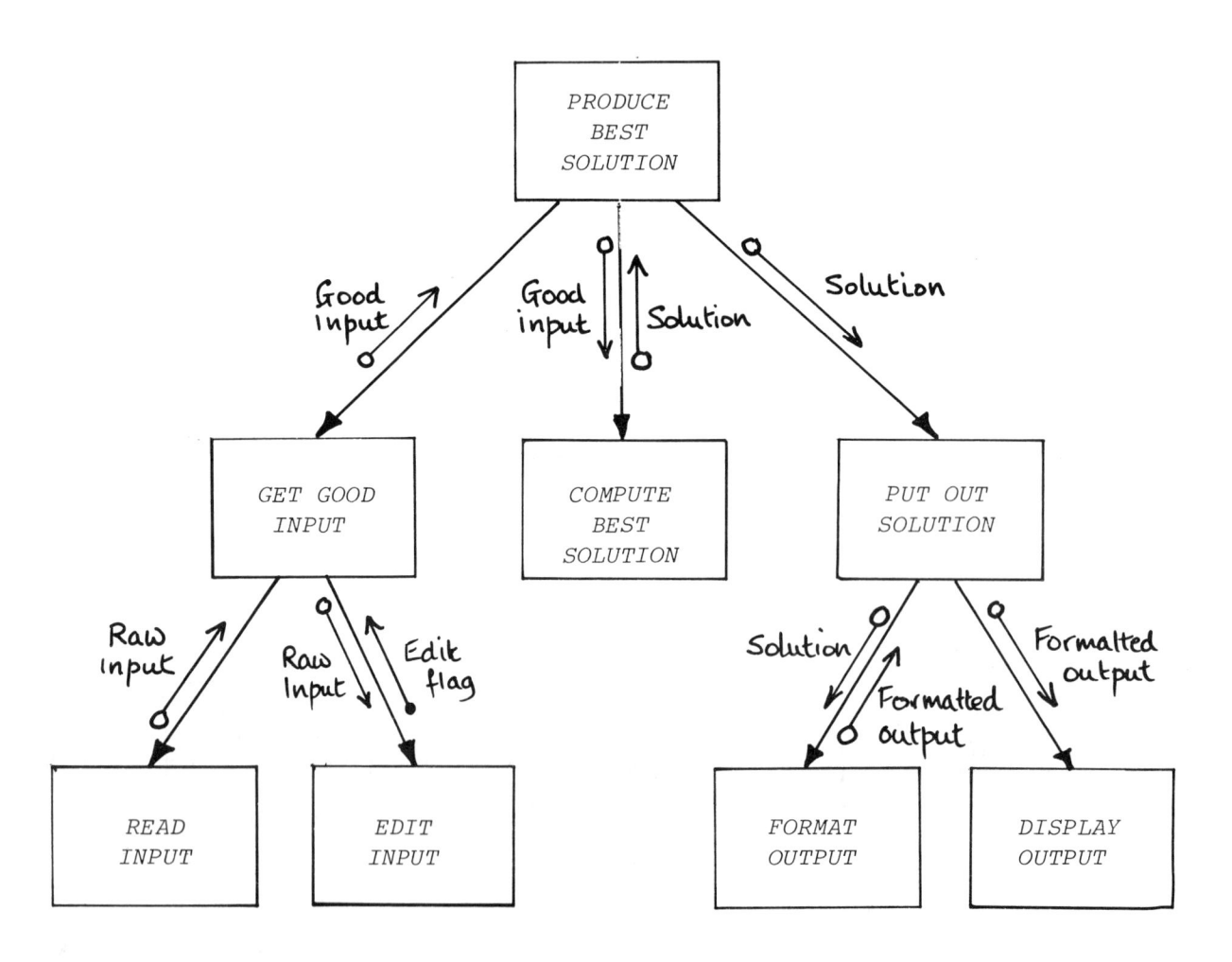

Figure 9.10

Structure chart for simple transform-centered system

9.2 STRUCTURED DESIGN FOR CHANGEABILITY

The "edit flag" is a control data element set by the module EDIT INPUT, perhaps to "Yes" if the input passes the edit or "No" if it does not.

The transform centered system usually is derived from a data flow in which all transactions follow the same, or closely similar, paths. Often, especially in commercial systems, this is not the case; the data flow shows the system handling several *different* types of transactions, as shown in Figure 9.11. Here four different types of transaction are shown (there might be more or less), each of which has to be edited in a separate way, and then used to update different master files, according to different logic, after which a log, or journal is printed showing the details of each transaction, together with the contents of each master record before and after updating.

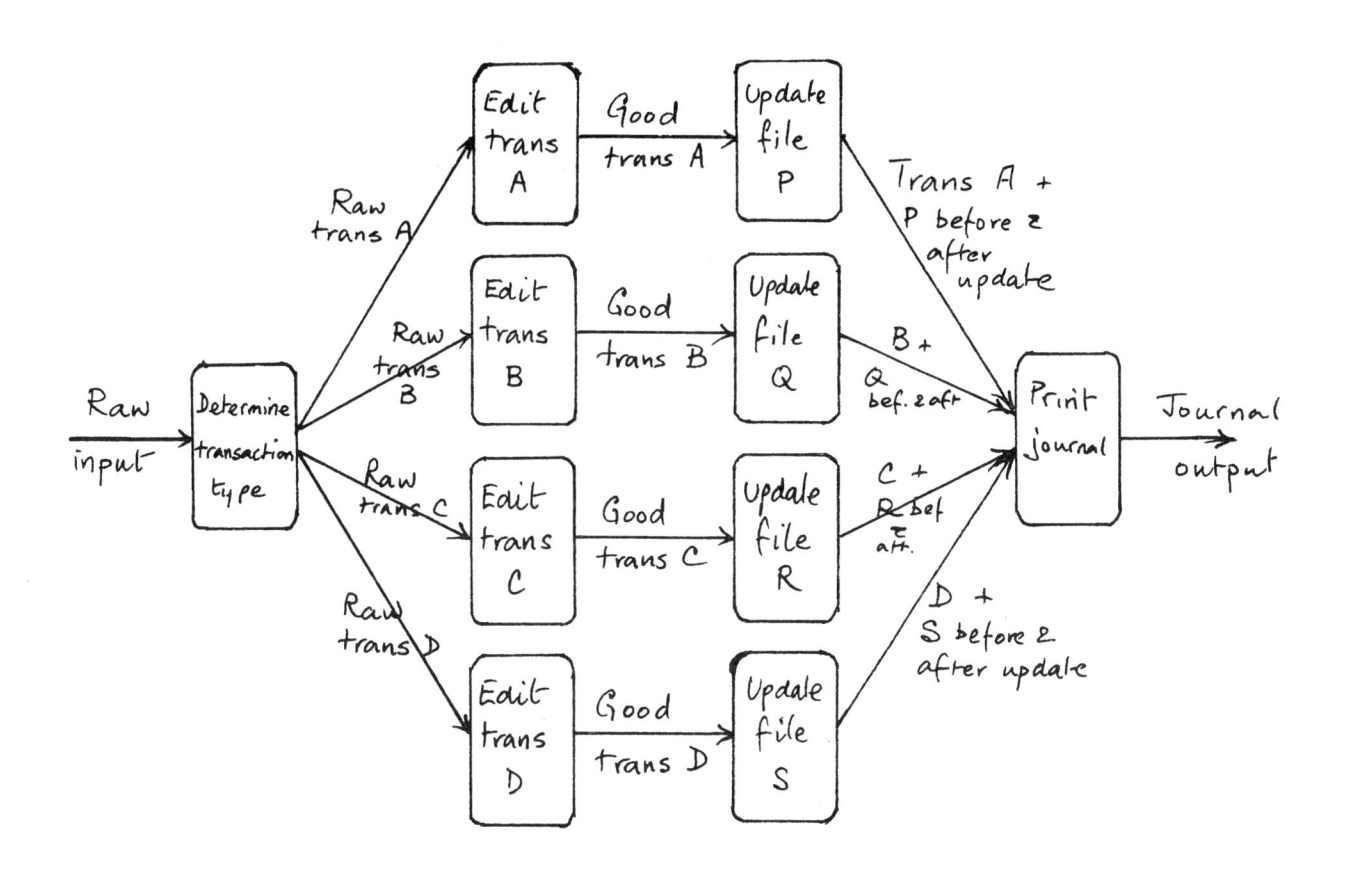

Figure 9.11

9. DERIVING A STRUCTURED DESIGN FROM THE LOGICAL MODEL

A changeable hierarchy corresponding to this data flow pattern is shown in Figure 9.12. In this type of system, the main executive module first invokes an "analyzer" module which returns with a transaction and its type; the main executive then invokes a "dispatcher" whose job is to route the transaction to an appropriate subsystem set up for each transaction type.

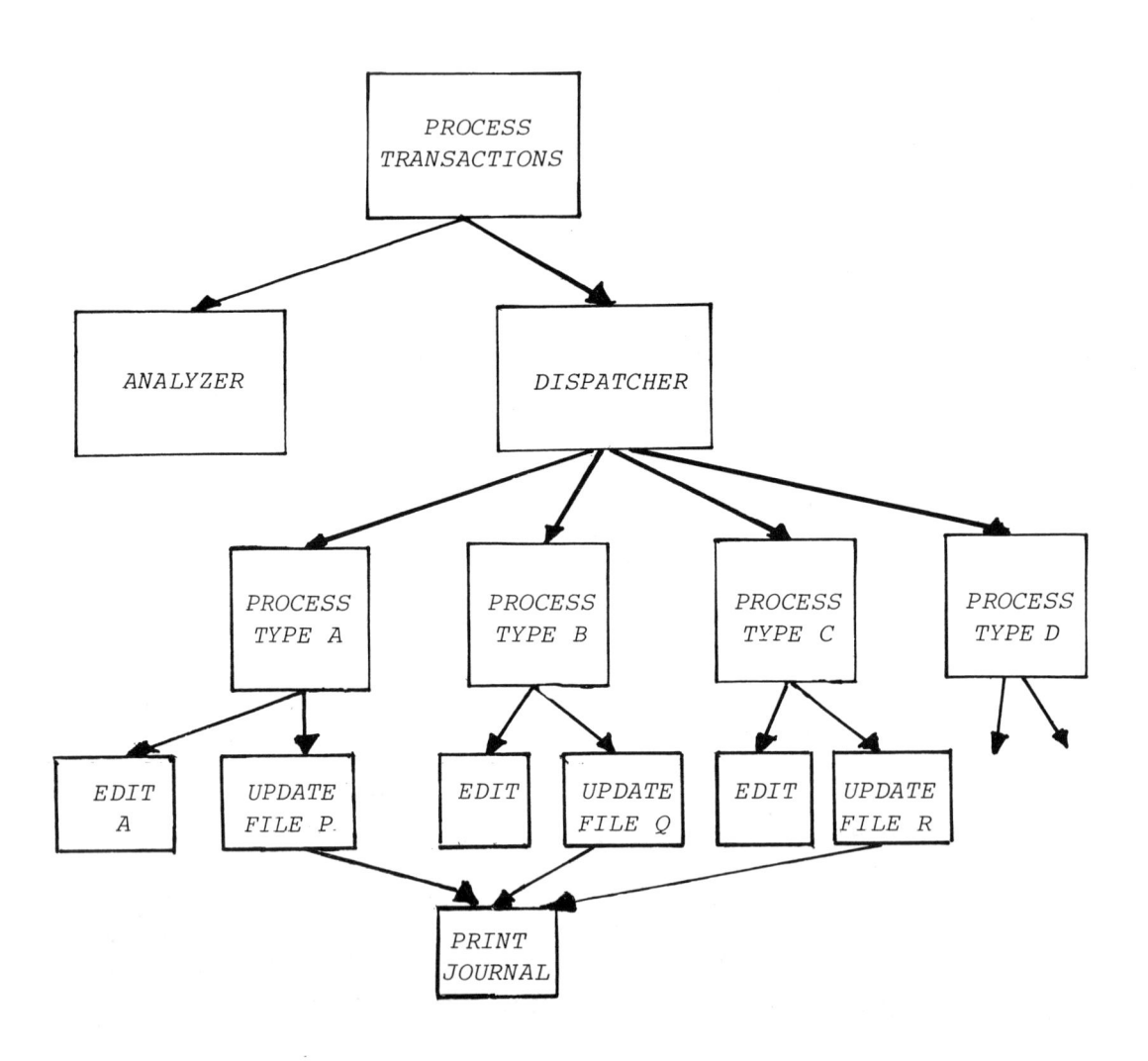

Figure 9.12

9.2 STRUCTURED DESIGN FOR CHANGEABILITY

This model of hierarchy is known as "transaction centered." As we shall see when we come to consider our design example, many system data flows imply a combination of both types of hierarchies, transforms and transactions, so the key issue is whether, at the point of transformation, there is one data path or many. By studying the detailed data flow diagram, the designer can determine the top-level structure (transform or transaction centered) and derive a "first-cut" hierarchy; the problem is then to ensure that the modules of the hierarchy and the communication between them, are as changeable as possible. To do this, we need some guidelines as to what leads to the most changeable coupling between modules, and what constitutes a well-formed module. Let us review each of these factors in turn.

9.2.3 MODULE COUPLING

As we noted before, one of the aims of design for changebility is to have the least possible coupling between the various modules, while still allowing the system to function. What types of coupling are there? Are some worse than others? Various studies of coupling have come up with slightly different types [9.2, 9.3, 9.10]; we will summarize the points on which there is general agreement.

1. *Data Coupling*

   It is clear that the most desirable form of coupling is where one module passes data to another as part of invocation or returning control. It is the best coupling because it is the least coupling.

   These two modules are data coupled. In general, a design in which few pieces of data are passed between modules is more changeable than one in which many pieces of data are passed.

   It is clearer and easier to follow the coupling when the data elements are passed as parameters in a module interface, rather than when the

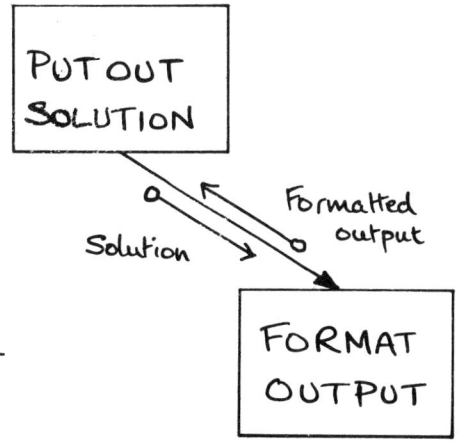

9. DERIVING A STRUCTURED DESIGN FROM THE LOGICAL MODEL

data is part of a global or common pool, which all modules can access. One of the changeability problems that arise with COBOL stems from the fact that all modules (sections or paragraphs) within a given COBOL program have access to the whole of the data division, especially working-storage. Thus each module within the program is in theory data coupled to every other module. It is often valuable to set aside an area of working-storage which belongs to each module, and is not used by any other module except for explicit passing of data as shown above. (This situation is also true for the FORTRAN COMMON statement but in FORTRAN alternative implementation is possible by passing explicit parameters.)

2. *Control Coupling*

Ideally, a system would be designed with only explicit data coupling from module to module. However, each time a worker module reads or writes or otherwise comes in contact with the outside world, it has to report back to its boss on what happened; it may have hit end-of-file, or found that a transaction is invalid, or found that an account number is not on file. Such "reporting back" involves passing a control variable, as we saw in Figure 9.10.

Control coupling has a more serious effect on changeability than data coupling, and must be kept to the absolute minimum necessary for the system to function. The more switches and flags, the more complex the maintenance programmer's job.

Control flags passed *down* the hierarchy, i.e. at invocation time, are indications that the invoked module is not "black-box"; it will execute in different ways depending on the control flags. This situation is undesirable, because it implies that the invoked module contains a mixture of functions.

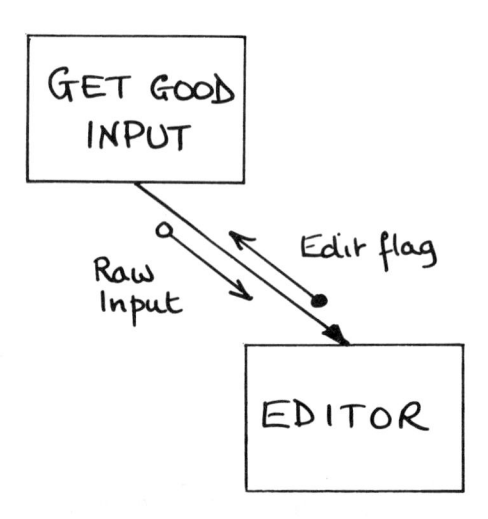

9.2 STRUCTURED DESIGN FOR CHANGEABILITY

3. *External/Content/Pathological Coupling*

These terms refer to severe coupling between modules in which one module refers to the inside of another module, either extracting some data defined within the second module, branching control to the inside of a second module, or modifying the way the second module executes.

If, for example, an output module reads a transaction counter which belongs to an input module, the coupling between the modules may not be apparent, since the transaction counter is not passed up and down the hierarchy. A maintenance programmer can cheerfully modify the READ module without realising the damage done to the output module.

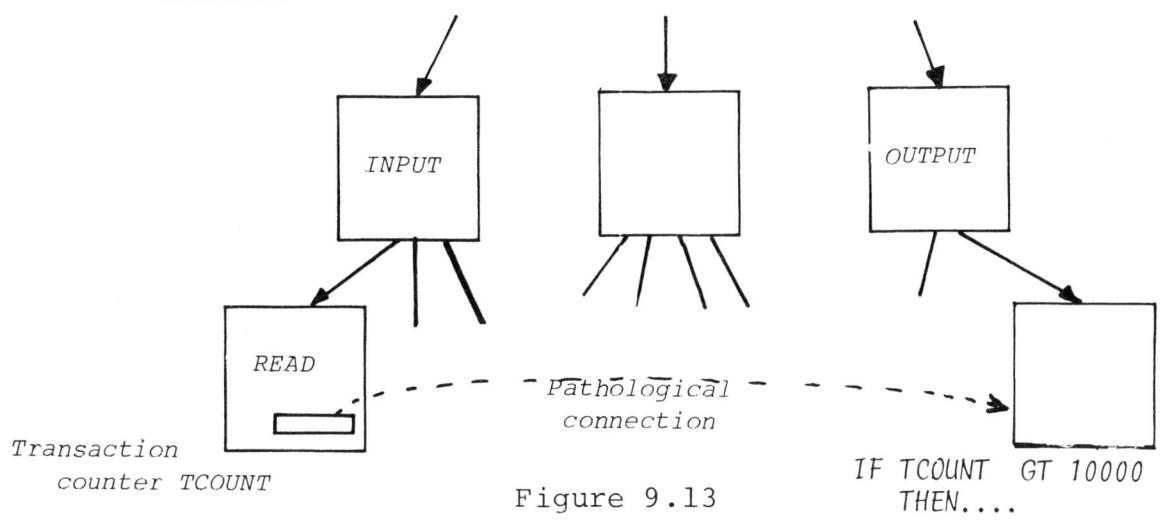

Figure 9.13

A Pathological connection

Where a situation occurs so rarely that it is not worth passing data or control all the way around the hierarchy, a severe coupling of this nature *may* be justified. Generally speaking, though, any form of coupling other than data and control should be avoided at all costs; that is where the endless ripples begin. This implies that when we are designing the modules, we must be sure to specify explicitly all the data or control information necessary for a module to function.

9. DERIVING A STRUCTURED DESIGN FROM THE LOGICAL MODEL

9.2.4 WELL-FORMED MODULES: COHESIVENESS, COHESION, BINDING

At the same time as minimizing coupling we want to be able to specify modules that are as well-chosen and well-formed as possible. The terms used to describe module quality-cohesion cohesiveness and binding--describe the extent to which all the parts of a module belong together. ("Binding" should not be used in this context, since it also describes the process of associating values with symbolic names.) A highly cohesive module, whose parts all contribute to a single function, is not likely to need much coupling to other modules; a poorly cohesive module, often a combination of unrelated parts, is likely to need high coupling to others. Thus low coupling and high cohesion go hand in hand, and vice versa.

Various studies identify six or seven types of module cohesion. We will describe the generally recognized types, from worst to best.

1. *Coincidental cohesion (worst)*

    The elements of the module cannot be seen as achieving any definable function: they are there by accident. When one organization introduced a standard that programs would be modular, with no module containing more than 50 statements, one programmer took a listing of a 2,000 statement program and a pair of scissors, cutting the listing every 50 lines, and making modules out of the pieces. Those modules were coincidentally cohesive.

2. *Logical cohesion (next to worst)*

    In this type of module several similar, but slightly different, functions are combined together making a more compact module than if each function had been programmed separately. A typical example is a module which edits all types of transactions, using common sections of code where appropriate, and branching round other parts of the code where not required for a particular transaction type. A logically cohesive module often requires a control switch to be passed to it from the module that invokes it, to tell it how to execute on any specific occasion. Control flags being passed down the hierarchy are thus often clues to the presence of logically cohesive modules.

    Modules of this type are often difficult to change, because the logic paths through them are so complex. They should be replaced by special purpose modules, one per function.

9.2 STRUCTURED DESIGN FOR CHANGEABILITY

3. *Temporal cohesion (moderate to poor)*

   Modules such as "initialization," "housekeeping," "wrap-up," contain a variety of functions whose only common element is that they are executed at the same *time*. Changeability is improved by isolating each separate function into its own module.

4. *Procedural cohesion (moderate)*

   This type of cohesion is found where the modules have been derived from a flowchart, and each "chunk" or procedure of the flowchart has been built into a module. Within each module several functions are carried out, but at least the functions are related by the flow of control between them.

5. *Communicational cohesion (moderate to good)*

   The subfunctions of a communicationally cohesive module all operate on the same data stream in addition to being procedurally cohesive. Some communicational module descriptions might be "Display result and write log record," "Read source statement and remove blanks," "Compute solution and print result."

6. *Functional cohesion (best)*

   A functionally cohesive module, which is our ideal, carries out one, and only one, identifiable function. One test for a functional module is to see whether it can be accused of being communicationally, or procedurally, or temporally, or logically cohesive. If it is innocent of these charges, it probably is functional. Functionally cohesive modules can usually be described by single phrases, with an active verb and a single object, of the type we have met before, such as "Compute best solution," "Edit inquiry," "Display answer."

   One guideline for determining the level of cohesion of a module is to write a single sentence beginning, "The full purpose of this module is to ..." and then examine that sentence.

   If the sentence cannot be completed, the module is probably only coincidentally cohesive.

   If the sentence has a plural object, and includes the word "all", e.g. "Edit all transactions" the module is probably logically cohesive.

9. DERIVING A STRUCTURED DESIGN FROM THE LOGICAL MODEL

If the sentence uses words to do with time or sequence, "First, next, then, after, otherwise, start," it may be temporally, procedurally, or communicationally cohesive.

If the sentence consists of a single active verb, with a non-plural object, the module is probably functionally cohesive.

One of the reasons for taking pains to describe the logical processes we identified during analysis in functional terms, is that doing so makes it easier for the designer to identify functional modules from the data flow diagram. Just as knowing about third normal form makes it more likely that the analyst will identify simple logical data stores, so knowing about functional cohesion makes it more likely that the analyst will identify simple processes.

9.2.5 SCOPE OF EFFECT/SCOPE OF CONTROL PROBLEMS

Given that we have designed a hierarchy using these guidelines for minimizing coupling and maximizing cohesion, there is another refinement to consider.

In some cases a module may contain logic which makes a decision about some event; based on that decision the module will invoke other modules. Figure 9.14 shows a simple case:

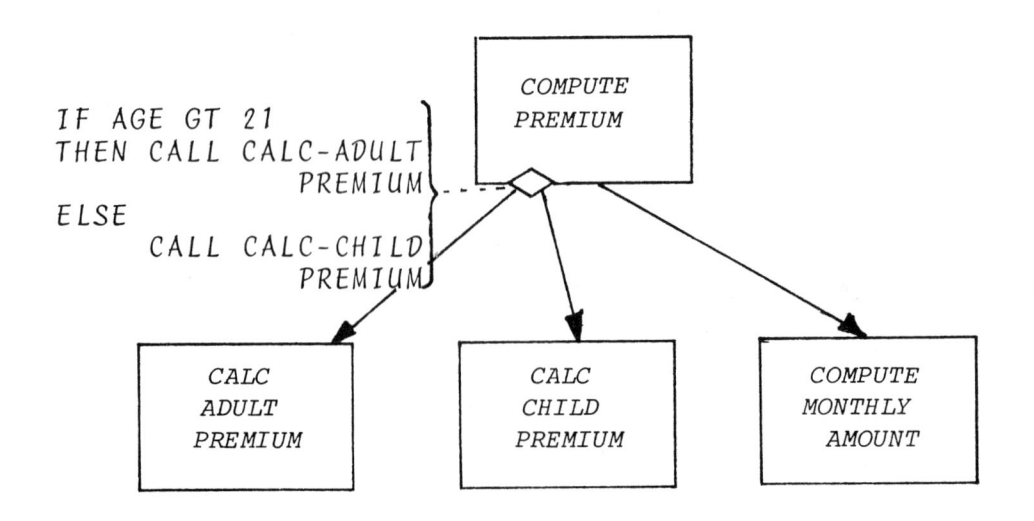

Figure 9.14

9.2 STRUCTURED DESIGN FOR CHANGEABILITY

The so-called "scope of effect" of the decision is the two modules below COMPUTE-PREMIUM; the decision has an effect on which one of them gets invoked. The so-called "scope of control" of a module is all of the modules that it calls, and all the modules called by them, and so on.

In Figure 9.14, the scope of control of COMPUTE-PREMIUM covers all three modules called by it. The scope of effect of the decision on age is thus within the scope of control of the module making the decision; this is as it should be. Just as no officer should issue orders to troops other than those under his command, no module should make a decision which affects modules outside its scope of control.

But suppose the design looked as shown in Figure 9.15.

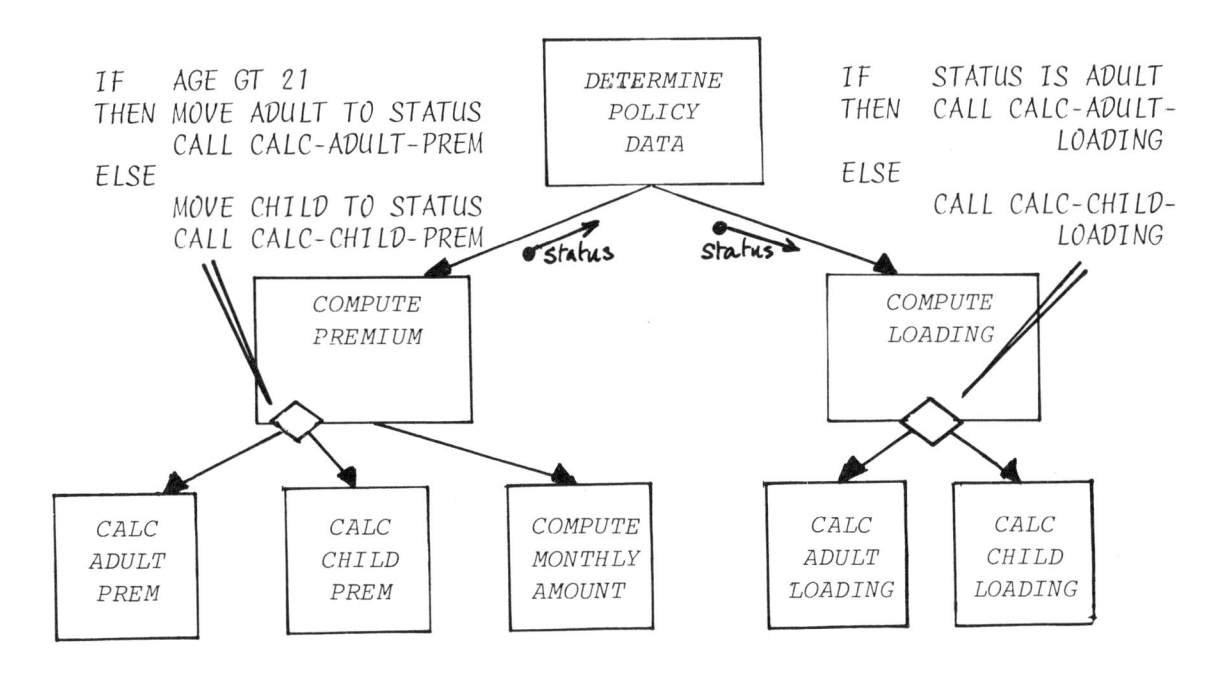

```
IF   AGE GT 21                                          IF   STATUS IS ADULT
THEN MOVE ADULT TO STATUS    DETERMINE                  THEN CALL CALC-ADULT-
     CALL CALC-ADULT-PREM    POLICY                          LOADING
ELSE                         DATA                       ELSE
     MOVE CHILD TO STATUS                                    CALL CALC-CHILD-
     CALL CALC-CHILD-PREM                                    LOADING
```

[Structure chart: DETERMINE POLICY DATA calls COMPUTE PREMIUM (passing status) and COMPUTE LOADING (passing status). COMPUTE PREMIUM has a decision diamond invoking CALC ADULT PREM, CALC CHILD PREM, and COMPUTE MONTHLY AMOUNT. COMPUTE LOADING has a decision diamond invoking CALC ADULT LOADING and CALC CHILD LOADING.]

Figure 9.15

Structure chart showing scope of effect outside scope of control

The small diamonds with invocation arrows coming out of them indicate that a decision is taken somewhere in the module, and which subordinate modules are invoked depends on that decision. We have spelled out the decision logic in each case.

301

9. DERIVING A STRUCTURED DESIGN FROM THE LOGICAL MODEL

Now the scope of effect of the original decision is far wider; it covers not only the modules called by COMPUTE-PREMIUM, but also both of those called by COMPUTE LOADING, since COMPUTE PREMIUM passes a control element, STATUS, to COMPUTE LOADING, which determines which lower module to call. The scope of effect of the decision no longer lies within the scope of control of the module containing the decision. This is undesirable, for several reasons; it creates control coupling between modules (via the STATUS control flag, in this case); it creates duplicate decisions, which may lead to maintenance problems; and it obscures the effect of the decision.

The structure should be redesigned to bring the scope of effect of the decision within the scope of the control of the module containing the decision, as shown in Figure 9.16.

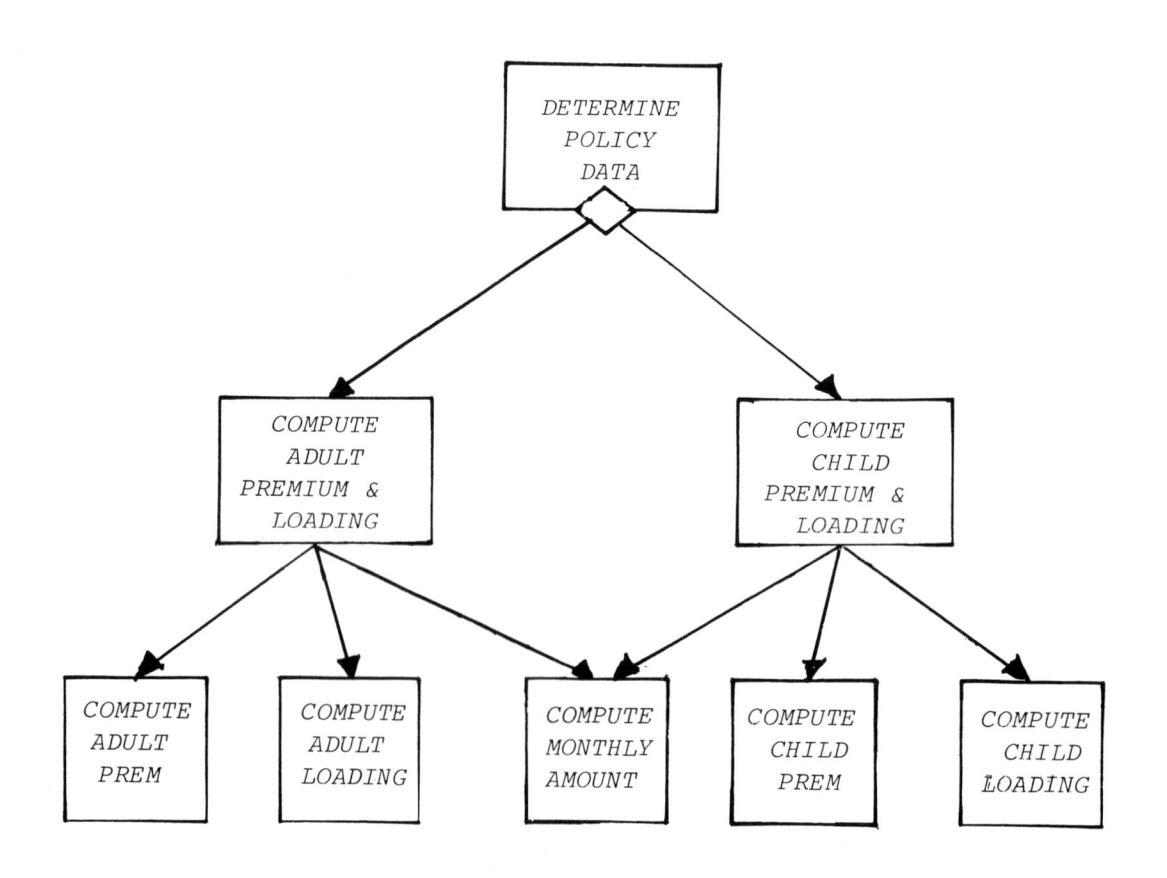

Figure 9.16

Structure chart showing scope of control within scope of effect

302

9.2 STRUCTURED DESIGN FOR CHANGEABILITY

In this modular structure, no control coupling is necessary, and the decision on age is made once only.

The designer should be on the look out for scope of control/scope of effect clashes as the design evolves. Their presence is often signalled by unnecessary control variables as we saw in this example.

We can summarize the main guidelines of Structured Design as in Figure 9.17:

GUIDELINES FOR PRODUCING CHANGEABLE MODULAR DESIGNS

- *Derive the structure from the data flow*

    - *transform-centered where input/output is homogeneous*
    - *transaction-centered where different types of data structures get processed differently*

- *Minimize inter-module coupling*

    - *data coupling as far as possible*
    - *control coupling as little as necessary*
    - *more severe coupling only under protest*

- *Maximize module cohesion*

    - *functional modules as far as possible*
    - *avoid logically and coincidentally cohesive modules at all costs*

- *Check for scope of control/scope of effect clashes*

Figure 9.17

9. DERIVING A STRUCTURED DESIGN FROM THE LOGICAL MODEL

9.3   THE TRADE-OFF BETWEEN CHANGEABILITY AND PERFORMANCE

Now that we have discussed the factors that make for a changeable design, we can consider the impact of changeability on performance, and the trade-off decisions that the designer has to make.

Many designers, encountering the techniques of Structured Design for the first time, fear that they will have a serious effect on the performance of any system designed for changeability. This concern mainly comes from two factors:

1. Designing a system as a set of small changeable modules, rather than a few large programs, may involve much more overhead caused by modules calling other modules.

2. Avoiding poorly cohesive modules may mean splitting one general purpose logically cohesive module, say, into several functional modules, resulting in programs requiring more memory.

In addition, though we have not discussed it specifically, if Structured Programming is used to code the modules it may lead to somewhat larger numbers of generated machine instructions than if more traditional methods of "tricky" coding for machine efficiency were used. (It is generally conceded that Structured Programming produces code which is easier to read and debug; studies indicate that structured code may sometimes be slightly larger and slower than traditional code [9.11], but sometimes smaller and faster! [9.12].)

These performance concerns cannot be swept under the carpet, although much of the worry about them is not justified. Indeed, the strange fact is that changeable designs can actually produce better performance than traditional designs. Consider the following issues:

1. We noted that one of the most important factors in performance of data-oriented system was unnecessary files, file passing and reading. Structured Design does not worsen these effects; on the contrary, the production of the hierarchy of logical modules encourages the designer to avoid intermediate files, and see how best to optimize file access.

9.3 THE TRADE-OFF BETWEEN CHANGEABILITY AND PERFORMANCE

2. We noted that the amount of generated code and the time spent in passing control between modules in main memory were of much less importance in data-oriented systems than file passing and accessing, except in the case where the calling of a module causes a library file read, or a paging operation in a virtual memory system. With this exception, producing a changeable design affects the least important of the performance factors.

3. It is difficult to know in advance where the performance bottlenecks in a system are going to be, especially in a complex operating environment. What we need to be able to do is to implement a system where the major performance issues have been resolved, exercise the system with realistic data and measure where the time is spent, whether in input/output, in computation, or in system overhead. Once we know *from measurement* where the bottlenecks are, we want to be able to change the system to remove the bottleneck. Typically the sort of changes that are needed to improve performance are re-writing of a time-consuming table search in assembly language, changing the overlay structure to ensure a frequently called section is always in main memory, or changing the details of file access. In the past, it has been difficult and expensive to do this "post-implementation optimization." Why? Because systems have been designed in such a way that they are hard to change without introducing a ripple-effect! The very fact that we are now designing our systems to be changeable, means that we can optimize their performance quickly and easily once we know what parts need improvement.

The trade-off between changeability and performance is thus a paradoxical one. If the designer worries too much about the fine details of the performance of the system during design, chances are that logical cohesion and severe coupling will be introduced into the modular structure through attempts to save memory and execution time. Those very features will make it difficult and expensive to improve performance of the system once it has been implemented.

So what is the moral? The designer needs to adopt a six step approach:

STEP 1 Derive the overall tentative physical design of **the** system and main files from the logical model with the

9. DERIVING A STRUCTURED DESIGN FROM THE LOGICAL MODEL

        least number of intermediate files, the fewest possible file passes, and as few seeks as possible to those remaining files.

STEP 2 Build in the necessary controls, trying to minimize any degrading effect on performance. (The data flow diagram is a useful tool to present to auditors for discussion of audit controls.)

STEP 3 Design the structure of logical modules for maximum changeability, paying no regard at this stage to the performance of the ultimate system (other than as dealt with in steps 1 and 2).

STEP 4 Study the logical modules in the hierarchy structure to estimate the size of the physical modules that will be needed to implement them. Note those modules that will be invoked frequently (on every transaction, or on every data element), and note those that will be invoked less frequently during execution, (e.g. error handling routines, initialization, or termination). Choose a physical "packaging" of the logical modules into program, subprograms, or sections, so that, as far as possible, frequently called modules will be in memory at the time they are called i.e. the number of seeks to program libraries will be minimized.

STEP 5 Implement the system as specified in Step 4 without further compromising the changeability of the design, except in cases where performance is so vital that it overrides the need for changeability. In these cases, compromise by combining modules and sharing resources, but do so to the least extent possible.

STEP 6 Exercise the system with representative data, and use an execution time monitor to measure which parts of the system take up the largest part of the execution time, or identify which modules are causing memory problems. Change those parts to make them faster or smaller as needs be. In a virtual memory environment rearrange the physical order of modules in the appropriate library to get the "working set" (those modules which are used frequently) all within the minimum number of pages. Consider fixing that working set in real memory if the operating system allows it.

Repeat Step 6 until the system meets its performance objectives.

The design of algorithmic-oriented systems can in principle *still* follow the steps outlined above. Particularly with very complex systems, the necessary testing and debugging will be aided by keeping the functions as isolated as possible until *all* the system functions can be demonstrated to work correctly, at which time optimization can begin.

9.4 AN EXAMPLE OF STRUCTURED DESIGN

In this section, we will apply the guidelines and techniques discussed earlier in the chapter, to the CBM Corporation's new system. We will assume that the methodology discussed in Chapter 8 has been followed, and that the users have opted for a medium budget alternative, in which orders for books are entered into the system via CRT terminals and validated on-line; files of shippable and non-shippable items are created during the day. Shipping notes are printed from these files overnight, along with invoices. Inventory levels are adjusted as orders are entered, but inventory control and reordering is handled by a separate subsystem which checks on stock available at the end of each day. Purchasing, accounts receivable, and accounts payable are separate batch subsystems. Some management reports are created in batch mode; query facilities are available into some of the files.

9.4.1 THE BOUNDARIES OF THE DESIGN

For the purposes of this exercise, we will design the order-entry subsystem whose boundary is defined in Figure 9.18, (which shows an extract from the overall data flow diagram). For simplicity, we will design an initial version which handles those orders which are not prepaid, i.e. those which ask for credit. As the data flow diagram shows, the entry of new customer details is outside the boundary of our design. In fact, when an order is detected as not being from an existing customer, or showing a change of address, it is routed to the order-entry supervisor who enters the details into the CUSTOMERS file, before returning the order to the system for normal processing. We thus have to design a system which handles the three processes:

9. *DERIVING A STRUCTURED DESIGN FROM THE LOGICAL MODEL*

    1. Edit orders (omitting prepayments for the time being)

    2. Verify credit is OK

    3. Verify inventory available and adjust stock level

Figure 9.19 (the four-page spread) shows the "explosions" of each of these processes, together with some detailed data flow in the manual process of taking the written or phone order and entering it into the CRT. These processes are numbered M1, M2, etc, since they are parts of a process which does not appear as such on the overall data flow diagram.

9.4.2 PHYSICAL FILE DESIGN CONSIDERATIONS

Before considering the building of the modular hierarchy, we will consider some of the factors in physical file specification, as they will clarify the DFD and the modular design.

1. DM/1: AUTHOR/TITLE-INDEX, in Process M - Manual Entry

    When we analyzed the data stores that described books, in Chapter 6, we noted that the unique identifier of a book was its ISBN. At the same time, customers usually order a book by specifying its title and author, with the publisher, where they know it. When we come to enter orders, we thus have an immediate access requirement to go from BOOK-TITLE or AUTHOR or both to find the ISBN. This could be done by providing secondary indexes into the BOOKS file and allowing access via the CRTs in the Order-Entry Department. The characteristics of the file would be:

    Number of records: About 1500 books on computers are known to CBM
    Number of accesses: Once per item ordered
    Volatility (rate of
    changes to the file): About 20 new books per month

    Where we see a small file with low volatility like this, requiring multiple accesses, we ask ourselves whether it is feasible to provide users with hard-copy of the file pre-sorted in various ways, rather than creating and maintaining software indexes. During the tentative design for the medium budget alternative, the designer decided to save money in this way.

9.4 AN EXAMPLE OF STRUCTURED DESIGN

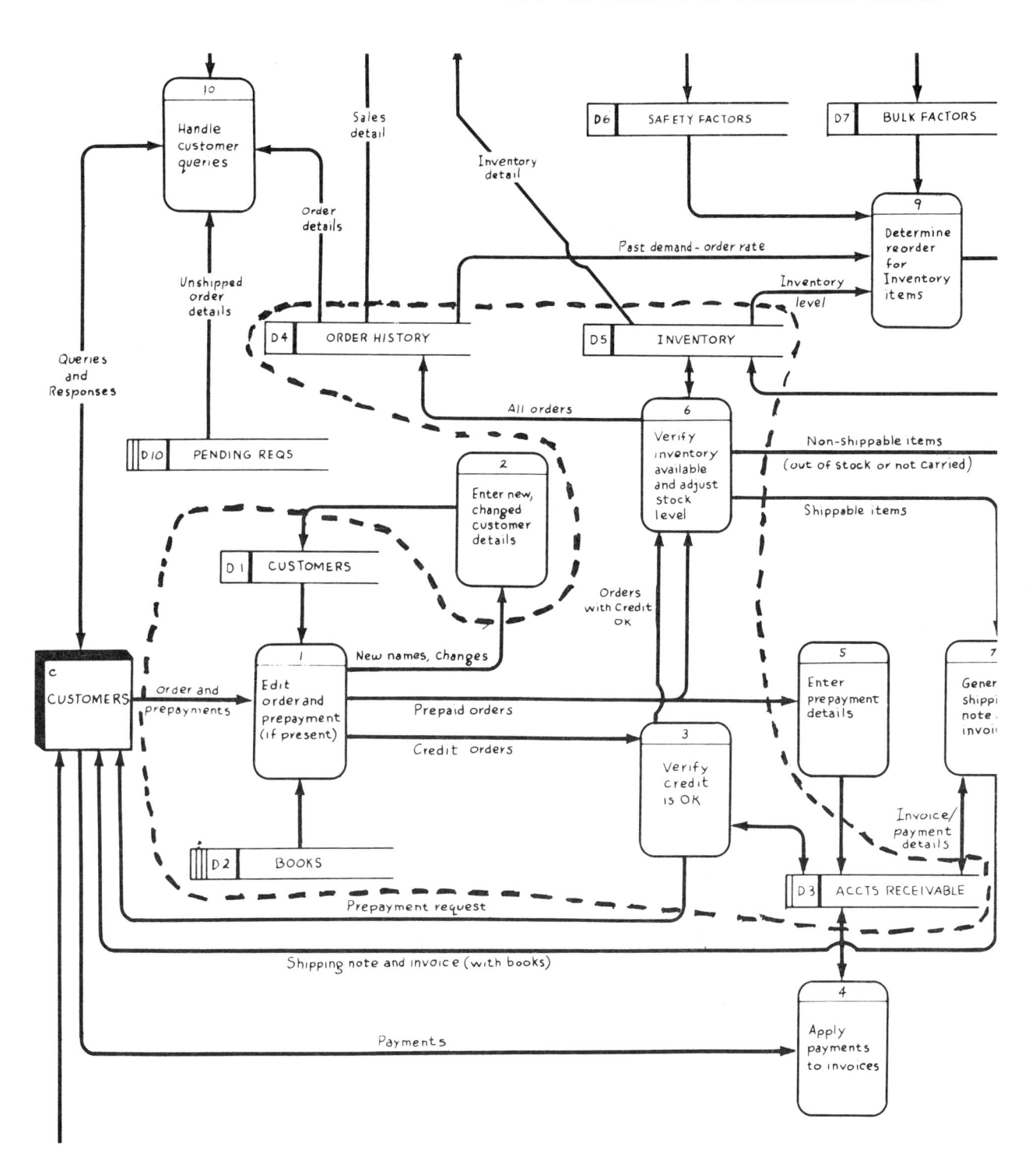

Figure 9.18

The order-entry subsystem

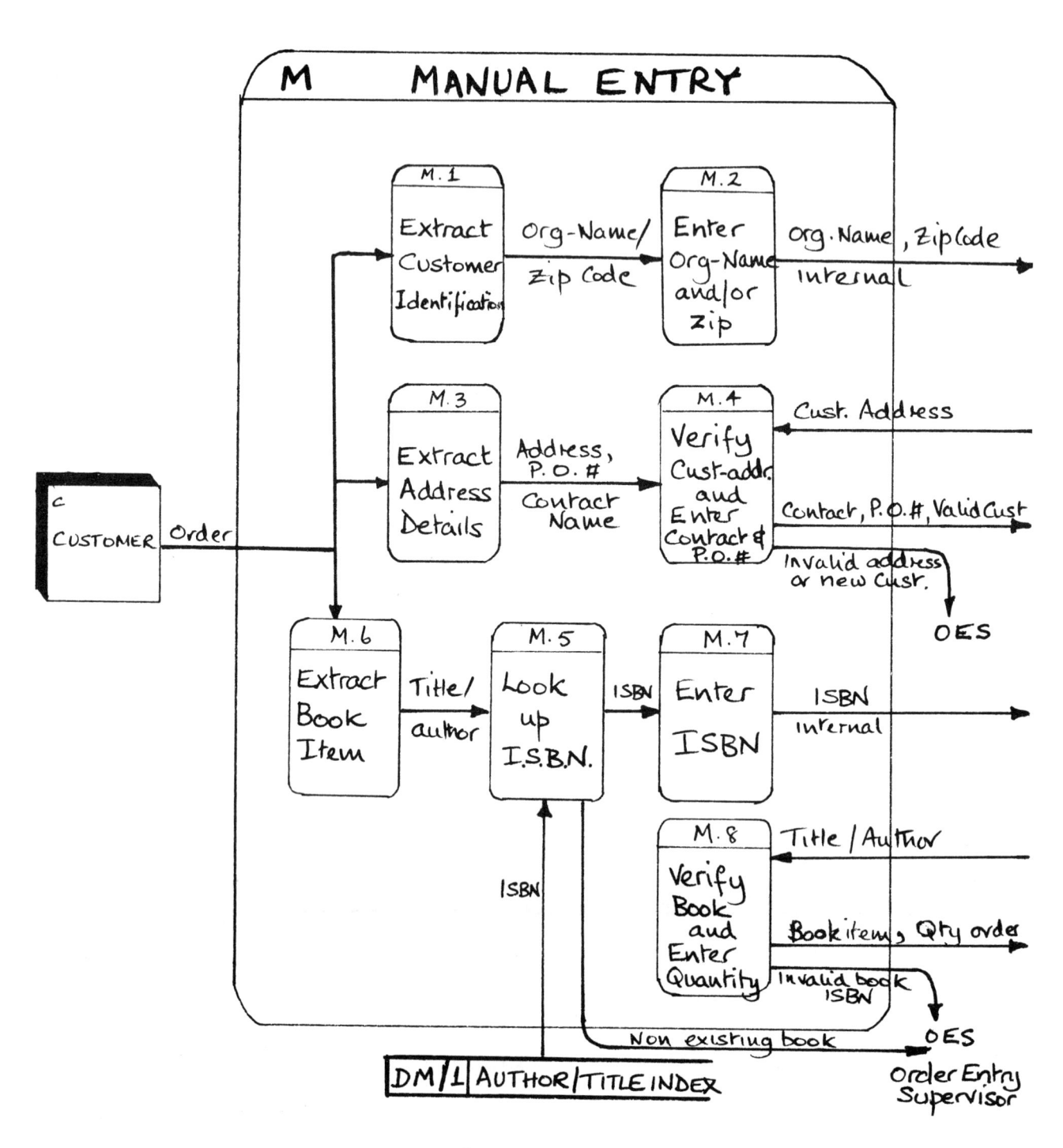

9. DERIVING A STRUCTURED DESIGN FROM THE LOGICAL MODEL

Figure 9.19 (part 1)

9.4 *AN EXAMPLE OF STRUCTURED DESIGN*

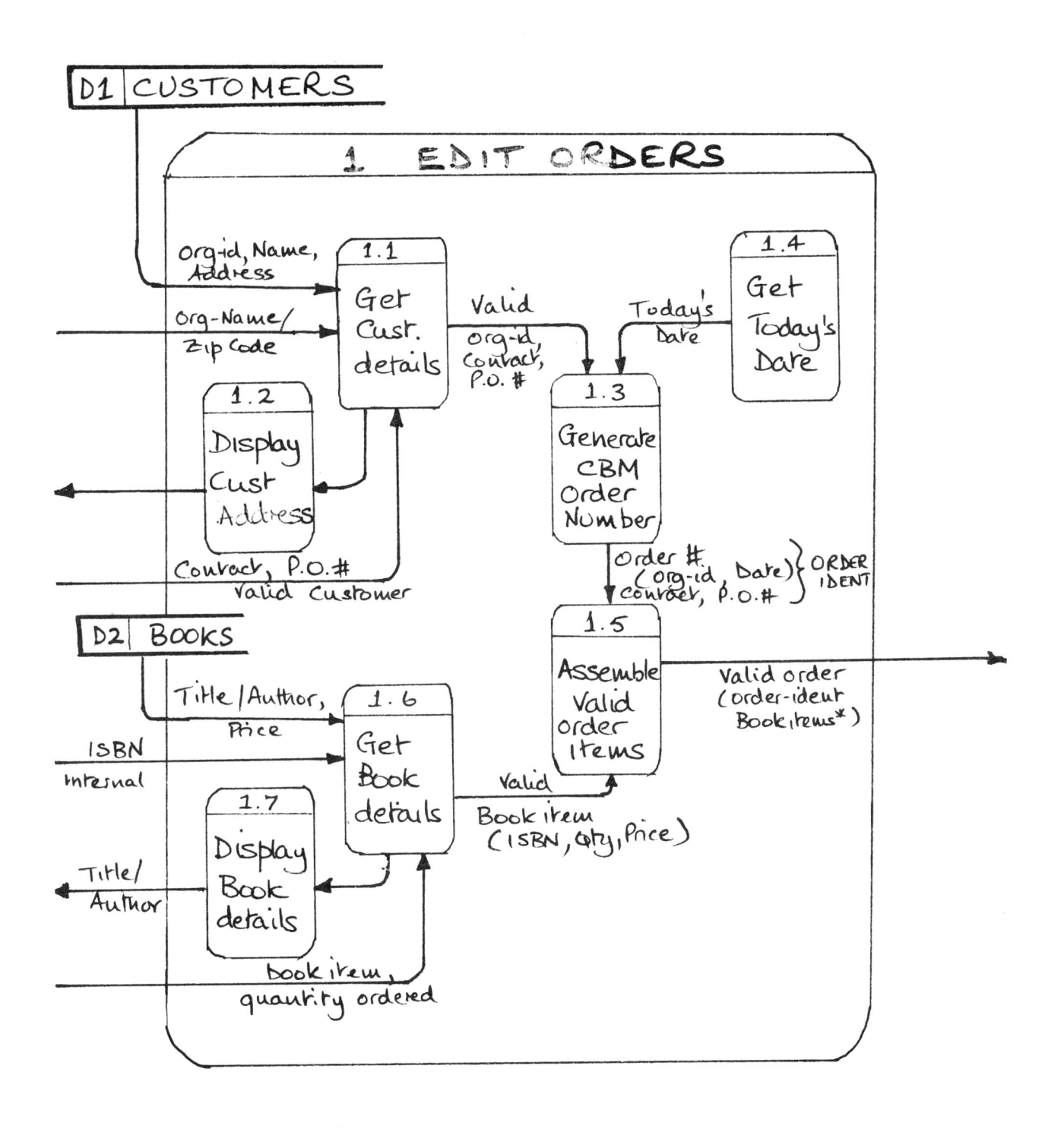

Figure 9.19 (part 2)

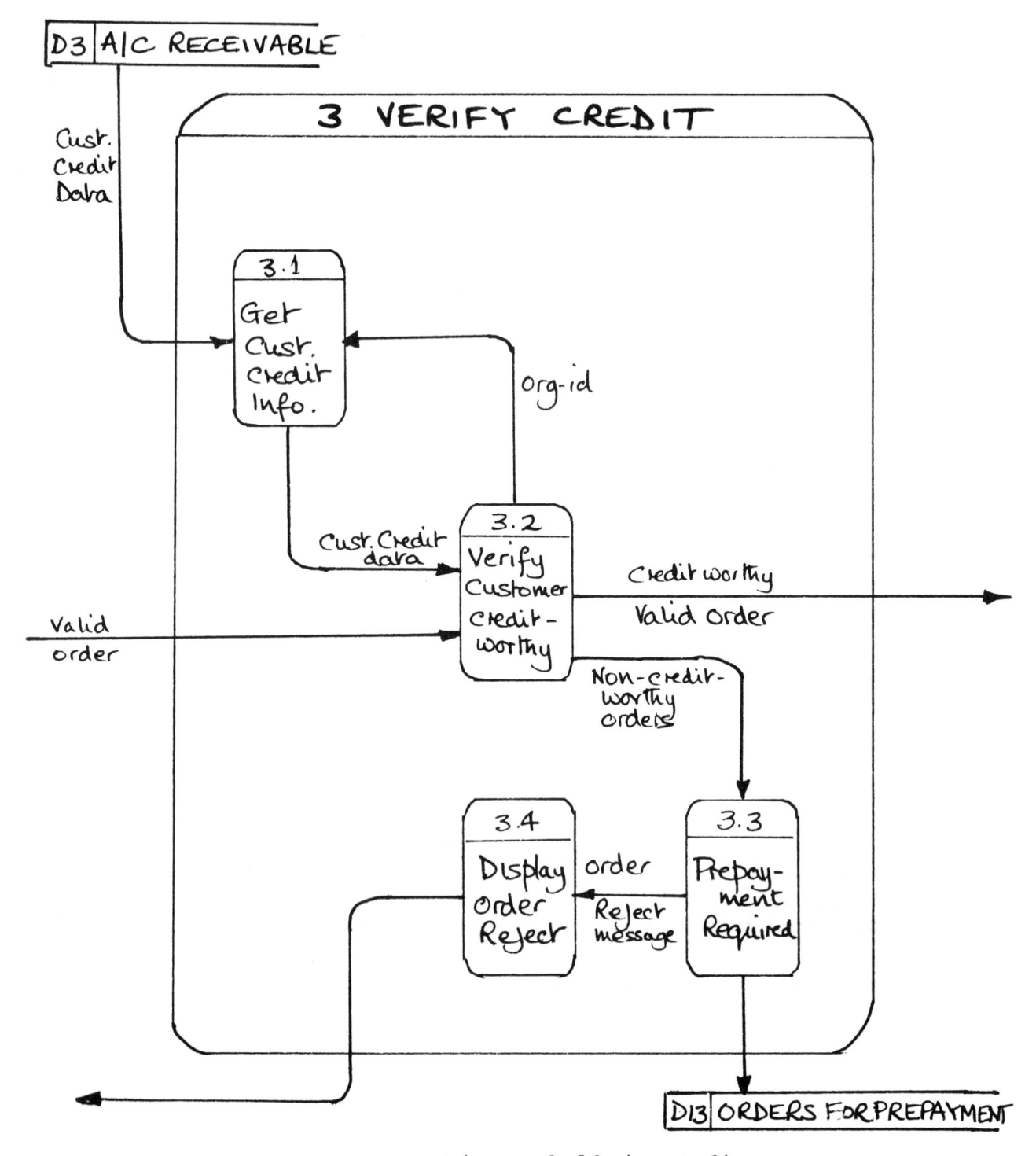

9. DERIVING A STRUCTURED DESIGN FROM THE LOGICAL MODEL

Figure 9.19 (part 3)

9.4 AN EXAMPLE OF STRUCTURED DESIGN

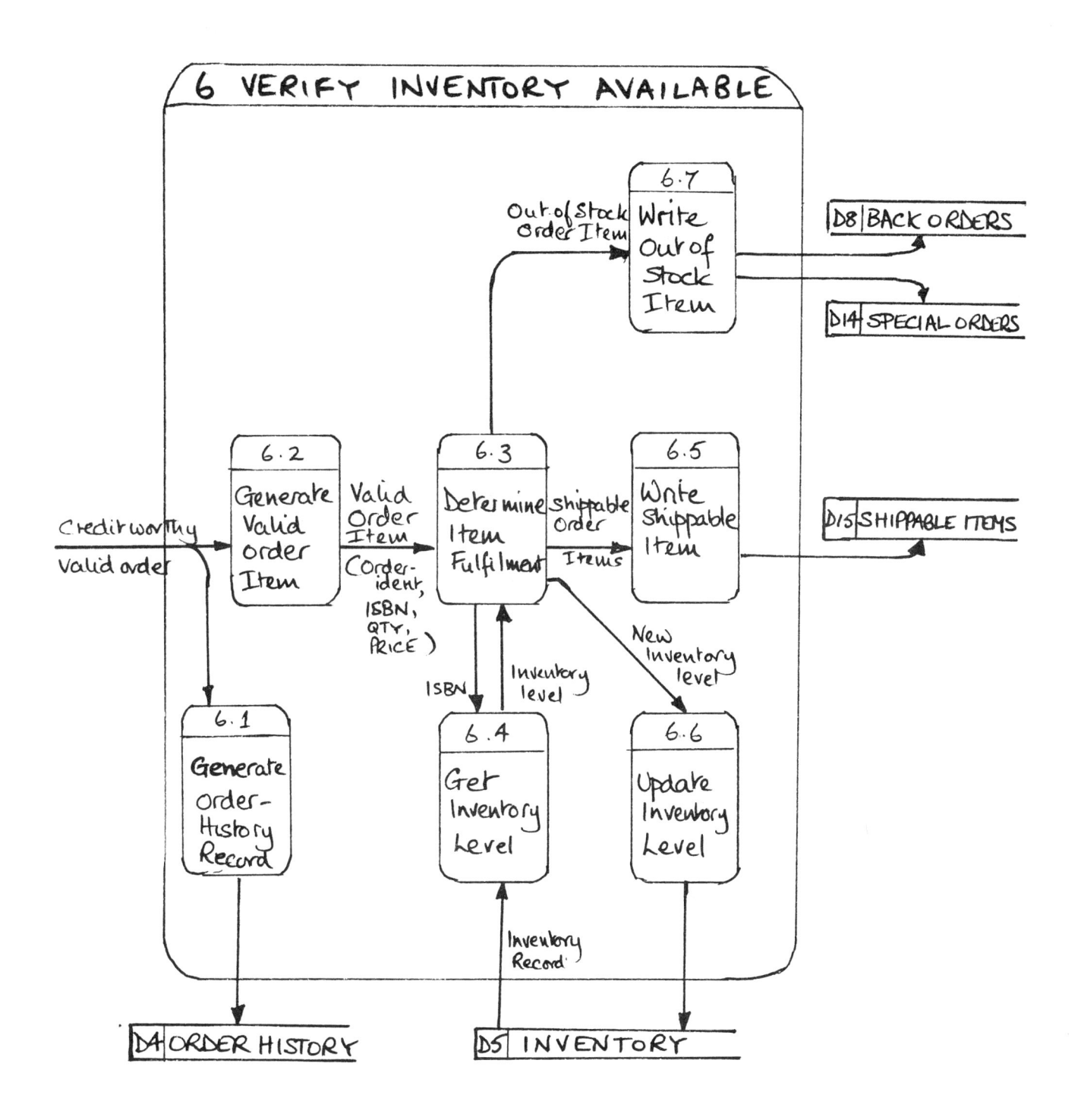

Figure 9.19 (part 4)

9. DERIVING A STRUCTURED DESIGN FROM THE LOGICAL MODEL

Each order-entry clerk is provided with a computer print-out listing all the known titles, with author, publisher and ISBN for each. Because of the various forms in which people remember book titles (this book might be referred to as "Structured Analysis," "Systems Analysis tools," "Structured Systems Techniques," and so on), the listing is provided as a Keyword - In - Context (KWIC) index, in which the title appears once for every important word in it, and the list of titles is sorted by each word. An extract from the S section of the KWIC-index looks like this:

```
                              .
                              .
                      SYSTEMS    ANALYSIS AND DESIGN USING NETWORK TECHNIQUES
                                         Whitehouse        ISBN..........
         STRUCTURED    SYSTEMS    ANALYSIS: tools and techniques
                                         Gane, Sarson      ISBN..........
     INTRODUCTION TO   SYSTEMS    SAFETY ENGINEERING
                                         Rodgers           ISBN..........
INTRODUCTION TO GENERAL SYSTEMS   THINKING
                                         Weinberg          ISBN..........
                              .
                              .
```

The Whitehouse book will also appear under ANALYSIS, DESIGN, NETWORK, and TECHNIQUES; the Weinberg book will also appear under GENERAL and THINKING, and so on. Thus, to locate any book, the order-entry clerk can pick the most significant word in the title, look that up in the KWIC index, and scan through all the books which have that word anywhere in the title, until the correct one is located. The KWIC index provides some help in advising customers as to the books on any given topic.

If the customer specifies the author, but is vague about the title, or the title cannot be found in the KWIC index, the order-entry clerk can use a second print-out which lists books in AUTHOR sequence.

The print-outs are produced once a month (from the BOOKS file), with the current date clearly printed on them; each week, any additions or changes which may have taken place are circulated as update sheets, to be added to the front of each print-out.

2. *D2: BOOKS, accessed by Process 1 - Edit Order*

Once the ISBN of each book has been determined and entered into the system, the details of the book need to be displayed to the order-entry clerk as a double check that the correct ISBN is in use. The BOOKS file holds the same information as the TITLE/AUTHOR index, with the addition of the current price for each book. It is a small file; 1,500 records averaging 100 characters per record, or 150,000 characters. Since, as we discovered in Chapter 6, both BOOKS and INVENTORY are organized by ISBN, the designer considered combining them and having just one physical file, but decided against it, since of the total of 1,500 or so titles, only 100 - 200 will be held in inventory. Also, the BOOKS file, which will be maintained overnight, will only be read during the day, not updated. Consequently, there will be no need to take back-up copies for security purposes. The INVENTORY data will be continually being updated and so will need to be copied for security at frequent intervals, since it would be very inconvenient to lose track of what is in stock. As a minor performance issue, it might be advantageous to put BOOKS and INVENTORY on separate disk drives, allocating the rest of each drive to low activity files such as BACK-ORDERS, SPECIAL-ORDERS, or ORDERS-REQUIRING PAYMENT. In each case the highly active file is small, and will take up only a few tracks. The read-write head will thus spend most of its time positioned over the active file, reducing seek time to a minimum.

If the management of CBM are successful in attracting 1,000 orders a day, with each calling for, say, 3 titles, and if many of the orders come in by phone, there could be activity during the morning peak period (10.30 A.M.-11.30 A.M.) of 700 - 800 items requiring an access to BOOKS and an access to INVENTORY. This amounts to one access every 4 seconds in that peak hour, which while not taxing, is not negligible.

9. DERIVING A STRUCTURED DESIGN FROM THE LOGICAL MODEL

3. *D1: CUSTOMERS, accessed by Process 1 - Edit Orders*

   We noted in Chapter 6 that the logical data store CUSTOMERS consisted of records describing both organizations and people within organizations. The unique definition of an organization is the key ORG-ID, consisting of five digits specifying the organization, plus two digits specifying a particular location, plus a check digit. There will be 10,000 - 20,000 customers of CBM in the new system, counting each location as a separate customer, with potential multiple contacts at each location. The designer has decided in this medium-budget system, not to store the contact, but to have the name of the person placing the order (and the customer's order number where it exists) entered with the details of the order. The data structure of the CUSTOMERS file will thus be:

   ```
           ORG-ID
               ORG-CODE
               LOCATION-CODE
               CHECK-DIGIT
           ORGANIZATION-NAME
           ORGANIZATION-ADDRESS
               STREET
               CITY
               STATE
               ZIP-CODE
           MAIN-PHONE
   ```

   We note that this decision means that it will not be possible to retrieve the phones of individual CONTACTS; should we need to call them we will have to go through their switchboard. The ORGANIZATION records require an average of 150 characters; each CONTACT record requires an average of 60 characters, and there are on average 1.8 contact per location. The designer has thus reduced the on-line disk requirement from about 5 million to 3 million characters (20,000 organizations x 150 = 3 million characters; 20,000 x 1.8 x 60 contacts = 2.16 million characters). A small saving for a small penalty; such is the stuff of physical design.

9.4 AN EXAMPLE OF STRUCTURED DESIGN

Most customers will not know their ORG-ID, and most orders will not carry it. (With each shipment, the analyst plans to send out order forms pre-printed with the customer's ORG-ID, but these will represent a minority of the orders received.) How is the order-entry clerk to find the ORG-ID for each customer, so that an order can be processed? We could take the approach that was just described for finding the ISBN, and give each clerk a regularly updated listing of CUSTOMERS. While this is feasible for 1,500 books, it becomes unwieldy for 20,000 customers. Consequently, the designer has decided to provide a secondary index capability into the CUSTOMERS file by ORGANIZATION-NAME and/or ZIP-CODE, as shown in the DIAD of Figure 9.20.

```
┌──────────┐       ┌──────────────┐       ┌──────────┐
│          │       │  CUSTOMERS   │       │          │
│ ORG-NAME │──────▶│ ──────────── │◀──────│ ZIP-CODE │
│          │       │  Org-id      │       │          │
└──────────┘       └──────────────┘       └──────────┘
                   ┌ ─ ─ ─ ─ ─ ─ ─┐
                   │ Org-name     │
                   ├ ─ ─ ─ ─ ─ ─ ─┤
                   │ Org-address  │
                   ├ ─ ─ ─ ─ ─ ─ ─┤
                   │ Main-phone   │
                   └ ─ ─ ─ ─ ─ ─ ─┘
```

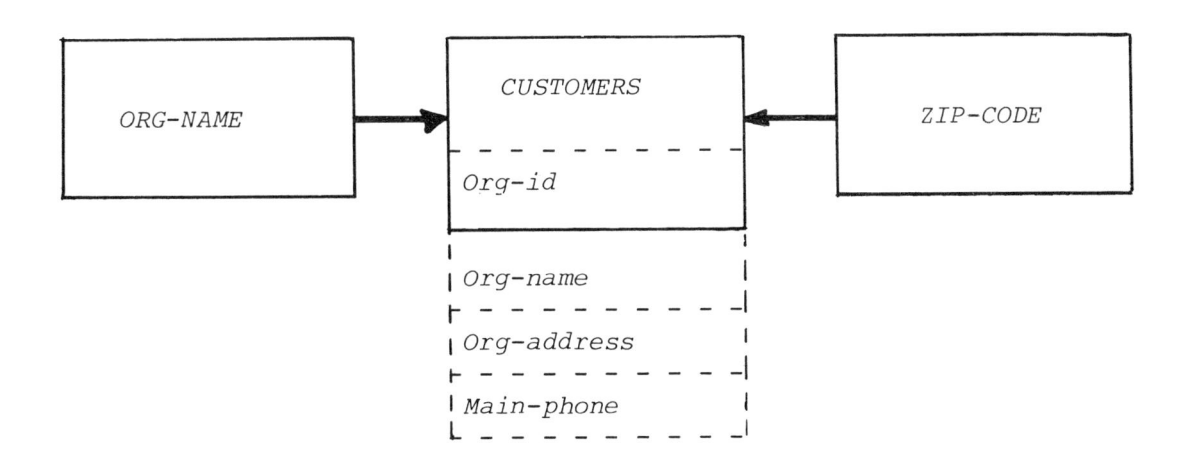

Figure 9.20

This means that the immediate accesses outlined in Figure 9.21 are possible for the order-entry clerk. Typically, the clerk will key in the name of the organization as it appears on the order, using certain standard abbreviations, plus the zip-code of the address on the order, and will retrieve the full organization record

9. DERIVING A STRUCTURED DESIGN FROM THE LOGICAL MODEL

ENTER	SYSTEM DISPLAYS
ZIP-CODE	All organization with addresses in that zip-code
ORGANIZATION-NAME	All organizations with that name
ORGANIZATION-NAME and ZIP-CODE	Only organizations with that name in that zip-code

Figure 9.21

showing the ORG-ID. Provided the address shown on the screen matches the address on the order, the details of the order can be entered. On the data flow diagram of Figure 9.19, the corresponding processes are M2, 1.1, 1.2, and M4.

4. *D3: ACCOUNTS RECEIVABLE, accessed by Process 3 - Verify Credit*

Though the static portion of ACCOUNTS RECEIVABLE could be combined with the CUSTOMERS record on logical grounds, the auditors of CBM require that accounting records be kept separate from other files, and only accessed from terminals in the Accounts Department. Management requires that 6 months history of invoices and payments be kept on-line; every 6 months the records for the period 7 months ago are to be written to tape and stored in the vault. Consequently, ACCOUNTS RECEIVABLE will be set up as a separate physical file from CUSTOMERS. (We note that a data base management system with proper security features, would enable us to satisfy the auditors, and still combine the data of CUSTOMERS with ACCOUNTS RECEIVABLE.)

ACCOUNTS RECEIVABLE will be a somewhat larger file. There will be 20,000 accounts with master information for each ORG-ID. We must allow for 1,000 orders a day, which creates 1.2 invoices on average (some orders will be filled in two or three parts with each part being invoiced separately), which will result in a maximum of 1,000 orders x 120 days (6 months) x 1.2 invoices, or

9.4 AN EXAMPLE OF STRUCTURED DESIGN

144,000 invoices on file at any one time. Some payments will cover more than one invoice; it is reasonable to expect about 100,000 payments. Allowing 100 characters for each header record and 30 characters for each invoice payment record, we get a file size of:

```
20,000 headers  x 100   = 2 million
144,000 invoices x 30   = 4.3 million
100,000 payments x 30   = 3 million
                          _____
TOTAL CHARACTERS:         9.3 million
```

The file will be accessed by the order-entry subsystem once for each valid order, and of course, by other subsystems concerned with invoices and payments.

5. *D15: SHIPPABLE ITEMS, created by Process 6 - Verify Inventory Available*

The main objective of the whole order-entry subsystem can be seen as that of producing this file, which has a record for each item that is valid, part of an order from a creditworthy customer, and in stock according to the computer records. This file will be used as the main input to the shipping subsystem, which will create shipping notes based on the actual quantities shipped from stock, and feed the invoicing subsystem with information on the fulfilment of each order.

The data structure of the file is:

```
ORDER-IDENTIFICATION
   ORDER-NUMBER
     ORG-ID
     ORDER-DATE (as received by the system)
   CONTACT-NAME
  [CUSTOMER-PURCHASE-NUMBER] entered only if
ISBN                                    available
QUANTITY-ORDERED
QUANTITY-SHIPPABLE
```

We should recognize that this is actually an intermediate file, created because we have broken the whole system into subsystems.

9. DERIVING A STRUCTURED DESIGN FROM THE LOGICAL MODEL

6. Other files created by the subsystem

> D13:ORDERS REQUIRING PREPAYMENT holds those orders from customers who are not creditworthy according to the policy logic in Process 3.2. It contains details of the orders they have placed, and is input to a subsystem which displays each order, with the customers credit history to the Accounts Department, for them to make a final determination as to whether credit should be extended or not.
>
> D8:BACKORDERS and D14:SPECIAL ORDERS have the same structure as SHIPPABLE ITEMS but reflect cases where the item is out-of-stock or for one of the 1,300 titles not held in inventory.
>
> D8:ORDER HISTORY contains a complete record of each valid order, whether shippable or not. It is used as the basis for analysis of demand for books, and in management reporting. It will be used for customer inquiries when that subsystem is implemented.

9.4.3 LOCATING THE CENTRAL TRANSFORM ON THE DATA FLOW DIAGRAM

Reviewing the data flow diagram of Figure 9.19 in the light of the physical files, it is at first sight hard to see how the flow corresponds to the simple model of input-transform-output that we described in Section 9.2.

The manual subsystem breaks each order into component book items, for entry and editing against the BOOKS file, Process 1.5 recombines the items into a valid order (with prices attached to each item), so that the total amount of the order can be calculated, and used in deciding whether this much credit can be extended to this customer. (When the design is extended to handle prepayments, we will need this total amount for each order, to check that the right amount of money has been sent with the order.) Once credit has been established, the order has to be broken into items again by Process 6.2, for checking against inventory.

Where does the main output stream, SHIPPABLE ITEMS first become visible as output? In the output data flow "Shippable Items," from Process 6.3. From this train of reasoning we come to the view that Process 6.3 represents the central func-

9.4 AN EXAMPLE OF STRUCTURED DESIGN

tion of the system, and the points of highest abstraction of input and output are on either side of it, as shown in Figure 9.22.

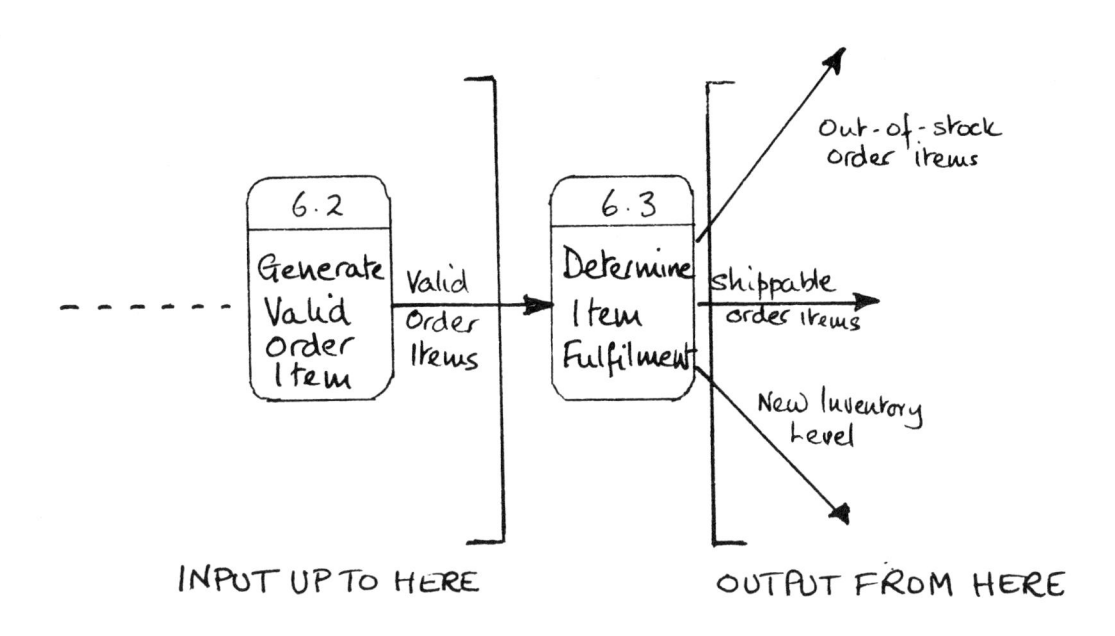

Figure 9.22
The central process of the order-entry subsystem

This may seem a little strange, since it implies that the majority of the system is in the input leg, with very little on the output side. Yet isn't that what one would expect on a system which does order entry? We now know that we can derive a top-level structure of the modular hierarchy, by setting up a module which calls for a valid order item, relates it to inventory, and writes records appropriately. Further, we see that we have at the top of the hierarchy a transaction center, rather than a transform center. The data flow splits into three different paths depending on the inventory situation. Figure 9.23 shows such a top-level structure:

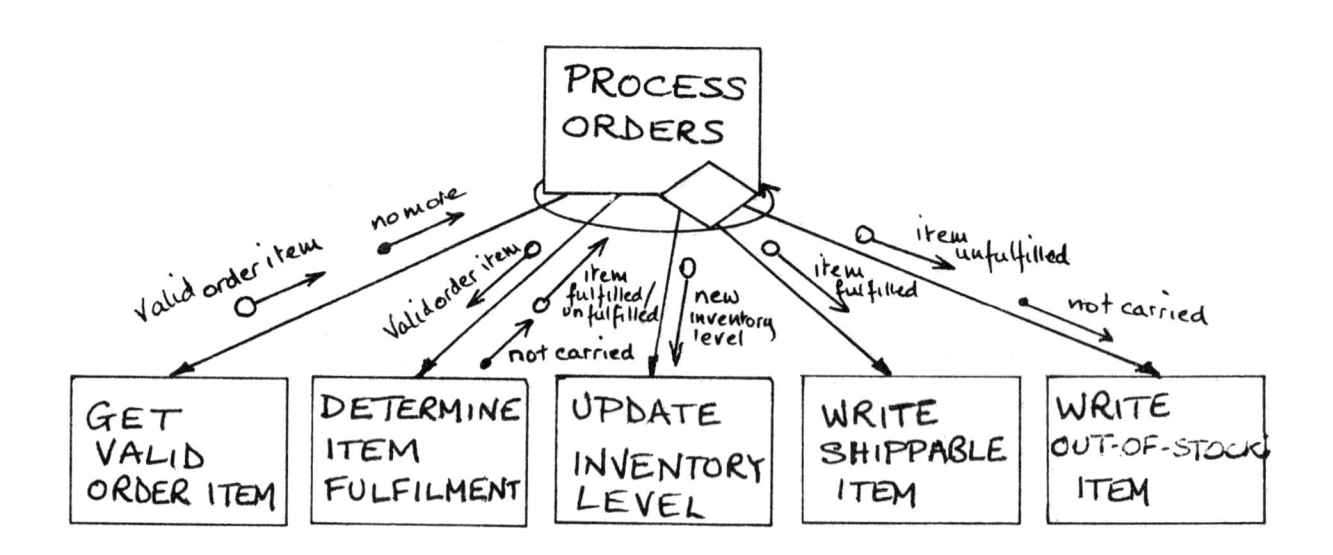

9. DERIVING A STRUCTURED DESIGN FROM THE LOGICAL MODEL

Figure 9.23

9.4.4 REFINING THE DESIGN FROM THE TOP DOWN

The module DETERMINE ITEM FULFILMENT plays the "analyzer" role, determining what type of transaction is being handled. The dispatcher function is contained within the main module PROCESS ORDERS, as shown by the decision symbol, calling modules which handle the case where the item is either shippable or out-of-stock or both (in the case where there are only enough of the items to fill part of the order). The curving arrow, crossing all the invocation arrows, indicates that a loop exists within PROCESS ORDERS, which invokes all of the lower modules each time an item is processed. UPDATE INVENTORY will not be called if the item is one that is not carried, or if the inventory is already zero (out-of-stock).

The next draft refinement of this design, as shown in Figure 9.24, specifies the functions of the output leg in more detail. We create a dispatcher module which invokes a separate module to deal with each of three circumstances: either

9.4 AN EXAMPLE OF STRUCTURED DESIGN

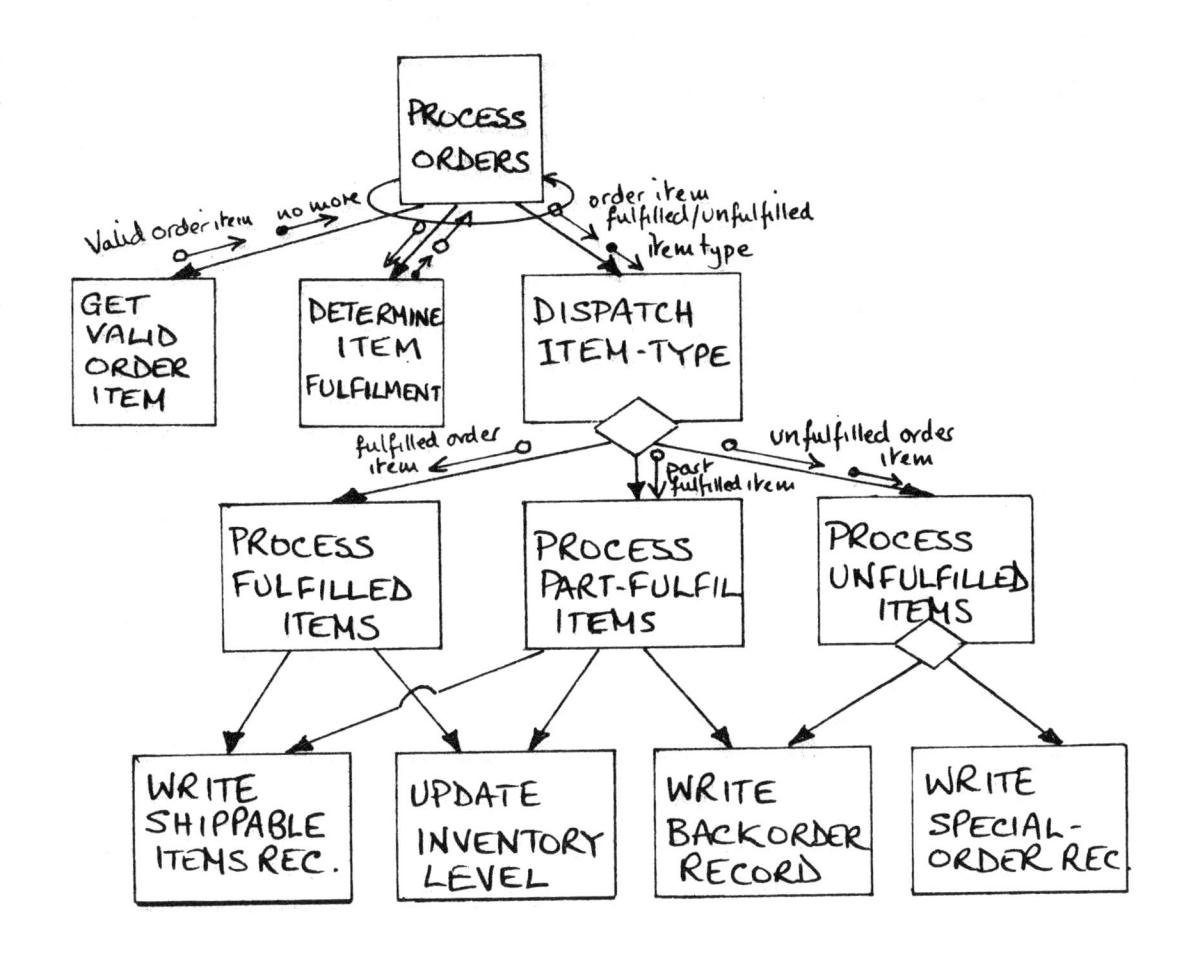

Figure 9.24

an order item can be fulfilled completely (the most frequent situation) or it can only be part fulfilled, or it cannot be fulfilled at all. The dispatcher is kept very simple--indeed trivial--because it just has to call the appropriate module; the logic for determining the various actions is built into those modules which are called. Indeed the dispatcher is so simple, that though we show it as a separate module on the structure chart, we can make up our minds that it will be implemented as part of the code for PROCESS ORDERS. The term for this is *"lexical inclusion"* of one module within another; two logical modules exist within one physical module. This is shown on the structure chart by a flat triangle on top of the module, as shown in Figure 9.25.

9. DERIVING A STRUCTURED DESIGN FROM THE LOGICAL MODEL

In Figure 9.25, the analyser has been further broken down (or "factored"), into a module which gets the inventory level for the item being processed, and a second module, DETERMINE QUANTITY FULFILLED/UNFULFILLED, which is called only if the item is one ordinarily held in stock.

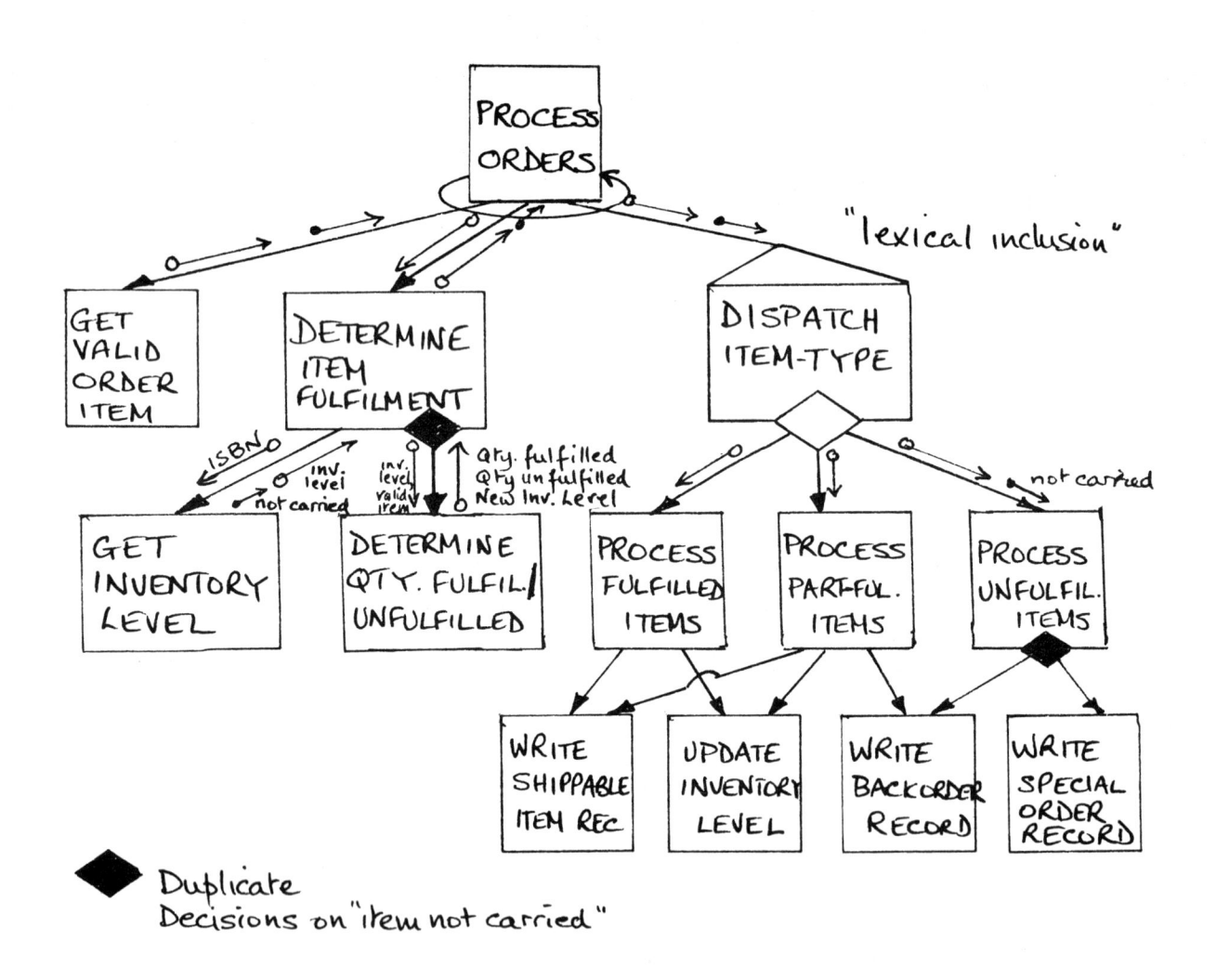

Figure 9.25

Now we perceive that we have a scope of control/scope of effect conflict on our hands. A decision is made in DETERMINE ITEM FULFILMENT as to whether the item is one carried in inventory or not (based on GET INVENTORY LEVEL finding no inventory record). But this same decision is made again in

9.4 AN EXAMPLE OF STRUCTURED DESIGN

PROCESS UNFULFIL ITEMS, in deciding whether to call WRITE BACK ORDER RECORD or WRITE SPECIAL ORDER RECORD.

How can we resolve this conflict, and remove the duplicate decision making? We could process special orders as part of DETERMINE ITEM FULFILMENT, but that would confuse the function of the module, creating a procedurally cohesive module "DETERMINE ITEM FULFILMENT AND WHEN THE ITEM IS A SPECIAL ORDER, CREATE THE SPECIAL ORDER." A better approach is to have the main control module handle the decision once only, calling a module which handles only special orders. This solution is shown in Figure 9.26, in which we have also incorporated the "analyzer" function and the DETERMINE QTY FULFILMENT function back up into PROCESS ORDERS, since we can see that they are both trivial.

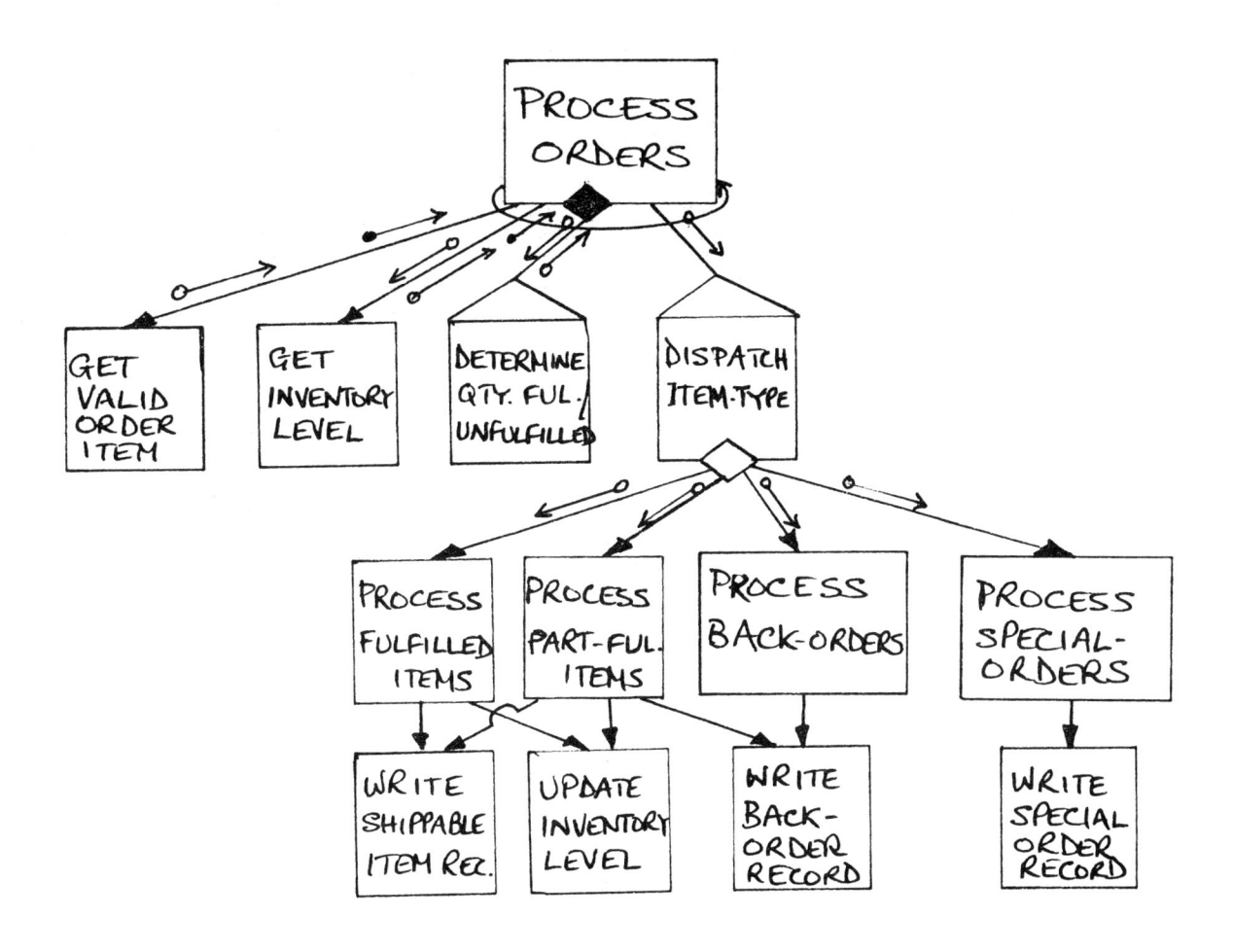

Figure 9.26

9. DERIVING A STRUCTURED DESIGN FROM THE LOGICAL MODEL

Have we made the main control module too complicated? It invokes six other modules--its so-called "span of control"--so we want to be careful that we don't build too much complexity into it. Let us express the logic we would need in PROCESS ORDERS using pseudocode. Figure 9.27 shows the result.

```
DO-WHILE    MORE INPUT

    GET VALID ORDER ITEM
    GET INVENTORY-LEVEL
    IF       NO INVENTORY RECORD
      THEN   DO  SPECIAL-ORDER-ROUTINE
    ELSE     (STOCK CARRIED)
      SO  IF       INVENTORY-LEVEL IS ZERO
            THEN   DO BACK-ORDER-ROUTINE
          ELSE     (STOCK IS HELD)
            SO     SUBTRACT ORDER-QTY FROM INVENTORY-LEVEL
                        GIVING NEW-INVENTORY-LEVEL
                   IF       NEW-INVENTORY-LEVEL GE ZERO
                     THEN   DO FULFILLED-ORDER-ROUTINE
                   ELSE     (PART ORDER FULFILMENT)
                     SO     SUBTRACT INVENTORY-LEVEL FROM ORDER-QTY
                                GIVING QTY-UNFILLED
                            DO PART-FULFILLED-ORDER-ROUTINE
END-DO
```

Figure 9.27
Pseudocode for PROCESS ORDERS

So far, we have been concerned mainly with the output leg of the system. It is time to turn to the more extensive input leg--everything subordinate to GET VALID ORDER ITEM--and design that. Figure 9.28 shows the first step; GET VALID ORDER ITEM has been factored into GET VALID ORDER and ISOLATE NEXT ITEM. When the last item on each order has been processed, ISOLATE NEXT ITEM will pass a "No more items" flag up to GET VALID ORDER ITEM, whereupon it will call for the next order, and then call ISOLATE NEXT ITEM again.

9.4 *AN EXAMPLE OF STRUCTURED DESIGN*

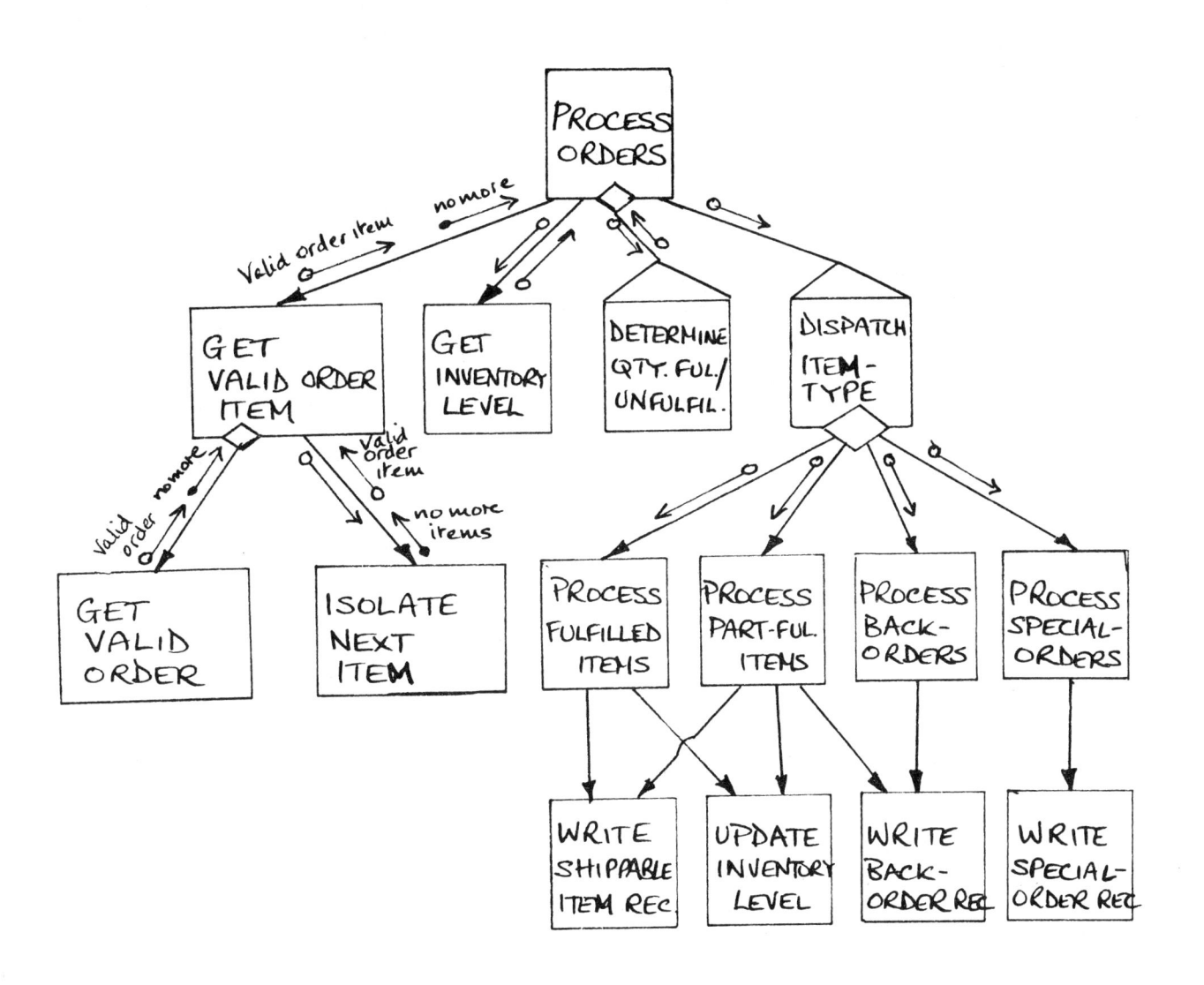

Figure 9.28

9. DERIVING A STRUCTURED DESIGN FROM THE LOGICAL MODEL

What is involved in GET VALID ORDER? We can see from the data flow diagram that several functions are involved; they will come under the control of GET VALID ORDER as shown in Figure 9.29, a subset of the structure.

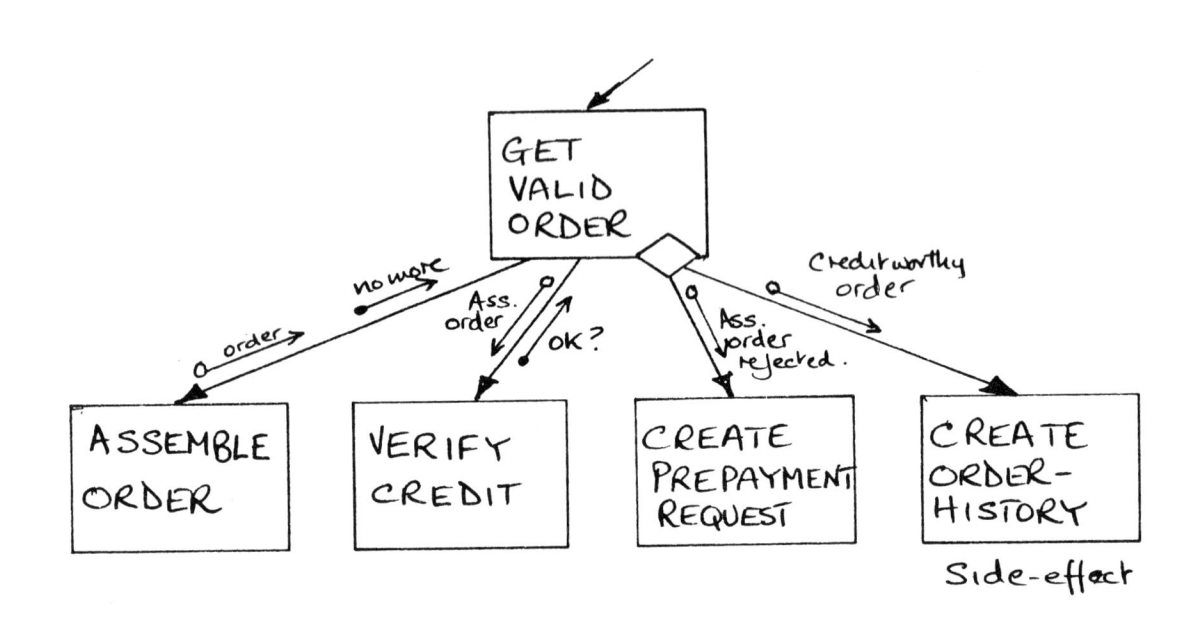

Figure 9.29

We see that CREATE ORDER HISTORY is one of these functions, being a subsidiary output that just happens to come out of the input stream at this point. Strictly speaking, CREATE ORDER HISTORY is nothing to do with GET VALID ORDER, or with GET VALID ORDER ITEM, for that matter. It is an example of a "side effect," something which a module is required to do "on the side" from its proper function. A maintenance programmer looking at the code for GET VALID ORDER ITEM would not know from the name of the module that the order history was created as part of one of its subfunctions. Can we redesign the structure so that CREATE ORDER HISTORY falls into a more

9.4 AN EXAMPLE OF STRUCTURED DESIGN

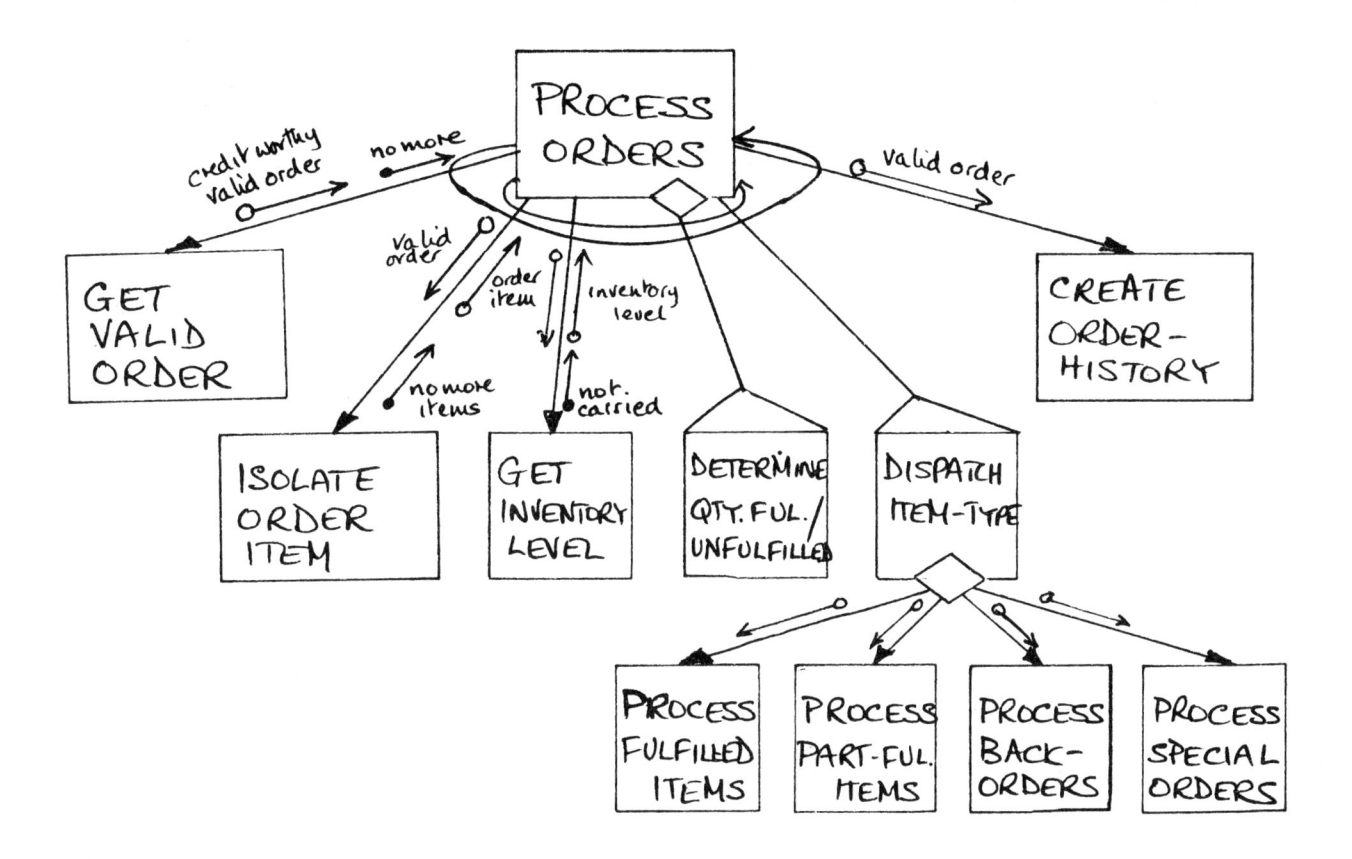

Figure 9.30
An attempt to resolve the side effect caused by CREATE ORDER HISTORY

natural and comprehensible place? We could move the logic of GET VALID ORDER ITEM up into PROCESS ORDERS and have CREATE ORDER HISTORY invoked from there, but that would involve two nested loops, as shown in Figure 9.30, one loop to handle orders, and one loop to handle items. This is an unduly complex design; even so, simplifying it by producing a module which manages the processing of items as shown in Figure 9.31, implies a delegation of the real work of managing the system, for the sake of handling what, after all, is merely a logging function.

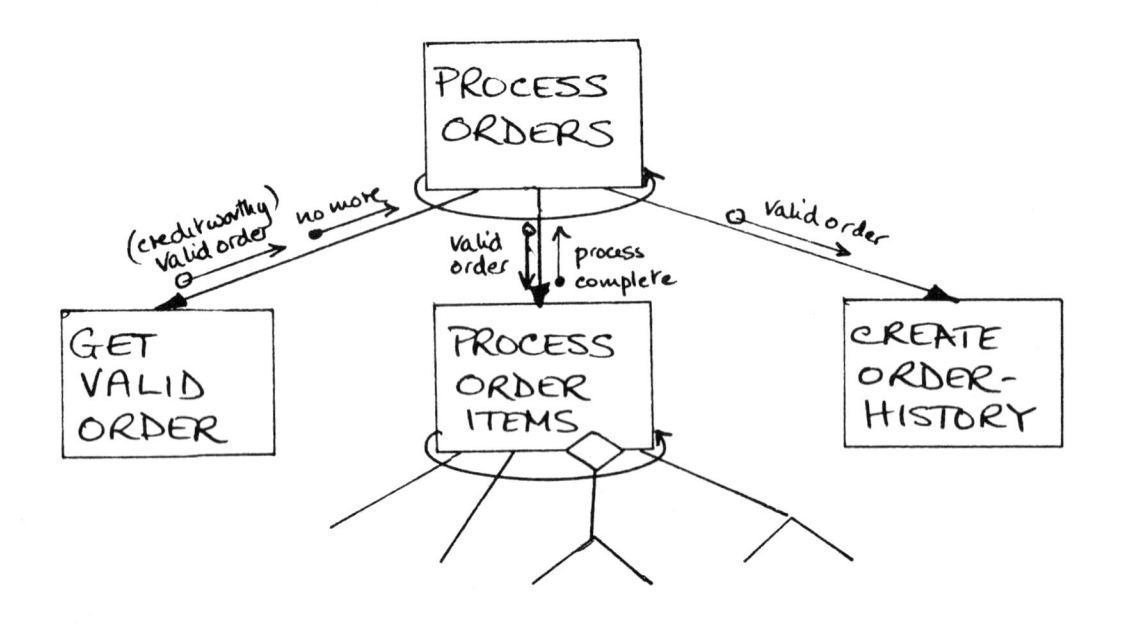

9. DERIVING A STRUCTURED DESIGN FROM THE LOGICAL MODEL

Figure 9.31

No, we shall do best to recognize that CREATE ORDER HISTORY is a side output, subordinate to the main function of the system, and accept that it will be invoked by GET VALID ORDER, thus making GET VALID ORDER a communicationally cohesive module. What we must do, to avoid unpleasant surprises for the maintenance programmer, is to document the existence and structural position of CREATE ORDER HISTORY at the point in GET VALID ORDER ITEM where GET VALID ORDER is invoked, and also in PROCESS ORDERS. We might also rename the GET VALID ORDER module to truly represent its function.

Now we are in a position to fill in the lower level modules of the input leg, and arrive at a complete design for the subsystem, shown in Figure 9.32. Note the use of the data dictionary extract table to hold abbreviations for the various data structures and data elements that are passed around the system.

This design, of course, can continue to be refined, for example, by expanding the lower level modules even further. However, apart from a few instances such as we have noted, it is composed of functionally cohesive modules, which are coupled largely by the passage of data, with a few control variables "reporting back" what has happened. To test how changeable the design is, it is interesting to consider the effect of various possible user changes. How many modules would need to be changed if the user wanted to alter the rules for granting credit? How many would have to be changed if all orders, valid and invalid, were to be written to the ORDER HISTORY file? What about if the user decides to eliminate part shipments, and either ship the full order, or nothing at all except a confirmation? We can also see that adding the logic to deal with prepayments is clearer than when we started out.

Assuming that we implemented the system as defined, what could we do if the response time was too great to be acceptable? Firstly, we could "package" the logical modules on the input leg into fewer physical modules by lexically including subordinates. If that structure still took too long to process, we could consider breaking the subsystem into two, at the point where a valid order is produced. The order-entry clerk would invoke a program which dealt with valid orders. Separately, we could invoke the top-level structure which would pick up valid orders from the ORDER HISTORY file, and process them against inventory. Since the order-entry clerks are not concerned with whether an order is actually shippable or not, this would reduce the number of file accesses and the number of modules invoked. The point is that the design can be optimized for performance in many ways.

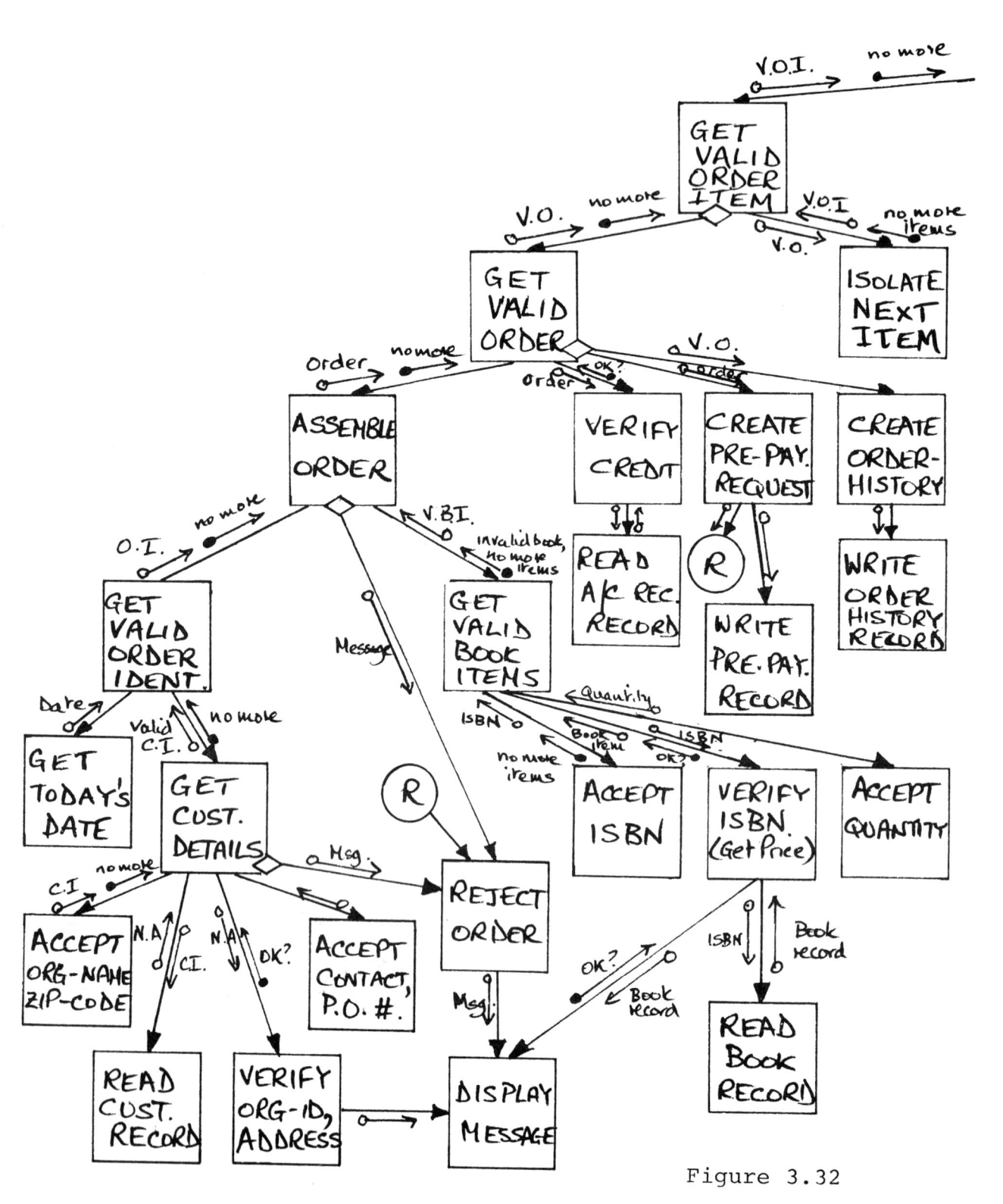

9. DERIVING A STRUCTURED DESIGN FROM THE LOGICAL MODEL

Figure 3.32

9.4 AN EXAMPLE OF STRUCTURED DESIGN

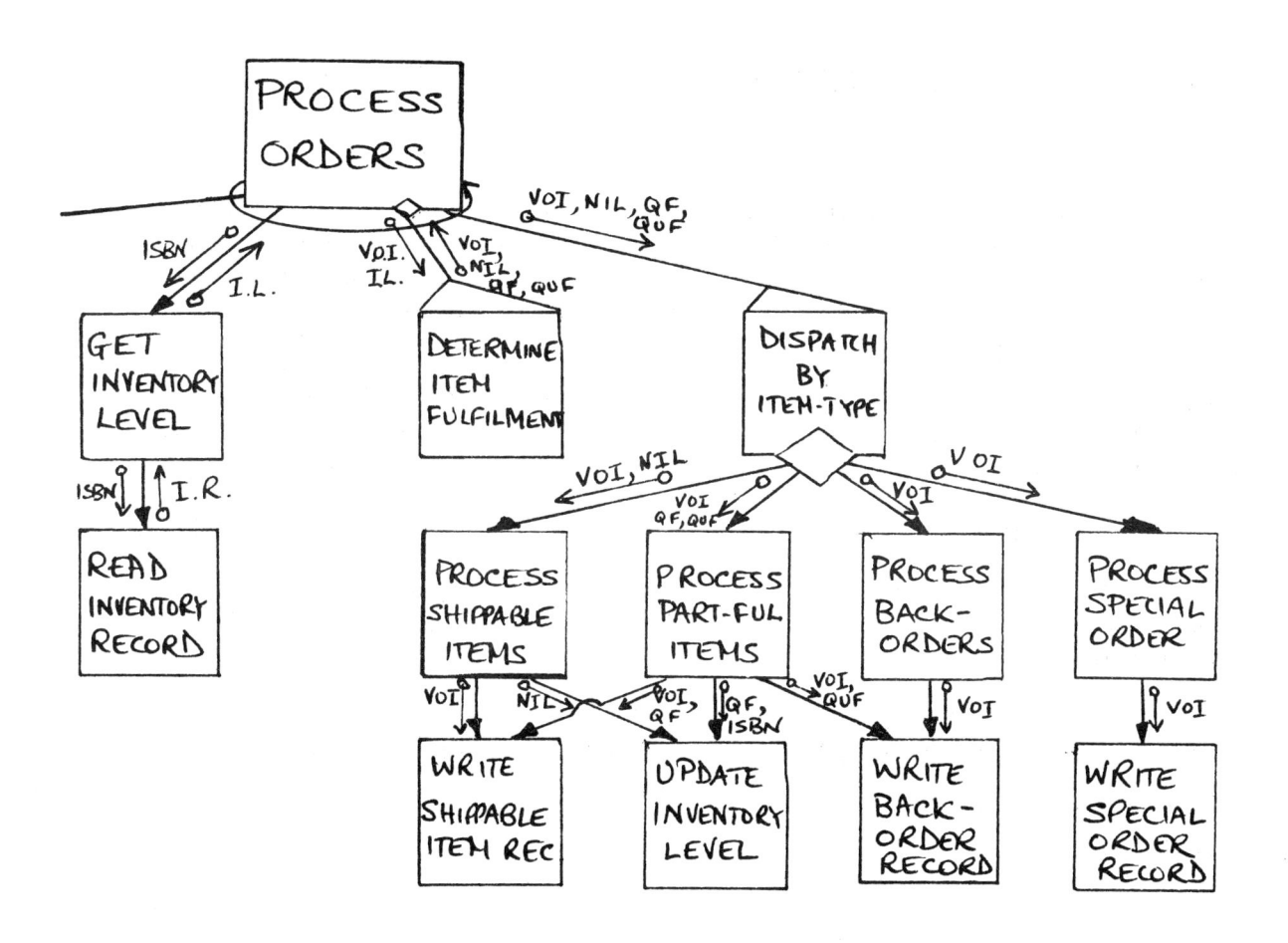

ABBREV.	NAME	DESCRIPTION
CI	Customer Identification	Organization name and/or zip-code
IL	Inventory Level	Current level of inventory
IR	Inventory Record	ISBN, Inventory-level, Quantity-on-order, reorder-level
ISBN	International Standard Book Number	10 digit book identifier
NA	Name and Address	Organization-name, Organization-address, org-id
NIL	New Inventory Level	Inventory-level minus Order-quantity
OI	Order Identity	Org-id, date, [contact, P.O.#]
QF	Quantity Filled	Quantity of book items to be shipped
QUF	Quantity Unfilled	Quantity of book items for back order
VBI	Valid Book Item	ISBN, price, quantity
VO	Valid Order	Order identity, valid book items*
VOI	Valid Order Item	Order identity, valid book item

333

9. DERIVING A STRUCTURED DESIGN FROM THE LOGICAL MODEL

9.5 TOP-DOWN DEVELOPMENT

Once the data flow diagram has been finished and the structure charts for all the subsystems have been drafted, they can be used to plan the implementation of the system *top-down*. That is, rather than define programs and build subsystems as complete units, and test that the units work together after they are ready (the "bottom-up" development approach), we can *start development by producing a crude "skeleton" version of the system, which accepts some simple input data, processes it in some very simplified way through as many subsystems as possible, and creates some very simple output*. After this skeleton version is working on our real machine, we can add more complexity to each subsystem and make *that* work. Let us see how this concept could apply to CBM, and then discuss why it has advantages over the traditional approach.

9.5.1 POSSIBLE TOP-DOWN VERSIONS OF THE CBM SYSTEM

Version 1

Our aim with Version 1 will be to get something working as quickly as possible which involves as many subsystems and as many of the interfaces between them, as possible. A realistic Version 1 might have the following functions:

- accept orders for one of 10 customers from a single CRT

- accept orders for only 10 titles, one item per order

- always accept that the customer credit is good

- always say there are 100 copies of each title in stock

- write a shipping note and invoice on plain computer paper, without updating ACCOUNTS RECEIVABLE

To deliver these functions would not take much effort, and would present a fairly simple task for the order-entry clerk! It would demonstrate that we could read from and write to the CRT screen, write out shippable items on the intermediate file and read the file again

9.5 TOP-DOWN DEVELOPMENT

as input to the shipping and invoicing subsystem. The VERIFY CREDIT module in Figure 9.32 would be a so-called "stub" consisting of two lines:

```
MOVE 'OK' TO CREDIT-STATUS.
EXIT.
```

Similarly, for all the other stub or dummy modules which would not be required for Version 1, just enough of their code would be required to allow the various programs to compile and execute. While the logic in the main control module PROCESS ORDERS would have to be complete, the module PROCESS PART FULFILMENTS could simply consist of the statements:

```
DISPLAY 'PART FULFILMENT MODULE INVOKED'.
EXIT.
```

and if this message came up, while testing the version with orders of less than 100 copies, we would know something had gone wrong.

Version 2

Once Version 1 has been developed and shown to work, the project team will implement the next version, which involves more subsystems and exercises more interfaces. Version 2 might have the following functions:

- accept orders for the same 10 customers (as Version 1, testing the CUSTOMERS file interface)

- accept orders for any number of books produced by a certain publisher provided they are normally held in stock

- always accept the customer's credit as good

- always say there are 100 copies of each title in stock

- write a correctly formatted shipping note and invoice

- set and change reorder levels for the INVENTORY file

- create a purchase order for 100 copies to the publisher of any book below reorder level

9. DERIVING A STRUCTURED DESIGN FROM THE LOGICAL MODEL

This will involve the implementation of parts of the inventory control and purchasing subsystems as well as more of the order entry and shipping subsystems, and will exercise more file and program interfaces. It will provide the users with realistic samples of three of the important documents that go outside the corporation, and do so at an early stage in development.

Subsequent Versions

Once Version 2 is demonstrated working as specified, Version 3 can be developed, followed by Version 4, and so on, each delivering more of the functions of the ultimate system. It is convenient to plan the implementation of the whole of the system in a series of such versions, each being delivered some 1 - 3 months apart, depending on the scale of the project. To have such a milestone every 30 - 60 days, and to be able to show tangible evidence of progress at regular intervals, is motivating to the project team, and reassuring to the users.

9.5.2 WHY DEVELOP TOP-DOWN?

Why approach implementation in this way? The principal reason is that top-down development avoids one of the most troublesome, timewasting, and costly problems that "normal" (bottom-up) implementation tends to run into, namely the "interface problem" or "making the parts fit together." The "normal" approach to implementation has been to divide the system into programs, specify the programs, write and test each program separately, assemble the programs into subsystems, and then, towards the end of the project, assemble the subsystems into a working system. The system is built up from the detailed lower level modules with the highest levels of control, such as JCL, being added last, hence the term "Bottom-up development." Especially if the different subsystems are developed by different teams, or different people, many projects have found that when they come to put the subsystems together there is a bug somewhere in the interface between two subsystems. Due to miscommunication of a change in specifications, or due to vague specifications in the first place, the interface to subsystem A produced by Team A does not fit the interface required by Subsystem B produced by Team B. All bugs are nasty, but interface bugs are the nastiest, for three reasons:

9.5 TOP-DOWN DEVELOPMENT

- they tend to involve more changes to existing code than, say, logic bugs. If a discount calculation is being made wrongly, we will have to fix one program; if the inventory control subsystem has been coded using the wrong record format for an order, and consequently cannot accept data from the order entry subsystem, we may have to recode, recompile and retest all the programs that use that record in one or other of the subsystems.

- they tend to be found one after another; not until the first interface bug has been put right can integration testing proceed to see if there are any others. This stretches out the testing period unpredictably.

- in "bottom-up" development interface bugs are found after most of the project budget and time has been spent, and when the deadline for delivery of the system is near or past. This means that there is always a risk that, just when the users are expecting delivery of the system, the project will be hung up for an unpredictable period of time, fixing an unpredictable number of interface bugs.

The project manager who has bad luck with interface bugs during bottom-up development, running into a series of them, is thus put in the position of repeatedly going to his users and saying "I know that four weeks ago I promised the system would be ready in two weeks time, and I know that two weeks ago I promised you the system would be ready today, but we've hit *another* problem,......" The users understandably get irritated, and quickly lose confidence in the project manager who has spent a substantial chunk of their money, and seems not to be in control of his operation.

Top-down development, being the opposite of bottom-up, creates the high level logic and all the important system interfaces first, *before* a great deal of detailed code has been written. Versions 1 and 2 are devised to exercise interfaces so that if any interface problems arise, they will be detected and corrected relatively early in the project, before much rework is required. This has a major effect on the smoothness of later implementation; the horrors of integration are largely banished. It makes it possible to show users tangible evidence of progress at regular intervals. Top-down development also is a great help to the analyst in "modelling" the system to the users, since it is possible to demonstrate an early version, say the entry of orders via a CRT, or inquiries on to the ORDER HISTORY file, and ask users whether that is what they have in mind. Though this demon-

9. DERIVING A STRUCTURED DESIGN FROM THE LOGICAL MODEL

stration usually cannot take place until some 65-70% of the way through the total project, when it is too late to make major changes, it is still very valuable in testing out the man-machine dialog, giving the users time to assimilate the nature of the new system, and discovering misconceptions which were not shown up by the logical model. Especially where users have no experience with on-line systems, showing an early version makes the previous discussions of requirements, (and the immediate access diagram), much more real. Often this will stimulate the users to ask for enhancements to the system which they would not otherwise have thought of. If these enhancements can be incorporated within the existing project budget, everybody wins.

9.5.3 THE ROLE OF THE ANALYST

We include this discussion of top-down development in a book on analysis, partly because of the value and interest of the technique, but mainly because of the impact which the approach has on the users, and the contribution which the analyst needs to make to a successful top-down project.

While Structured Design and Structured Coding are hidden from the user community (except insofar as they produce faster, better systems), top-down development makes the whole project look different. In a normal project, the **users** are involved in the study and the agreement of the functional specification after which a deathly hush descends while the project team return to their lair to design, code and test the system. The users may well hear nothing but status reports about their system for many months, until acceptance testing and user training begins. In a top-down development, a much shorter period elapses after functional specification, before the users are invited to "Come and take a look at Version 1." Thereafter, demonstrations and acceptance of versions takes place at (regular) intervals throughout the last third of the project duration.

In this situation of much greater visibility, the analyst has a number of roles to play:

1. *The analyst should help in devising the top-down versions for the implementation plan*

    As we saw earlier in this section, the data flow diagram and the structure charts of the subsystems are the prime tools for planning early versions. One guideline is to imagine the logically simplest of all transactions, and work out the minimum amount of implementation necessary to process that transaction through as many of the interfaces as possible. The analyst is in a good position to know what is the logically simplest transaction; the designer will know the amount of implementation necessary to process it. Once the early versions are established, the analyst can see whether it is possible to create later versions which provide some useful, though partial, service for the users. For instance, it might be possible to start providing inquiry capability on some files, *before* the full version of the system is complete. It might be possible to start producing mailing lists from the CUSTOMERS file, *before* the full order entry system is in place. The analyst knows which features are of value to the users, and can contribute that to the implementation plan.

2. *The analyst should explain the top-down development concept to the user community to ensure sympathetic collaboration*

    As pointed out at the start of this section, top-down development is much more visible to the users. The analyst should take pains to explain the plan to all the users who will be impacted by each version, explaining what each version will do, but also what it *won't* do. If this explanation and orientation is not done well, two dangers arise. The first is that the users will be disillusioned by the early versions that they see, because of their limited function, and will lose confidence in the system. The second is the opposite danger, that they will be so thrilled with Version 1, they will want to start using it for their business the following Monday.

    The analyst can explain the version approach by comparison with the construction of a house, saying, "Version 1 is comparable to seeing the framework of the house only, with no walls or roof or interior. It enables you to reassure us that we're building the house in the right place, and that it's about the right size, but you

9. DERIVING A STRUCTURED DESIGN FROM THE LOGICAL MODEL

wouldn't complain because the rain came in, or expect to move in on Monday."

The same comparison is useful for explaining the impact of changes the users might want to make. Changing a report format is like wanting the living room walls painted a different color; it's nice if we know beforehand, and anyhow it's one of the last things we do in construction (and top-down development), but if you want a change after the work is done, it's no big deal. Some changes are comparable to changing a bedroom into another bathroom -- serious structural work involving time and expense. Some changes are comparable to the user saying, "I like the house just as it is, but could you move it 10 feet further from the roadway?"

3. *The analyst should ensure management support for top-down development so that the hardware is available when needed*

Typically Version 1 is implemented about 60-70% of the way through a project, as we mentioned earlier. Thus on a 12 month project starting in January, the logical model may emerge in March/April, the structure charts in May/June, and Version 1 could be delivered by the end of July, with perhaps five more versions to follow at 4 week intervals. If the system is to have on-line functions, this means that at least one terminal, with communication facilities and access to the required hardware, needs to be available by June at the latest. The managers signing the checks for the hardware, though, expect it not to be needed until November! The analyst must "sell" the benefits of this development strategy to management, and ensure that there are no obstacles to getting the necessary hardware and software. Where the hardware is being developed at the same time as the system (say for a special purpose terminal), it is important to get hold of a prototype at the earliest stage possible, and to test the interfaces to it.

Against this requirement for having some hardware earlier, can be weighed the fact that the testing load in the latter part of the project tends to be less with top-down development than with bottom-up development. This is because by developing top-down, we avoid the rework and retesting load that is typical of the integration phase of a bottom-up project; and which often results in the use of more and more test time up to the delivery of the system.

9.5 TOP-DOWN DEVELOPMENT

4. *The analyst must exert pressure for frequent full integration of subsystems*

    Once coding and testing has started top-down, there is frequently a temptation to "water-down" the concept. For instance, the programmer working on the order-entry subsystem may prefer to develop his subsystem and test it alone, after demonstrating that Version 1 works. He may develop the subsystem top-down, but if he does not regularly get together with all of the other people who are developing other subsystems, and prove that all the interfaces still function, the advantages of top-down development will be largely lost. Similarly, if some piece of hardware is not available, and that hardware interface is simulated, or not tested, along with the others, the risk of integration problems is increased.

    The analyst, as a representative of the users, can exert pressure to encourage thorough top-down testing, ensuring that the project team take the time to exercise every interface in the final system as early and often as possible. The people who are most reluctant to integrate their subsystem with the rest of the system are probably those who most need to do so.

5. *The analyst should see that the personnel subsystem is developed top-down and exercised along with the versions of the application system*

    Though we have given a lot of attention to software/software and software/hardware interfaces, the man/system interface is just as critical and just as likely to present integration problems. One of the benefits of top-down development is that the user manual and user training can be evolved along with the various versions of the system. To take the case of the hypothetical version of the CBM system which we discussed, the documentation and training needed to enter one of 10 books for one customer is clearly not very extensive, yet Version 1 provides an opportunity to train someone to interact with a CRT terminal even if they have never seen one before, in a very simple, risk-free situation. Later versions allow the documentation and training to gradually develop in complexity and be proven with the tests of the subsystem.

    One corollary of this concept is that each version must be exercised, with a representative clerical user at the terminal, using documentation produced by the people

9. DERIVING A STRUCTURED DESIGN FROM THE LOGICAL MODEL

responsible for the personnel subsystem. It is no good to have the version exercised by a senior programmer, using codes and procedures which only he/she knows!

6. *The analyst should act as the user's representative in accepting each version of the system*

    The implementation plan should specify the deliverables of each version, in terms of strongly-stated objectives. On the day fixed for the delivery of each version the analyst, and appropriate user representatives, should supervise the exercising of the system to see that, in fact, the version deliverables have been met. Naturally this exercise gets more elaborate as later versions supply more functions, until the final version receives a full acceptance test and is cut over into production.

    The analyst acts as technical adviser to the user in this process, and should ensure that everyone in the user community who can provide useful feedback on a version should have the chance to do so. He should co-ordinate the user's feedback and present it to the project team. In doing this he may get reinforcement for the comments he has made on points 4 and 5, ensuring that all interfaces are exercised properly, including the human ones.

9.5.4 SUMMARY

Top-down development is an exciting approach. Properly planned and carried out, it can avoid many of the problems that have plagued us for years in systems development. In the Walston and Felix study of productivity within IBM [9.13] projects which did not use top-down development had an average productivity of 196 lines of code per man-month; projects which did use it averaged 321 lines per man-month, an improvement of some 60%. Perhaps the most important aspect, at least for us as analysts, is the increased involvement of the users as the system is built, both in terms of the valuable comments and help they can give as the system evolves, and in terms of their confidence that the right system is being built for them, and making tangible progress.

REFERENCES

9.1 W. P. Stevens, G. J. Myers, and L. L. Constantine, "Structured Design," *IBM Systems Journal*, Vol. 13, No. 2, 1974.

9.2 G. J. Myers, *"Reliable Software through Composite Design,"* Petrocelli/Charter, 1975.

9.3 E. Yourdon, and L. L. Constantine, *"Structured Design,"* Yourdon inc., 1975.

9.4 E. Yourdon, *"Techniques of Program Structure and Design,"* Prentice-Hall, 1975.

9.5 J. Martin, *"Security, Accuracy, and Privacy in Computer Systems,"* Prentice-Hall, 1973.

9.6 F. P. Brooks, *"The Mythical Man-Month,"* Addison-Wesley, 1975.

9.7 B. Boehm, "Software Engineering," *IEEE Transactions on Computers*, Dec. 1976.

9.8 H. Mills, "Software Development," *IEEE Transactions on Software Engineering*, Dec. 1976.

9.9 F. M. Haney, "Module Connection Analysis: A Tool for Scheduling Software Debugging Activies," *Proceedings FJCC*, Vol 41, 1972.

9.10 R. C. Tausworthe, *"Standardized Development of Computer Software,"* Prentice-Hall, 1977.

9.11 R. J. Weiland, "Experiments in Structured COBOL," in "Structured Programming in COBOL - Future and Present," ACM Publications, 1975.

9.12 P. Kraft and G. M. Weinberg, "The Professionalization of Programming," *Datamation*, Oct. 1975.

9.13 C. E. Walston and C. P. Felix, "A Method of Programming Measurement and Estimation," *IBM Systems Journal*, Vol. 16, No. 1, 1977.

9. DERIVING A STRUCTURED DESIGN FROM THE LOGICAL MODEL

EXERCISES AND DISCUSSION POINTS

1. Identify a system that uses a large number of intermediate files. Roughly what proportion of the average run-time of that system is spent reading and writing those files, or waiting for the operator to mount volumes?

2. What is the longest and average seek time for the disk drives in your installation? What is the longest and average time for executing a machine instruction on your machine?

3. If you have access to a program execution monitor, such as PPE or SUPERMON, use it on a program with which you are familiar to see where the program spends most time. Were you surprised by the results?

4. What proportion of the total professional time in your installation is spend on maintenance activities?

5. Take a system you know of that has been in production for several years and assess the number of man-hours per year spent on maintaining it. How long do you expect the system to continue in use before it is finally replaced? What will the total maintenance effort over the system life-time be, compared with the original development effort?

6. If you have access to a system written in modular fashion, draw a structure chart for the system, and assess the coupling and cohesion types involved. How do your findings correlate with the changeability of the system?

7. Some general purpose modules are logically cohesive, such as general editing functions, general I/O handlers, etc. Can you think of some examples of functionally cohesive modules which are still general purpose in that they can be used in a number of places in a system and in different systems? What would be the benefits of having a library of functionally cohesive, black-box, general purpose modules?

8. In your view, should changeability ever be reduced for the sake of performance? If so, can you define the circumstances?

9. EXERCISES AND DISCUSSION POINTS

9. Starting with the detailed data flow diagram of Figure 9.19 and the structure chart of Figure 9.32, upgrade the design to handle prepayments, verifying that the prepayment is for the correct amount, and writing a record to the ACCOUNTS RECEIVABLE file, connecting the amount with the order identity.

10. Making reasonable assumptions, design a purchasing subsystem for CBM.

11. On the structure chart of Figure 9.32, mark the modules that would have to be implemented in order to deliver the hypothetical Version 1 as specified in Section 9.5.1. Categorize each module as F (needing full implementation), P (needing part implementation), or S (needing to be implemented only as a stub).

12. Produce a top-down implementation plan for a recently-delivered system with which you are familiar. What would the likely benefits have been had this plan been followed?

Chapter 10

Introducing structured systems analysis into your organization

In this chapter we consider the steps that need to be taken to introduce the tools and techniques discussed in this book into a systems development organization, and then consider the benefits that may reasonably be expected together with the problems that people have experienced.

10.1 STEPS IN IMPLEMENTATION OF STRUCTURED SYSTEMS ANALYSIS

10.1.1 REVIEWING THE GROUNDRULES FOR CONDUCTING PROJECTS

The amount of work involved in this area will vary greatly, depending on whether or not a formal systems development methodology is in place and in use, and if so, what that methodology prescribes for the analysis phase.

In response to problems in systems development, some organizations have adopted formal sets of procedures for systems development, either working them out themselves, or adopting one of the "packaged" methodologies developed by consulting firms. Each methodology specifies, much as we did in Chapter 8, the sequence of activities to be followed in

10. INTRODUCING STRUCTURED SYSTEMS ANALYSIS INTO YOUR ORGANIZATION

developing the system, the products to be developed at each stage, and the management controls to be applied. Typically, it may specify the conduct of a feasibility study, the contents of the feasibility study report, and the management group who should review the feasibility study and authorize further work. Following the feasibility study, a general design phase and detailed design phase may be specified, followed by turn by coding, unit testing, subsystems testing, and system testing.

If such a methodology is in place, it needs to be reviewed carefully to see whether what is prescribed or forbidden conflicts with the use of Structured Systems Analysis techniques. The following questions are among those that need to be considered:

> Does the present methodology prescribe or encourage premature physical design? If so, how can the methodology be modified to allow the building of the logical model prior to physical design?
>
> Does the methodology encourage "over-documentation" i.e. the exhaustive writing down of narrative detail? How can the methodology be modified to allow the graphical tools of Structured Systems Analysis to take the place of narrative wherever feasible? Though the use of the data flow diagram, the data dictionary, and the other tools do not remove the need for *all* narrative in the specification, they can greatly reduce the volume of narrative descriptions needed, provided their use is acceptable in the context of the methodology.
>
> Does the present methodology allow top-down development? While not specifically to do with the analysis phase, this issue has its impact on the analyst, as we saw in Section 9.5. Some methodologies require all of the detailed design to be complete before any coding can start, and only envisage the delivery of one version of the system (the final one), using the completion of unit testing and subsystems testing as management checkpoints. Of course, this makes the use of top-down development somewhat difficult, since the top-down approach enables coding of the high-level modules in a system to begin before the detailed design of the low level modules is complete, and delivers a series of working versions, rather than completed testing phases.

10.1 STEPS IN IMPLEMENTATION OF STRUCTURED SYSTEMS ANALYSIS

This last point raises a more fundamental and subtle issue, that of the "straight-line" approach to projects as opposed to the "spiral" approach. In the past we have tended to assume that a well-managed development project went in a "straight line" from feasibility study through analysis through design into testing, acceptance, and operation. Figure 10.1 shows this diagramatically.

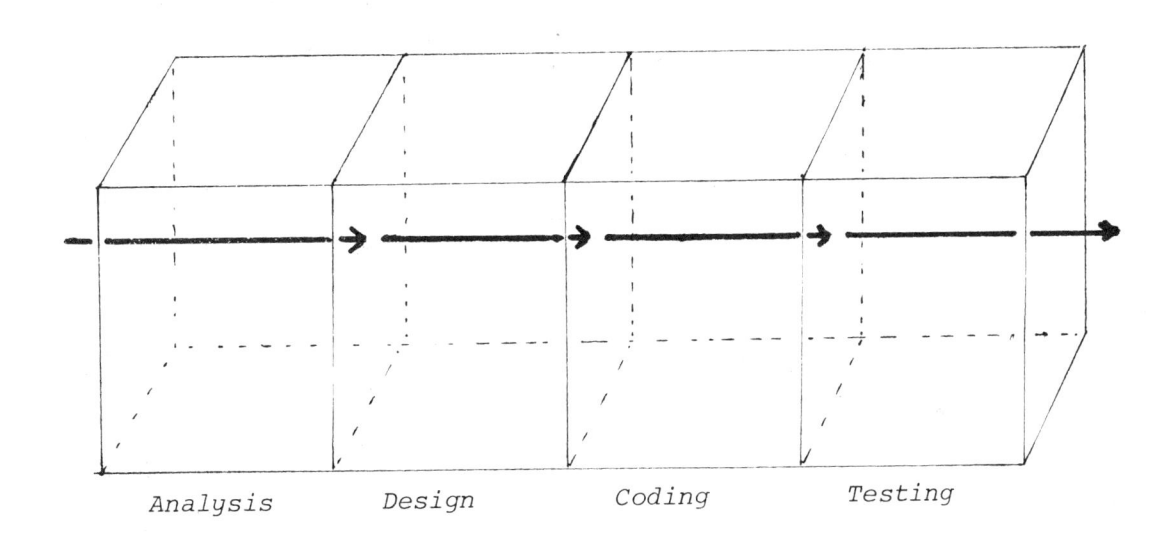

Figure 10.1
The "ideal" of project progress

Clean and manageable as the straight-line may be, it does not appear to correspond to the realities of system development. Even a well-managed project, staffed by competent people, needs to proceed iteratively, doing some analysis, then a little design, then back for more detailed analysis, then more design, then coding Version 1 perhaps, then more design and so on. The path of such a project can be pictured as a "spiral" as shown in Figure 10.2.

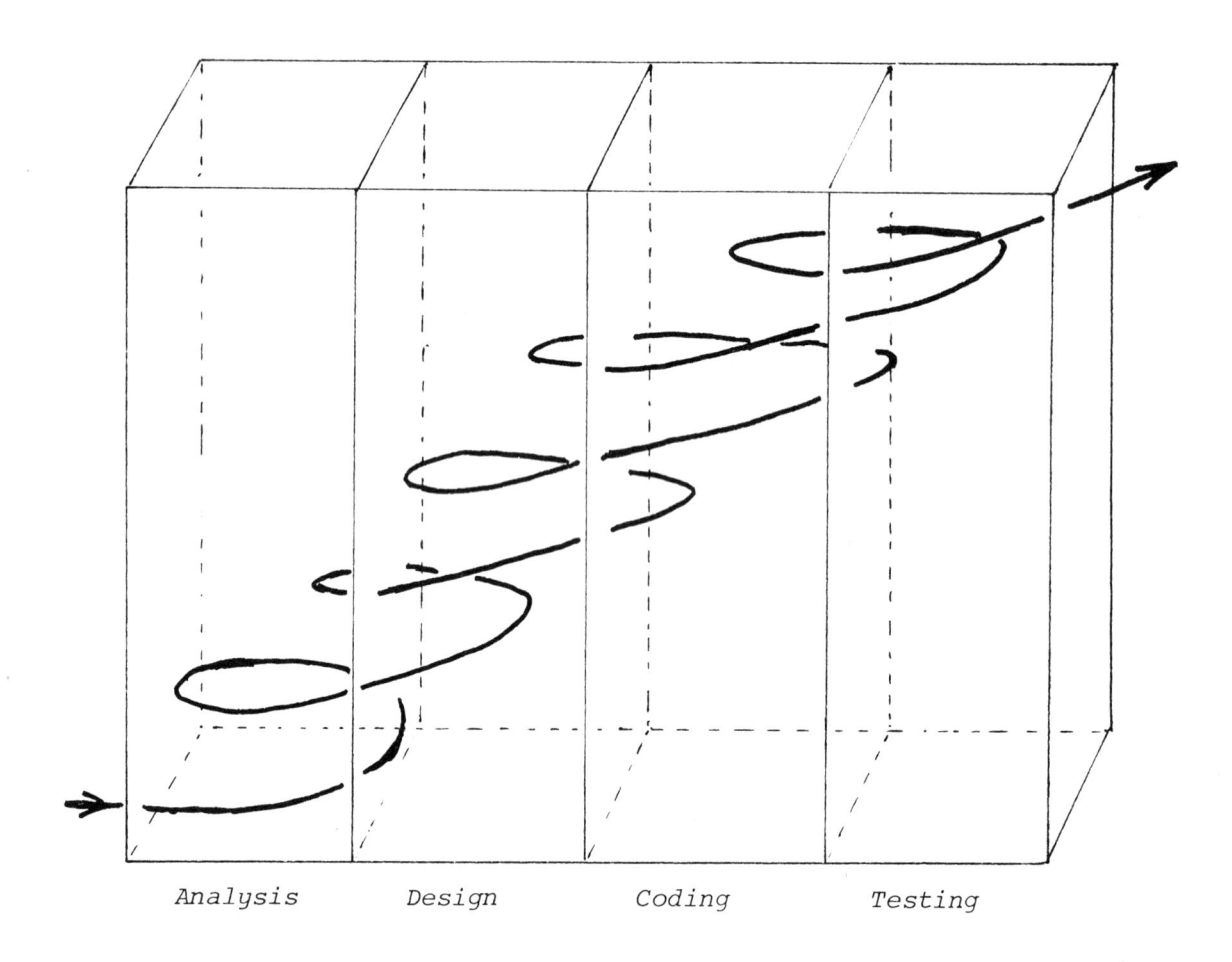

10. INTRODUCING STRUCTURED SYSTEMS ANALYSIS INTO YOUR ORGANIZATION

Figure 10.2
The reality of "spiralling" projects

The spiral concept was built into our discussion of a structured methodology in Chapter 8. We described the production of an overall data flow diagram before the production of detailed data flows and the production of a tentative design followed by successive refinement. This approach, in our view, reflects the reality of the difficult problems we face in systems development; as well as top-down development, we are doing top-down design and top-down analysis.

10.1 STEPS IN IMPLEMENTATION OF STRUCTURED SYSTEMS ANALYSIS

In each case and at each level we build a skeleton, first logical then physical, see how well the skeleton works, and then go back to put the flesh on the bones.

How can we exercise management control over a spiral activity? Conventional methodologies typically identify project milestones as "analysis finished," or "design completed." Interim status reporting requires the analyst or project manager to estimate the degree of completion of the activity, such as "analysis 40% complete," "90% of coding done." These criteria mean little, especially when the project is going in a spiral. Management control has to stop being based on *activities* and be based on *deliverables;* instead of saying "How far are you into analysis?" managers say, "Show me the latest version of the data flow diagram," or "Show me the structure chart." Though we discussed the structured methodology in Chapter 8 in terms of a series of phases, that was for convenience only; what is really being produced in a structured project is a set of deliverables of increasing refinement leading up to the delivery of each of the top-down versions of the system. Here, of course, the deliverables are much more definite than in the traditional project, and the management control is correspondingly tighter; if the implementation plan specifies that Version 3 will be delivered on December 31st, consisting of the processing of two transactions producing seven reports, and if by 5 P.M. on that day the project leader cannot demonstrate that facility to his manager, the project is behind schedule. No question of "90% complete" arises; the version can either be delivered, or it cannot. The length of time it takes to deliver the version, after the specified version deadline, is an exact measure of how far the project is behind schedule, and how much the final deadline is likely to slip.

Thus the groundrules of management control of a project need to be changed in the structured environment to put emphasis on delivery of products--data flow diagrams, structure charts, version functions--rather than the completion of activities.

10. INTRODUCING STRUCTURED SYSTEMS ANALYSIS INTO YOUR ORGANIZATION

10.1.2 ESTABLISHING STANDARDS AND PROCEDURES FOR THE USE OF THE DATA DICTIONARY AND OTHER SOFTWARE

A decision may have to be made to develop or acquire an automated data dictionary capability, if one does not already exist. If data dictionary techniques are to be manual, forms and control procedures will have to be worked out. If a text-editing capability is available, it may be adapted to give a measure of data dictionary capability as described in Chapter 4. The clerical people who use the text-editor will need to be trained in the data dictionary conventions.

It is likely that more extensive software will become available for the support of analysis; it would be very convenient to have a data flow diagram plotter for the maintenance and easy updating of diagrams, and software for reading and checking process logic in Structured English form. One pioneering effort in this direction is the ISDOS project at the University of Michigan [10.1], which has developed software allowing analysts to express data flows, data structures, and process logic in an English-like language called Problem Statement Language (PSL). The PSL statements are processed by the Problem Statement Analyzer (PSA) which checks the PSL statements for correctness and consistency and builds a data base of information about the project from which reports and charts can be produced.

10.1.3 TRAINING ANALYSTS IN THE USE OF THE TOOLS AND TECHNIQUES

In this book we have tried to set out the tools and techniques of Structured Systems Analysis as simply and yet as realistically as possible. To use the tools fluently takes study and practice. While the rules and conventions can be learnt fairly easily, say from a one-week workshop, the hardest change seems to be in starting to think at the logical level, rather than in terms of a physical implementation. We have noticed a parallel phenomenon in our seminars on Structured Design; the hardest aspect is the viewing of a system as a hierarchy, rather than as a sequential flowchart showing the sequence of events. Both in analysis and in design, we are asking people to think about problems at a higher level of abstraction than before, and this can take time and persistance.

10.1 STEPS IN IMPLEMENTATION OF STRUCTURED SYSTEMS ANALYSIS

As well as fluency with the logical tools, analysts need to become familiar with the new support software or with the data dictionary manual procedures.

If the groundrules for projects are to change, analysts will have to explain them to the user community, so that analysts will need to be briefed on the new methodology and think through its implications for users.

If top-down development is to be used, the analysts must be thoroughly briefed on the concept, and the implementation plan for each project with which they are concerned, since they have to play such a significant part in the greater user involvement that top-down development implies. Just as we have gone into Structured Design in some detail in this book, so each analyst should understand the principles of Structured Design and be able to criticise a design in the light of those principles, even if he/she does not produce the design. This may be achieved by study of this book and the references on Structured Design, or by a formal workshop.

10.1.4 ORIENTING USERS TO THE NEW APPROACHES

Since the new techniques and approaches improve communication with users, and involve them more in setting the direction of the project, they are, in general, welcomed by the user community. At the same time, the new ideas represent a change in the rules of "How you get something developed round here," and, as such, may meet with distrust unless their implications and benefits are clearly explained.

The user briefing, usually done as each "structured" project starts up, should cover the following points:

- the notation of the data flow diagram (and possibly the "decision tree")

- the concept of presenting a "menu" of alternative systems for the user decision makers to consider (if this is relevant to the project in hand)

- the concept of top-down development

- a statement of the involvement needed on the part of the users

- a reassurance that the new techniques *replace* the existing

10. INTRODUCING STRUCTURED SYSTEMS ANALYSIS INTO YOUR ORGANIZATION

approach rather than imposing yet more effort, and that the project will generate less written material for users to review rather than more!

The more involved tools--the data dictionary, the data immediate access diagram, structured English--should be left until each user has to review a product which uses them. The analyst should be aware of who has been briefed on what, and be prepared to explain each tool before showing it to a user for the first time.

Similarly, the concept of top-down development should be explained again when the analyst presents the implementation plan to users.

Should users be trained to draw up data flow diagrams of their own, and write their policies down in structured English? We come down heavily in favor of this, provided each individual user wants to do it. In other words, we cannot *require* users to express their needs in a logical model, but we should *encourage* them to do so, where they are prepared to commit the time and effort. We have had a few non-technical users attend our seminars on Structured Analysis. Though they have found the technical details of physical systems beyond their grasp, without exception they have found the logical model an easy concept to pick up, and a valuable one to use in thinking about the systems that they need. The data flow diagram and the decision tree appear to be the easiest techniques for non-technical people to acquire. Paradoxically, it may be their very ignorance of technical detail that makes it easier from them to start thinking at a logical level!

Where specific executives are assigned as User Liaison Representatives, it is desirable that they should be trained in all of the tools and techniques of Structured Systems Analysis. This will give them the ability to think more precisely about their business and its requirements, to communicate them to the analyst and designers in standard ways, and to be informed critics of the logical models produced by the data processing people.

10.2 BENEFITS AND PROBLEMS

With Structured Coding or top-down development, it is possible to quantify some of the benefits that result; improved productivity in lines of code per day, more manageable use of test time, and so on.

With Structured Design, the benefits are just as real, but harder to quantify. We can ask a group of maintenance programmers to rate the changeability of a system using Structured Design compared with one not using Structured Design; in theory we could measure the maintenance cost of a group of such systems and compare it with that of a group of unstructured systems. One unpublished study suggests that a system using Structured Design is as much as *seven* times easier and cheaper to change than traditional designs [10.2]. Other studies tend to confirm this dramatic result, but are anecdotal; Bill Inmon [10.3] commented that, in a Structured Design system, "The largest change required four days, and the next largest, less than a day."

With Structured Systems Analysis, the benefits are even harder to quantify. Indeed, in some senses, if the work of analysis were perfectly done, the only result would be an absence of problems! We have a number of subjective and anecdotal comments which appears to be fairly representative of people's experience with the Structured Systems Analysis tools. See for example [10.4]. These benefits are summarised in the next section.

10.2.1 BENEFITS FROM USING STRUCTURED SYSTEMS ANALYSIS

1. Users get a much more vivid idea of the proposed system from logical data flow diagrams than they do from narrative and physical system flowcharts. Because they understand it, they are more positive towards the project. The probability of building a system which, though excellent, does not meet the user's needs, is sharply reduced.

10. INTRODUCING STRUCTURED SYSTEMS ANALYSIS INTO YOUR ORGANIZATION

2. Presenting the system in terms of logical data flow shows up misunderstandings and contentious issues much earlier than is normally the case. The comment has been made that "with narrative specifications, everyone draws their own mental data flow diagram." Once these mental data flows are put down on paper, and made public, many differences between people's private ideas about the system become obvious. For example, a written narrative might specify that "An ORDER HISTORY file will be created containing details of all orders processed." As the statement stands it might be perfectly acceptable to users. However, when they come to walk through the data flow diagram, it will be very clear, from the place in the diagram where the order history data flow originates, exactly what will be captured in ORDER HISTORY. Is it only orders actually shipped? Or is it all orders, whether shippable or not? Or does it include all orders, whether shippable or not, including those rejected for credit?

   This merciless exposing of vagueness through the data flow diagram means that more discussion takes place over the data flow than over a typical narrative specification. It is, however, very productive discussion. Making changes on a piece of paper is very cheap compared with making them in the code.

3. The interfaces between the new system and existing clerical and/or automated systems are shown very clearly by the data flow diagram, and the need to document the details of the data flows in the data dictionary forces clear definition of those interfaces at an early stage. Some organizations specify that a data flow diagram should be drawn, not only for the system under study, but also for each other system--clerical or automated--with which it interfaces. This exercise, though involving a considerable amount of work, shows up duplication of function, and points to opportunities for including clerical functions within the new system.

4. The use of the logical model does away with a certain amount of duplication of effort which takes place in traditional projects. Typically the user representative and the analyst would, in the past, work together to produce a narrative specification of the system. Once the narrative specification was agreed, the design/programming group would take it, and effectively re-analyze it, doing much of the work of data and logic

definition over again. It is noticeable that the tools of Structured Systems Analysis are equally valuable to both users and technicians. Once the users agree the data flow, the immediate access analysis, and the policy logic, those documents can be used directly as inputs to physical design. This advantage is particularly noticeable to the data base designer, who previously had to hunt through the narrative specification to extract the data items and access requirements. Now he is presented with a data dictionary, relations in 3NF, and an immediate access analysis.

5. The use of the data dictionary to hold project glossary items saves time by quickly resolving those cases where people call the same things by different names, or where one term means different things depending on context. These uses of words are taken for granted by people in the user community, since they are a part of their daily life, but they can be quite baffling to analysts.

In a nutshell, the benefits reduce to only two:

- showing clearly what you're going to build so that everyone can be sure you're building the right system

- working out the alternatives and details with as little waste of time as possible

Put that way, they sound almost trivial, yet how much time and money has been wasted in the last twenty years because we have not been able to do these two things?

10.2.2 POTENTIAL PROBLEMS

The benefits of the new analytical tools are not free, of course; there are some costs and potential problems associated with their introduction. Partly these are the problems associated with any change, partly they are a result of the greater formality and discipline of the logical tools.

1. Orientation of users and training of the analysts is required, as discussed in the previous section. Since the introduction of Structured Systems Analysis is perceived as "changing the rules," everyone must be clearly told what the new rules are, and how they improve the game.

10. INTRODUCING STRUCTURED SYSTEMS ANALYSIS INTO YOUR ORGANIZATION

2. The effort, formality, and degree of detail required, especially in building the data dictionary, is often resisted. This is a question of making an investment of effort during analysis, for the sake of a smoother project later, of doing things right the first time, so that they will not have to be done over. Partly the resistance arises from users because previous projects have not required such a clear definition of terms and meanings; the project team has suffered so that users could go on with their old sloppy terminology. Resistance may be due to attempts to define excessive detail too early; as we commented in Chapter 8, a finely balanced decision has to be made as to the level of detailed documentation, especially of the current system functions that are not going to be included in the new system. But the fact has to be faced; doing good analysis takes effort from both the users and the analysts. The compensation is that it is more productive effort.

3. There has been some uneasiness on the part of programmers that getting detailed specifications of logic in Structured English will "take all the fun out of programming; make me a mere coder." These fears die away when the the designer and programmers see that Structured System Analysis gives them a much larger job to do, by giving them the logical model to work from. Our discussion of the design trade-offs in Chapter 9 should have made it clear just how much work is left to be done after the logical model is finalized. We think the problem arises because, until programmers have some experience of working with logical models, they do not appreciate the difference between the external logic of credit validation, say expressed in Structured English, and the internal logic of the physical module. External logic is given to them by the analyst; the internal logic is all theirs to design.

4. Lastly, a question arises in some organizations after their first positive experience with Structured Systems Analysis. "What a shame we didn't have these tools on the XYZ project; we've been working on it for six months and we haven't finished the analysis. Can we use the tools of Structured Systems Analysis on XYZ, now that we are part way through?" The lesson of experience seems to be that it pays to use the structured techniques for analysis, design, and development, starting at any point in a project. Even if it appears to the users of the XYZ project that we are going to junk the last six months

10.2 BENEFITS AND PROBLEMS

work and begin again (which is not true) they will be quickly won over when they see the improvement in progress. To quote one Insurance company analyst, "Once you've seen what these new methods can do, you're loathe to continue using the old methods."

10. INTRODUCING STRUCTURED SYSTEMS ANALYSIS INTO YOUR ORGANIZATION

REFERENCES

10.1 D. Teichroew and E. Hershey, "PSL/PSA: A Computer-Aided Technique for Structured Documentation and Analysis," *IEEE Transactions on Software Engineering*, Jan. 1977.

10.2 Larry Constantine, unpublished lecture notes.

10.3 Bill Inmon, "An Example of Structured Design," *Datamation*, March 1976.

10.4 W. James Kain, "The Practice of Structured Analysis," *paper presented at Life Office Management Association Systems Forum*, Atlanta, March 1977.

Glossary

ALIAS

    a name or symbol which stands for something, and is not its proper name.

ARGUMENT

    a value which is used as the input to some process, often passed through a module-module interface.

ATTRIBUTE

    a data element which holds information about an entity.

CANDIDATE-KEY

    an attribute or group of attributes, whose values uniquely identify every tuple in a relation, and for which none of the attributes can be removed without destroying the unique identification.

COINCIDENTAL COHESION

    used to describe a module which has no meaningful relationship between its components (other than that they happen to be in the same module). The weakest module strength.

COMMUNICATIONAL COHESION

    used to describe a module in which all the components operate on the same data structure. A good, but not ideal module strength.

CONTENT COUPLING

    a severe form of coupling, in which one module makes a direct reference to the contents of another module.

GLOSSARY

CONTINUOUS DATA ELEMENT

> one which can take so many values within its range that it is not practical to enumerate them, e.g., a sum of money.

CONTROL COUPLING

> a form of coupling in which one module passes one or more flags or switches to another, as part of invocation or returning control.

DATA ADMINISTRATOR (DATA BASE ADMINISTRATOR)

> a person (or group) responsible for the control and integrity of a set of files (data bases).

DATA AGGREGATE

> a named collection of data items (data elements) within a record.
>
> See also: GROUP

DATA BASE

> "a collection of interrelated data stored together with controlled redundancy to serve one or more applications; the data are stored so that they are independent of programs which use the data; a common and controlled approach is used in adding new data and in modifying and in retrieving existing data within a data base."
>
> James Martin

DATA DICTIONARY

> a data store that describes the nature of each piece of data used in a system; often including process descriptions, glossary entries, and other items.

DATA DIRECTORY

> a data store, usually machine-readable, that tells <u>where</u> each piece of data is stored in a system.

DATA ELEMENT (DATA ITEM, FIELD)

the smallest unit of data that is meaningful for the purpose at hand.

DATA FLOW DIAGRAM (DFD)

a picture of the flows of data through a system of any kind, showing the external entities which are sources or destinations of data, the processes which transform data, and the places where the data is stored.

DATA IMMEDIATE-ACCESS DIAGRAM (DIAD)

a picture of the immediate access paths into a data store, showing what the users require to retrieve from the data store without searching or sorting it.

DATA ITEM

See DATA ELEMENT.

DATA STORE

any place in a system where data is stored between transactions or between executions of the system (includes files--manual and machine readable, data bases, and tables).

DATA STRUCTURE

one or more data elements in a particular relationship, usually used to describe some entity.

DECISION TABLE

a tabular chart showing the logic relating various combinations of conditions to a set of actions. Usually all possible combinations of conditions are dealt with in the table.

DECISION TREE

a branching chart showing the actions that follow from various combinations of conditions.

GLOSSARY

DEGREE (OF NORMALIZED RELATION)

the number of domains making up the relation. (If there are 7 domains, the relation is 7-ary or of degree 7).

DESIGN

The (iterative) process of taking a logical model of a system, together with a strongly-stated set of objectives for that system, and producing the specification of a physical system that will meet those objectives.

DFD

See DATA FLOW DIAGRAM.

DIAD

See DATA IMMEDIATE ACCESS DIAGRAM

DISCRETE DATA ELEMENT

one which takes up only a limited number of values, each of which usually has a meaning.

See also: CONTINUOUS DATA ELEMENT.

DOMAIN

the set of all values of a data element that is part of a relation. Effectively equivalent to a field, or data element.

EE

See EXTERNAL ENTITY.

ENTITY

1. external entity: a source or destination of data on a data flow diagram

2. something about which information is stored in a data store e.g., customer, employees

EXTERNAL COUPLING

a severe form of intermodule coupling in which one module refers to elements inside another module, and such elements have been declared to be accessible to other modules.

EXTERNAL ENTITY (EE)

See ENTITY.

FACTORED

a function or logical module is factored when it is decomposed into subfunctions or submodules.

FIRST NORMAL FORM (1NF)

a relation without repeating groups (a normalized relation), but not meeting the stiffer tests for second or third normal form.

FUNCTIONAL

1. functional cohesion: used to describe a module all of whose components contribute towards the performance of a single function.

2. functional dependence: a data element A is functionally dependent on another data element B, if given the value of B, the corresponding value of A is determined.

GROUP (ITEM)

a data structure composed of a small number of data elements, with a name, referred to as a whole.

See also: DATA AGGREGATE.

HIPO (HIERARCHICAL INPUT PROCESS OUTPUT)

a graphical technique similar to the structure chart showing a logical model of a modular hierarchy. A HIPO overview diagram shows the hierarchy of modules: details of each module's input processing and output are shown on a separate detail diagram, one per module.

GLOSSARY

IMMEDIATE ACCESS

retrieval of a piece of data from a data store faster than it is possible to read through the whole data store searching for the piece of data, or to sort the data store.

INDEX

a data store that, as part of a retrieval process, takes information about the value(s) of some attribute(s) and returns with information that enables the record(s) with those attributes to be retrieved quickly.

INVERTED FILE

one in which multiple indexes to the data are provided; the data may itself be contained within the indexes.

IRACIS

acronym for Increased Revenue, Avoidable Costs, Improved Service.

KEY

a data element (or group of data elements) used to find or identify a record (tuple).

LEXICAL

to do with the order in which program statements are written. Module A is *lexically included* within module B if A's statements come within B's statements on the source listing.

LOGICAL

1. non-physical (of an entity, statement, or chart): capable of being implemented in more than one way, expressing the underlying nature of the system referred to.

2. logical cohesion: used to describe a module which carries out a number of similar, but slightly different functions--a poor module strength.

MODULE

1. *a logical module:* a function or set of functions referred to by name.

2. *a physical module:* a contiguous sequence of program statements, bounded by a boundary element, referred to by name.

NORMALIZED (RELATION)

a relation (file), without repeating groups, such that the values of the data elements (domains) could be represented as a two-dimensional table.

ON-LINE

connected directly to the computer so that input, output, data access, and computation can take place without further human intervention.

PATHOLOGICAL (CONNECTION)

a severe form of coupling between modules where one module refers to something inside another module.

See also: CONTENT COUPLING

PERSONNEL SUBSYSTEM

the data flows and processes, within a total information system, that are carried out by people: the documentation and training needed to establish such a subsystem.

PHYSICAL

to do with the particular way data or logic is represented or implemented at a particular time. A physical statement cannot be assigned more than one real-world implementation.

See also: LOGICAL.

PRIMARY KEY

a key which uniquely identifies a record (tuple).

GLOSSARY

PROCEDURAL COHESION

used to describe a module whose components make up two or more blocks of a flowchart. Not as good as communicational or functional cohesion.

PROCESS (TRANSFORM, TRANSFORMATION)

a set of operations transforming data, logically or physically, according to some process logic.

PSEUDOCODE

a tool for specifying program logic in English-like readable form, without conforming to the syntactical rules of any particular programming language.

RELATION

a file represented in normalized form, as a two-dimensional table of data elements.

RELATIONAL DATA BASE

a data base constructed out of normalized relations only.

SCOPE-OF-CONTROL (OF A MODULE)

all of the modules which are invoked by a module; and all those invoked by the lower levels, and so on. The "department" of which the module is "boss."

SCOPE-OF-EFFECT (OF A DECISION)

all those modules whose execution or invocation depends upon the outcome of the decision.

SEARCH ARGUMENT

the attribute value(s) which are used to retrieve some data from a data store, whether through an index, or by a search.

See also: ARGUMENT.

SECOND NORMAL FORM (2NF)

a normalized relation in which all of the non-key domains are fully functionally dependent on the primary key.

SECONDARY INDEX

an index to a data store based on some attribute other than the primary key.

SEGMENT

a group of (one or more) data elements; the unit of data accessed by IMS software.

Compare GROUP, DATA AGGREGATE.

SIDE EFFECT

the lowering of a module's cohesion due to its doing some subfunctions which are "on the side," not part of the main function of the module.

SPAN OF CONTROL

the number of modules directly invoked by another module. This should not be very high (except in the case of a dispatcher module) or very low.

STRUCTURE CHART

a logical model of a modular hierarchy, showing invocation, intermodular communication (data and control), and the location of major loops and decisions. See Figure 9.32.

STRUCTURED DESIGN

a set of guidelines for producing a hierarchy of logical modules which represents a highly changeable system.

See also: DESIGN

GLOSSARY

STRUCTURED ENGLISH

a tool for representing policies and procedures in a precise form of English, using the logical structures of Structured Coding.

See also: PSEUDOCODE.

STRUCTURED PROGRAMMING (CODING)

the construction of programs using a small number of logical constructs, each one-entry, one-exit, in a nested hierarchy.

THIRD NORMAL FORM (3NF)

a normalized relation in which all of the non-key domains are fully functionally dependent on the primary key and all the non-key domains are mutually independent.

TIGHT ENGLISH

a tool for representing policies and procedures with the least possible ambiguity.

See also: STRUCTURED ENGLISH.

TSO

Time Sharing Option: a feature of IBM software which allows the entering and editing of programs and text through on-line terminals.

TOP-DOWN DEVELOPMENT

a development strategy whereby the executive control modules of a system are coded and tested first, to form a "skeleton" version of the system, and when the system interfaces have been proven to work, the lower level modules are coded and tested.

TUPLE

a specific set of values for the domains making up a relation. The "relational" term for a record.

See also: SEGMENT

VOLATILITY

a measure of the rate at which a file's contents change, especially in terms of addition of new records and deletion of old.

Index

Action, 116
And/or ambiguity, 119-121
Adjectives, 80
   undefined, 121
Alias, 77
Alternation notation, of data
   structure, 84
Area code, 80
Aron, J., 267
Attribute, 216-217

Binding, 298
Black-box, of module, 287-288
Boehm, B., 9
Bottom-up development, 334
Boundary, of system, 18-21, 246-247
Brooks, F., 7

Candidate key, 179-180
Case structure, 149-150
Changeability, 274, 283-285
Clerical users, 245-246
CODASYL, 72
Codd, E., 178, 180
Cohesion, 298-299
Commissioners of systems, 244
Conditions, 116
Conditional sentence
   structures, 118
Consistency checking, 96
Constantine, L., 273
Continuous data element, 78
Control, 274, 281-283
Coupling, 295-297
CULPRIT, 235

Data Administrator, 91-93
DATA CATALOGUE, 112
DATA DICTIONARY, 112
Data dictionary, 21-23, 72-111
   reports, 93-94
   cross-referencing, 95
   organization-wide, 108
   distributed processing, 109-111
Data element, 73
   in data dictionary, 76-82
   continuous, 78
   discrete, 78-79
Data flow
   description, 86-87
   symbol, 39-42
Data flow diagram, 12-18
   guidelines for drawing, 50-52
Data immediate access diagram, 30-33, 226-229
Data store
   description, 87-88
   contents, 27
   symbol, 45
Data structure, 73-74, 83-85
   simplifying contents of, 174-175
DATAMANAGER, 98-107, 112
   cross-referencing, 104
Date, C., 178
Decision structure, 148-149
Decision table, 125-126, 135-144
   vs. decision tree, 145
   vs. other tools, 162
Decision tree, 24-25, 123-123, 126-133
   vs. decision table, 145
   vs. other tools, 162
Degree, of relation, 178
Design, definition, 273
Discrete data element, 78-79
Distributed system, 256
Distribution with inventory, 52-65

INDEX

Distribution without inventory, 21
Domain, 178

EASYTRIEVE, 235
Encoding, 81
English, limitation, 5
Entity, 216-217
  external, 38-39
  external, description of, 89-90
Errors, 18, 49
Estrin, G., 37
Explosion, 17, 46-48
Extended entry, of decision table, 141-144

Felix, C., 342
First normal form, 180-181
Flowcharts, 6-8
Freshness, of data, 205
Functional specification, 9, 33

GE, 118-119
Gilb, T., 264
GIS, 216
Glossary entries, in data dictionary, 90-91
GT, 118-119

Hierarchical records, 209-213

IBM DB/DC DATA DICTIONARY, 112
Immediate access, 29
IMS, 72, 212-213
Indifference, 139-140
Inmon, B., 355
Interface bug, 336-337
Inverted file, 208-209
IQF, 214-215

IRACIS, 241, 247-248
ISBN, 23
ISDOS, 352
Isolation of function, 287
Iteration notation, of data structures, 84

Jackson, M., 84
Join, 186

KWIC index, 314

LE, 118-119
Lexical inclusion, 323
LEXICON, 112
Logic, internal vs. external, 24
LT, 118-119

Machine-readable, data definitions, 96-97
MARK IV, 235
Martin, D., 37
Martin, J., 178, 262, 264
Materials flow, 66-67
Module, 286
Myers, G., 273

Normalization, 176-180
  procedure, 184

Optional notation, of data structures, 84

Performance, 274, 275-281
Personnel subsystems, 264

INDEX

*Predictability of information
    requests, 204
Problems of analysis, 2-5
Process
   description, 87-89
   symbol, 43-44
Projection, 186
Pseudocode, 156-158*

Questionnaire, 231-233

*Relation, 178
Repetition structure, 150-152
Request evaluation, 240
Ripple-effect, 287
Ross, D., 37*

*Scope of control, 300-302
Scope of effect, 300-302
Second normal form, 182
Secondary index, 206-208
Security, 234
Segment, 210-211
Sequential structure, 146-147
Side effect, 328
Span of control, 326*

*Spiral approach, 349-350
Structure chart, 292
Structured Design,
    guidelines, 303
Structured English, 25-26, 152-156
   compared with other tools, 162
System objectives, 251*

*Tight English, 158-161
   compared with other tools, 162
Third normal form, 182-184, 189-200
Tools of Structured Analysis,
    relationship, 34
Top-down development, 334
Transaction center, 293-294
Transform center, 292
Tuple, 178*

UCC TEN, 112

*Walston, C., 342
Weinberg, G., 264
Whitehouse, G., 37*

Yourdon, E., 273